Molecular Regulation
of Nuclear Events
in Mitosis and Meiosis

CELL BIOLOGY: A Series of Monographs

EDITORS

D. E. Buetow
*Department of Physiology
and Biophysics
University of Illinois
Urbana, Illinois*

I. L. Cameron
*Department of Cellular and
Structural Biology
The University of Texas
Health Science Center at San Antonio
San Antonio, Texas*

G. M. Padilla
*Department of Physiology
Duke University Medical Center
Durham, North Carolina*

A. M. Zimmerman
*Department of Zoology
University of Toronto
Toronto, Ontario, Canada*

Recently published volumes

Gary L. Whitson (editor). NUCLEAR-CYTOPLASMIC INTERACTIONS IN THE CELL CYCLE, 1980

Danton H. O'Day and Paul A. Horgen (editors). SEXUAL INTERACTIONS IN EUKARYOTIC MICROBES, 1981

Ivan L. Cameron and Thomas B. Pool (editors). THE TRANSFORMED CELL, 1981

Arthur M. Zimmerman and Arthur Forer (editors). MITOSIS/CYTOKINESIS, 1981

Ian R. Brown (editor). MOLECULAR APPROACHES TO NEUROBIOLOGY, 1982

Henry C. Aldrich and John W. Daniel (editors). CELL BIOLOGY OF *PHYSARUM* AND *DIDYMIUM.* Volume I: Organisms, Nucleus, and Cell Cycle, 1982; Volume II: Differentiation, Metabolism, and Methodology, 1982

John A. Heddle (editor). MUTAGENICITY: New Horizons in Genetic Toxicology, 1982

Potu N. Rao, Robert T. Johnson, and Karl Sperling (editors). PREMATURE CHROMOSOME CONDENSATION: Application in Basic, Clinical, and Mutation Research, 1982

George M. Padilla and Kenneth S. McCarty, Sr. (editors). GENETIC EXPRESSION IN THE CELL CYCLE, 1982

David S. McDevitt (editor). CELL BIOLOGY OF THE EYE, 1982

P. Michael Conn (editor). CELLULAR REGULATION OF SECRETION AND RELEASE, 1982

Govindjee (editor). PHOTOSYNTHESIS, Volume I: Energy Conversion by Plants and Bacteria, 1982; Volume II: Development, Carbon Metabolism, and Plant Productivity, 1982

John Morrow. EUKARYOTIC CELL GENETICS, 1983

John F. Hartmann (editor). MECHANISM AND CONTROL OF ANIMAL FERTILIZATION, 1983

Gary S. Stein and Janet L. Stein (editors). RECOMBINANT DNA AND CELL PROLIFERATION, 1984

Prasad S. Sunkara (editor). NOVEL APPROACHES TO CANCER CHEMOTHERAPY, 1984

Burr G. Atkinson and David B. Walden (editors). CHANGES IN EUKARYOTIC GENE EXPRESSION IN RESPONSE TO ENVIRONMENTAL STRESS, 1985

Reginald M. Gorczynski (editor). RECEPTORS IN CELLULAR RECOGNITION AND DEVELOPMENTAL PROCESSES, 1986

Govindjee, Jan Amesz, and David Charles Fork (editors). LIGHT EMISSION BY PLANTS AND BACTERIA, 1986

Peter B. Moens (editor). MEIOSIS, 1987

Robert A. Schlegel, Margaret S. Halleck, and Potu N. Rao (editors). MOLECULAR REGULATION OF NUCLEAR EVENTS IN MITOSIS AND MEIOSIS, 1987

In preparation

Monique C. Braude and Arthur M. Zimmerman (editors). GENETIC AND PERINATAL EFFECTS OF ABUSED SUBSTANCES, 1987

E. J. Rauckman and George M. Padilla (editors). THE ISOLATED HEPATOCYTE: USE IN TOXICOLOGY AND XENOBIOTIC BIOTRANSFORMATIONS, 1987

Molecular Regulation of Nuclear Events in Mitosis and Meiosis

Edited by

Robert A. Schlegel

Department of Molecular and Cell Biology
The Pennsylvania State University
University Park, Pennsylvania

Margaret S. Halleck

Department of Molecular and Cell Biology
The Pennsylvania State University
University Park, Pennsylvania

Potu N. Rao

Department of Medical Oncology
The University of Texas System Cancer Center
M. D. Anderson Hospital and Tumor Institute
Houston, Texas

1987

ACADEMIC PRESS, INC.
Harcourt Brace Jovanovich, Publishers

Orlando   San Diego   New York   Austin
Boston   London   Sydney   Tokyo   Toronto

13945243
DLC

4-1-87 JM

ACADEMIC PRESS, INC.
Orlando, Florida 32887

United Kingdom Edition published by
ACADEMIC PRESS INC. (LONDON) LTD.
24–28 Oval Road, London NW1 7DX

Library of Congress Cataloging in Publication Data

Molecular regulation of nuclear events in mitosis and
    meiosis.

    (Cell biology) — use
    Includes index.
    1. Mitosis. 2. Meiosis. 3. Molecular biology.
I. Schlegel, Robert A. II. Halleck, Margaret S.
III. Rao, Potu N. IV. Series. [DNLM: 1. Meiosis.
2. Mitogens. 3. Mitosis. 4. Molecular Biology.
QH 605.2 M718]
QH605.M69 1987   574.87'623   86-17216
ISBN 0–12–625115–0 (alk. paper)

PRINTED IN THE UNITED STATES OF AMERICA

87 88 89 90   9 8 7 6 5 4 3 2 1

Contents

6 Regulation of Chromatin Condensation and Decondensation in Sea Urchin Pronuclei

Dominic Poccia

7 Regulation of Mitosis by Nonhistone Protein Factors in Mammalian Cells

Ramesh C. Adlakha and Potu N. Rao

8 Mitosis-Specific Cytoplasmic Protein Kinases

*Margaret S. Halleck, Katherine Lumley-Sapanski,
and Robert A. Schlegel*

9 Antibodies to Mitosis-Specific Phosphoproteins

Frances M. Davis and Potu N. Rao

10 Mitosis-Specific Protein Phosphorylation Associated with Premature Chromosome Condensation in a *ts* Cell Cycle Mutant

Takeharu Nishimoto, Kozo Ajiro, Frances M. Davis,
Katsumi Yamashita, Ryosuke Kai, Potu N. Rao,
and Mutsuo Sekiguchi

11   Chromatin Structure and Histone Modifications
through Mitosis in Plasmodia of *Physarum
polycephalum*

*Hideyo Yasuda, Reinhold D. Mueller,
and E. Morton Bradbury*

Preface

In the past several years the molecular mechanisms involved in the control of the early events of amphibian oocyte maturation have been elaborated in some detail. The later events which culminate in germinal vesicle breakdown and chromosome condensation, the cytological events distinctive of meiotic maturation, are much less well understood. One point is, however, perfectly clear: transplantation of cytoplasm from a mature oocyte into an immature recipient can bypass normally requisite early events and initiate the processes which immediately precede final nuclear transformation. Studies of the regulation of the meiotic cell cycle dramatically converged with similar studies of the mitotic cell cycle upon demonstration that the signal which induced these nuclear events in oocytes could also be furnished by cytoplasmic extracts prepared from developing blastomeres and that this activity cycled in relation to the mitotic cell cycle. Only some five years ago this message was brought home most forcefully when cytoplasmic extracts prepared from culture cells arrested in mitosis were shown capable of inducing maturation when injected into immature oocytes. Such activity has now been revealed to be universal in oocytes undergoing maturation or in somatic cells in mitosis; extracts from maturing amphibian oocytes can induce maturation in immature starfish oocytes, and extracts from synchronized yeast can induce maturation in amphibian oocytes. Furthermore, in reciprocal experiments, cytoplasmic extracts from mature oocytes were shown to induce mitotic events in somatic nuclei, and, soon thereafter, extracts from somatic culture cells were shown capable of inducing condensation of interphase chromatin of somatic nuclei.

This convergence of meiotic and mitotic cell cycles has lent new impetus to identifying and comparing the molecular species responsible for the biological activity in each type of extract. Partial purification and characterization of the active factors from both meiotic and mitotic sources have been accomplished, and these pursuits continue. Concomitantly, efforts are under way to purify and characterize factors which are inhibitory to meiotic and mitotic processes. In the case of meiotic maturation, factors are present in follicular fluids which maintain arrest of oocytes prior to initiation of maturation, and other factors, cytoplasmic in nature, are required to maintain matured oocytes in metaphase arrest until fertilization. Reversal of this arrest upon fertilization and its dominance over sperm pronuclei are also of relevance in this context. In the case of cells traversing mitosis, factors inhibitory to nuclear membrane breakdown and chromosome condensation can be found in cells in the early portion of the G_1 phase of the cell cycle.

With the development of these various assays for biological activities and with the partial purification of the substances responsible, the time is ripe for investigating the possible modes of action of the molecular species involved. Reversible modifications of the proteins which compose the nuclear lamina and chromatin have been known for some time to occur concomitantly with breakdown of the nuclear membrane and condensation of chromatin into discrete chromosomes. Whether these modifications are causal has been a question of some controversy. Perhaps the modification most closely linked with mitotic nuclear events is phosphorylation. Thus, the possible role of protein phosphorylation/dephosphorylation in the regulation of these events is currently being actively pursued by identifying the protein kinases present at mitosis as well as their phosphoprotein substrates.

This book seeks to bring together in one volume the related studies of investigators in the various fields which this area of research encompasses. Authors describe their relevant background work and present their most up-to-date findings. Beyond this, however, they provide their views on how the various systems and factors described relate to one another and formulate the direction and priorities they anticipate for their continuing research. In this way, the goals achievable in the not-too-distant future may begin to take shape.

Robert A. Schlegel
Margaret S. Halleck*
Potu N. Rao

*Present address: Department of Pharmacology, The University of Texas Medical School, P.O. Box 20708, Houston, Texas 77225.

Molecular Regulation
of Nuclear Events
in Mitosis and Meiosis

1

Development of Cytoplasmic Activities That Control Chromosome Cycles during Maturation of Amphibian Oocytes

YOSHIO MASUI AND ELLEN K. SHIBUYA

Department of Zoology
University of Toronto
Toronto, Ontario
Canada M5S 1A1

I. INTRODUCTION

A. Maturation and Activation

The characteristic feature of female meiosis in animals, which distinguishes it from male meiosis and mitosis, is that female meiosis is arrested at a certain phase of the chromosome cycle, and an external stimulus is required to release it from the arrest. In most species, primary oocytes are arrested at prophase of the first meiosis (prophase I) while enclosed in the ovarian follicles, but in some marine animals, meiosis resumes when spawned eggs are fertilized. Chromosomes arrested at prophase I are partially condensed and contained in a large nucleus called the germinal vesicle (GV). Each chromosome is composed of two chromatids and is

1

MOLECULAR REGULATION OF NUCLEAR EVENTS
IN MITOSIS AND MEIOSIS

associated with its homologue. Normally, the resumption of prophase-arrested meiosis is brought about by gonadotropin stimulation of the follicles enclosing the oocytes, or by penetration of the sperm if the oocytes have already been spawned. The resumption of prophase-arrested meiosis and its subsequent processes have been called (meiotic) maturation. During maturation the oocyte undergoes a sequence of morphological changes including germinal vesicle breakdown (GVBD), followed by chromosome condensation to a metaphase state, and segregation of one of the haploid sets of chromosomes into the first polar body.

However, in many species, meiosis is arrested again before its completion, either at metaphase I, as seen in many invertebrates including insects, or at metaphase II, as seen in cephalochordates and vertebrates (see Masui, 1985, for review). Only in the echinoderms and coelenterates (*Cnidaria*) is the chromosome cycle arrested after meiosis is completed, and a pronucleus is formed. In all these cases, oocytes are induced to resume chromosome cycles by fertilization, completing meiosis and/or initiating mitosis. The process induced by fertilization is called (egg) activation and can be induced by other stimuli as well.

B. Early Studies of Oocyte Maturation

Our classic notion of maturation was derived exclusively from studies of the oocytes of marine invertebrates. Until the methods of inducing maturation of vertebrate oocytes *in vitro* were developed, oocytes of marine animals were the only source of information about the mechanism of maturation, since their maturation processes are easily observed and controlled *in vitro*. Thus, Wilson (1903) and Yatsu (1905), using oocytes of *Cerebratulus,* and Delage (1901), using starfish oocytes, discovered that maturation was a prerequisite for the oocyte to undergo cell division and that the contribution of nucleoplasm from the GV was essential for this process. Since then, numerous studies have been done concerning conditions for initiating maturation of marine eggs (Tyler, 1941; Brachet, 1951; see Heilbrunn, 1952, for review). At that time, it was generally accepted that the reaction that induces maturation first occurs in the cortex and that the most important factor in this process is calcium ions. Indeed, Heilbrunn (1952) stated that "thus in the marine egg cell response to stimulation involves a liquefaction of the cortex with liberation of calcium."

Pincus and Enzman (1935) were the first to observe maturation of vertebrate oocytes *in vitro,* i.e., spontaneous maturation of mammalian oocytes isolated from follicles. Rugh (1934) found that intraperitoneal injections of macerated pituitary glands caused not only oocyte maturation, but ovulation as well in the amphibian *Rana pipiens*. A year or so later, Shapiro (1936) and Zwarenstein (1937) found that ovarian follicles of *Xenopus laevis* could be induced to ovulate *in vitro* by culturing in Ringer's solution containing progesterone. On the other hand, Heilbrunn *et al.* (1939) succeeded in inducing oocyte maturation and ovulation *in vitro* by incubating dissected ovaries of *R. pipiens* in Ringer's solution containing homogenized frog pituitaries. This technique was widely used by Wright (1945), Nadamitsu (1953), and Tchou-Su and Wang Yu-Lan (1958) for further investigation of conditions for ovulation in amphibians. Wright (1961) also reported a synergistic effect of pituitary hormone and progesterone on ovulation in *R. pipiens*. Although in this early research less attention had been devoted to oocyte maturation than ovulation, the importance of oocyte maturation had been well recognized. Brachet (1951), commenting on the physiological significance of maturation, stated that "this field, so interesting both from the point of view of embryology and cellular physiology, remains to be explored." Tchou-Su and Wang Yu-Lan (1958) were the first researchers to place an emphasis on studying the cytological aspects of maturation as well as ovulation of amphibian oocytes *in vitro*. However, it was not until Dettlaff *et al.* (1964) performed a series of microsurgical experiments including nuclear transplantation on toad oocytes that nucleocytoplasmic relations during maturation were analyzed in vertebrate oocytes.

Over the years, animal oocytes have become a unique cell system particularly well suited to studies of nucleocytoplasmic interactions. Oocytes have provided us with several advantages over other cells in manipulating chromosome cycle events. First, because of their large cell size and capacity to survive mechanical injuries, oocytes are suitable for various microsurgical operations such as enucleation and nuclear transplantation. Second, by releasing oocytes from their meiotic arrest using artificial stimulation, we can obtain a fairly large quantity of cells with highly synchronous chromosome cycles. Third, mainly because of a large store of nutrients, fully grown oocytes can be handled and cultured with relative ease.

This chapter describes the progress in research of oocyte maturation during the past 20 years, with major emphasis on the problems of amphibian oocyte maturation. We discuss amphibian oocytes as a model system for the study of cytoplasmic control of chromosome behavior, and some relevant aspects of oocyte maturation in other animals are discussed in comparison with those in amphibians.

II. OOCYTE MATURATION

A. Hormonal Control of Oocyte Maturation

1. Dettlaff's Study

The paper published by Dettlaff *et al.* (1964) renewed interest in the study of oocyte maturation among embryologists and endocrinologists using amphibians. These authors, using the toad species *Bufo bufo* and *B. viridis,* reported that maturation could be induced by pituitary hormone in oocytes from which follicular investments had been removed, and oocytes induced to mature in this manner could respond to an activation stimulus, such as pricking with a glass needle, by undergoing surface contractions. Moreover, when used as recipients for nuclear transplantation, these oocytes were able to cleave, but failed to acquire this capability for cell division if the GV had been removed. Surgical breakage of the GV and mixing its contents with cytoplasm in oocytes without hormone treatment did not cause maturation, but "a small amount of karyoplasm with some cytoplasm taken at the onset of GV dissolution" from hormone-stimulated oocytes could cause maturation when injected into untreated oocytes. From these observations the authors speculated that "unknown changes in nuclear properties precede changes in the cytoplasm" and that "substances inducing cytoplasmic maturation are formed in the karyoplasm only just prior to dissolution of the GV membrane." Further, since the cytoplasm lacking the GV does not respond to the action of the hormone, the gonadotropic hormones must affect oocyte maturation through the oocyte nucleus. However, these authors were also cautious about implicating cytoplasmic factors, stating that "karyoplasm (with cytoplasm) taken from the oocyte at the onset of GV dissolution can stimulate maturation of oocytes at the initial stage, i.e., it substitutes for the action of gonadotropic hormone."

Two years later, using *R. temporaria* oocytes, Dettlaff (1966) showed that actinomycin D and puromycin could prevent hormone-induced maturation, and concluded that "gonadotropins induce DNA-dependent synthesis of specific mRNAs" and "after some time protein synthesis starts, including the synthesis of enzymes participating somehow in the rupture of the germinal vesicle membrane." Independently, Smith *et al.* (1966) studied protein synthesis by microinjecting [³H]leucine into *R. pipiens* oocytes and demonstrated considerable increase in labeled amino acid incorporation into oocyte proteins during maturation. These studies taken together suggested that the induction of oocyte genomic activities by gonadotropin leads to oocyte maturation.

2. *Relative Roles of Pituitary, Follicles, and Progesterone*

The implication of the findings by Dettlaff and associates was so important that it was necessary to repeat their experiments. Thus, Brachet (1967) confirmed that both RNA and protein synthesis inhibitors could inhibit oocyte maturation induced by gonadotropins in several amphibian species including *X. laevis*. However, Gurdon (1967) found that human chorionic gonadotropin (hCG) did not induce maturation of *Xenopus* oocytes if it was directly injected into the oocytes, while the same hormone externally applied to the isolated follicles was effective, thus casting a doubt on the theory of direct hormonal action on the oocyte genome. Schuetz (1967a), on the other hand, tested in the effects of various gonadal steroids on *R. pipiens* follicles cultured *in vitro*, and found that progesterone was the most potent steroid for inducing maturation. Schuetz (1967b) also found that manual removal of follicle cells reduced pituitary-induced GVBD by 50% but that it did not affect progesterone-induced GVBD and that the former could be inhibited by both actinomycin D and puromycin, but the latter only by puromycin. Based on these results, Schuetz suggested the presence of "a common mechanism concerned with protein synthesis" that mediated pituitary and steroid-induced GVBD and also "the presence of a pituitary-stimulated intermediary process prior to the initiation of protein synthesis." However, Masui (1967) pointed out that after manual removal of follicular investments, *R. pipiens* oocytes were not totally devoid of adhering follicle cells unless ovarian pieces were treated with Ca^{2+}-free medium prior to divestment of the follicular envelopes. He also showed that once all follicle cells were removed, oocytes were unable to respond to pituitary hormone, but still responded to progesterone. The oocytes induced to mature by progesterone were capable of cleavage upon nuclear transplantation. Similar results were obtained independently by Smith *et al.* (1968) with oocytes from which follicle cells were completely removed by pronase treatment. Masui (1967) also observed that pituitary hormone could induce maturation in follicle-free oocytes if oocytes were packed with follicle cells previously removed from the oocytes, suggesting that "pituitary hormone first affects follicle cells that in turn secrete some diffusible factor that acts on oocytes to cause maturation." The hypothesis that "the follicle cells secrete a progesterone-like substance" in response to pituitary hormone to induce oocyte maturation was substantiated by Fortune *et al.* (1975) in *X. laevis* and Schatz and Ziegler (1979) in *R. pipiens* using the radioimmunoassay technique of steroid determination, which had recently become available. The notion developed earlier that pituitary hormone-induced GVBD requires RNA synthesis, whereas progesterone-induced GVBD requires only protein synthesis, was also corroborated by Merriam (1972) and Wasserman and Masui (1974). These workers carried out experiments in which *X. laevis* ovaries were treated with various inhibitors, including actinomycin D, α-amanitin, ethidium bromide, puromycin, and cycloheximide. Also, the observation by Masui (1973b) that the resistance of progesterone-induced GVBD to X-ray irradiation is higher than that of pituitary-induced GVBD supports the above notion.

3. Hormonal Control in Nonamphibian Species

It is important to note that studies of oocyte maturation in starfish were also being done at this time and that results of experiments by Schuetz and Biggers (1967) and Kanatani and Shirai (1967) exactly paralleled those obtained with the frog. These results indicated that a gonad-stimulating substance secreted from the radial nerves causes follicle cells to secrete a diffusible factor that in turn acts on oocytes to induce maturation. This substance was later identified as 1-methyladenine (1-MA) by Kanatani *et al.* (1969). A similar pathway for the transmission of a hormonal signal which induces oocyte maturation was described in the sturgeon, *Acipenserides,* by Dettlaff and Skoblina (1969) and in the trout by Fostier *et al.* (1973). In mammals, the involvement of pituitary gonadotropin [luteinizing hormone (LH)] in the induction of maturation in follicle-enclosed oocytes was shown by Ayalon *et al.* (1972), although it remains unclear how follicle cells act on the oocytes to induce maturation in response to the hormonal signal. Therefore, it may be assumed that in all animals the signal for oocyte maturation is transmitted from the nervous system to the oocyte via the mediation of ovarian follicle cells. The only exception to this rule was reported by Sundararaj and Goswami (1977) in the Indian catfish, *Heteropneutes fossilis.* In this animal pituitary hormone stimulates the interrenal gland rather than the ovarian follicles to secrete a corticosteriod that induces oocytes to mature. However, according to a recent report by Van der Hurk and Richter (1980), this does not appear to be the case in the African catfish (*Clarias lazera*).

Retrospectively, the conclusions about the mechanism of oocyte maturation reached by Dettlaff *et al.* (1964) proved, at least in part, to be incorrect. Nevertheless, the importance of the contribution made by these authors as pioneers cannot be overemphasized. It was this paper that initiated a paradigm for research in the study of oocyte maturation.

B. Mode of Progesterone Action

1. Sites of Action

To determine the site in the oocytes where the initial reaction leading to maturation takes place, substances known to have maturation-inducing activity were injected directly into oocytes, avoiding any superficial contact. Smith and Ecker (1969) first reported such an experiment in which progesterone was injected into *R. pipiens* oocytes. Smith and Ecker (1971) and Masui and Markert (1971) later confirmed the finding that progesterone failed to induce maturation if injected into an oocyte even at a dose higher than required to induce maturation when applied externally. This suggests a surface localization of progesterone action sites. Results of experiments conducted by Ishikawa *et al.* (1977), using progesterone derivatives conjugated with agarose, and Godeau *et al.* (1978), using polyethylene oxide–progesterone conjugates, confirmed this notion. These authors showed that the polymer-conjugated steroids could induce maturation in *Xenopus* oocytes even though their penetration into the oocytes was prevented.

Contrary to these observations, Tso *et al.* (1982) reported that a direct injection of progesterone dissolved in paraffin oil into *Xenopus* oocytes could induce maturation. Using [^{3}H]progesterone, these authors found that the concentration of the steroid in the external medium (due to leakage from the oocytes) was less than that found to be effective in inducing maturation. Hence, these authors concluded that the hormone injected into oocytes can reach the internal receptor sites to induce maturation. However, it has been shown by Schuetz (1972) and Smith and Ecker (1971) that amphibian oocytes have a strong capacity to absorb progesterone from the external medium. Therefore, the conclusion by Tso *et al.* may be acceptable only if the oocytes did not reabsorb any progesterone that had leaked out. Proof for this, however, was not provided. Therefore, the possibility exists that more progesterone leaked out of the oocytes into medium than was reported, but some of it was reabsorbed by the oocytes before the authors' measurement was carried out. In this case, the external progesterone level may have reached a dose sufficient to induce oocyte maturation by external application. This possibility could have been checked by coculturing control oocytes with those injected with progesterone. In fact, this test was carried out in our laboratory some time ago, and the control oocytes were found to mature along with the hormone-injected ones. The hypothesis that assumes the internal localization of progesterone action sites would meet difficulties in explaining not only the induction of maturation by polymer-conjugated steroids but also the effectiveness of other maturation-inducing agents that are known to act very likely on cell surface receptors. These include β-adrenergic blocking agents and amphiphilic cations, as shown by Schorderet-Slatkine *et al.* (1977).

2. Role of Ca²⁺ Release

Amphibian oocyte maturation is known to be induced by various chemicals other than progesterone, most of which are known to cause Ca^{2+} mobilization in the cell (Baulieu *et al.*, 1978; see Masui and Clarke, 1979, and Morrill *et al.*, 1981, for review). The first evidence that Ca ion mobilization across the cell membrane causes maturation of *Xenopus* oocytes was provided by Wasserman and Masui (1975a). They demonstrated that exposure of oocytes to ionophore A23187 at high external Ca or Mg ion levels could induce maturation. Wasserman *et al.* (1980) and Moreau *et al.* (1980) observed a release of Ca ions in *Xenopus* oocytes following progesterone stimulation by recording chemiluminescence resulting from aequorin previously injected into the oocytes. However, recent measurements by Robinson (1985), using a Ca^{2+}-sensitive electrode, did not detect any significant change in Ca ion levels in *Xenopus* oocytes following progesterone stimulation. The discrepancy between the recent and previous results may reflect the fact that different kinds of Ca ion probes can reach different regions in the cells. It may be that while aequorin can monitor only Ca ions localized near the cell surface, the electrode can record Ca ion levels somewhat deeper in the cytoplasm. Therefore, progesterone appears to cause a release of Ca ions localized in the cortex of the oocyte rather than in the internal cytoplasm. If this is the case, in all probability it is the Ca ions released in the cortex of the oocyte that are responsible for initiating maturation. This inference has also been given support by Moreau *et al.* (1976) using *Xenopus* oocytes. They observed that Ca ions injected by ionophoresis into the cytoplasmic region located within 0.2 mm of the surface caused maturation, whereas the ions injected into deeper regions had no effect. It may be that the initial step of maturation in amphibian oocytes is the interaction of progesterone with the reaction sites located near the surface, thus releasing Ca ions within the cortex. These Ca ions then induce the subsequent events of maturation.

The initial reaction of oocytes to maturation-inducing agents in nonamphibian species has only been characterized in the starfish. Kanatani and Hiramoto (1970) microinjected 1-MA into starfish oocytes without effect, which suggests its surface action. However, roles of Ca ions in the initiation of maturation still remain unclear. Starfish oocytes, unlike amphibian oocytes, have been unresponsive to Ca^{2+}-mobilizing agents (see Moreau et al., 1985, for review) and even the initial release of Ca ions in response to hormone has recently been questioned (Eisen and Reynolds, 1984).

C. Acquisition of Developmental Capacities by Oocyte Cytoplasm during Maturation

1. Cytoplasmic Activities Controlling Nuclear Behavior

Changes in cytoplasmic activities during maturation can be monitored by transplanting nuclei into the oocyte. Dettlaff et al. (1964) demonstrated for the first time that amphibian oocytes which had undergone GVBD and activation became capable of cleavage in response to transplanted blastula nuclei. Gurdon (1968), on the other hand, found in X. laevis that adult brain nuclei could be induced to undergo nuclear membrane dissolution and chromosome condensation to a metaphase state after transplantation into maturing oocytes, whereas those transplanted into immature oocytes remained at interphase. By injecting cilia basal bodies isolated from protozoa into maturing oocytes of X. laevis, Heidemann and Kirschner (1975) found that the ability to assemble microtubules appeared in cytoplasm after GVBD. The ability to respond to activation stimuli develops near the final stage of maturation. Smith and Ecker (1969) and Skoblina (1969) showed, in R. pipiens and R. temporaria, respectively, that progesterone-stimulated oocytes responded to pricking with a glass needle by undergoing cortical granule breakdown. Elinson (1977) found that if R. pipiens eggs were precociously inseminated, they were not activated following penetration by several sperm. However, when these polyspermic eggs completed maturation, they became capable of activation and cleavage in response to electrical stimulation or reinsemination. Also, oocytes acquire the ability to decondense sperm chromatin and promote DNA synthesis in response to activation stimuli, as shown by Katagiri and Moriya (1976) and Skoblina (1974). They injected a suspension of sperm nuclei demembranated with detergent into oocytes at various stages of maturation. Evidently, oocytes have acquired all the cytoplasmic capacities throughout maturation to support the cell cycle in the zygote.

2. Roles of the GV

Early studies of some marine animals disclosed that anucleate fragments of oocytes, unlike nucleate fragments, separated prior to GVBD could not develop the ability to cleave if they were fertilized (see Wilson, 1925, for review). Since then, the importance of the role of the GV in maturation has been well recognized. However, studies confirming this notion in other animals did not appear until 1964, when Dettlaff and associates performed experiments with toad oocytes (see Section II,A,1). In this experiment, they injected blastomere nuclei into hormone-treated oocytes from which the GV had been removed and found that these oocytes were unable to cleave. Confirming this with *R. pipiens,* Smith and Ecker (1969) demonstrated further that the oocyte's lost ability to cleave due to enucleation could be recovered by reinjection of GV contents. Importantly, they noted that the GV contents from hormone-treated and untreated oocytes were equally potent, indicating the absence of hormonal effects on the activities of the GV material. The presence of the GV is also required for the development of the oocyte's capacity to induce sperm chromatin decondensation in response to activation stimuli. In amphibian species, Skoblina (1974), Katagiri and Moriya (1976), and Lohka and Masui (1983) showed that enucleated oocytes could not develop this capacity after progesterone stimulation, but that reinjection of GV contents restored this ability. The GV factor for sperm chromatin decondensation was found dispersed in the cytosol during maturation (Lohka and Masui, 1983). The requirement for the GV factor in developing the oocyte's ability to induce sperm chromatin decondensation was also demonstrated in the mouse by Bałakier and Tarkowski (1980) by fertilizing enucleated oocyte fragments. However, the oocyte's capacity to undergo cortical granule breakdown in response to activation stimuli was shown to develop independently of the GV in oocytes of *R. pipiens* and *R. temporaria* by Smith and Ecker (1969) and Skoblina (1969), respectively.

It is important to note, however, that the dependence of the oocyte on the GV for acquisition of its developmental capacities during maturation does not imply dependence on genomic functions. Masui (1973a) found that *R. pipiens* oocytes that had been irradiated with X rays before maturation was initiated were able to develop into androgenetic haploid tadpoles after fertilization. These tadpoles were indistinguishable from gynogenetic haploid tadpoles produced by insemination with X-ray irradiated sperm. This result implies that all the developmental capacities of the oocyte necessary for producing tadpoles have been provided during maturation without the involvement of maternal chromosomes.

III. MATURATION-PROMOTING FACTOR (MPF)

A. Development of MPF during Oocyte Maturation

1. *MPF in* Rana pipiens

As previously mentioned, the first step of oocyte maturation involves the interaction of the oocyte surface with a maturation-inducing agent applied externally. The signal that triggers all the subsequent changes in internal cytoplasmic activities of an oocyte required for maturation is received only by the cortex. Therefore, it must be translated into a cytoplasmic message inside the cell. This conjecture was tested in *R. pipiens* by Masui and Markert (1971) who, using a graduated micropipette, transferred cytoplasm from progesterone-treated oocytes into untreated recipient oocytes. Although essentially the same experiment was reported by Smith and Ecker (1971) in the discussion of the mode of steroid interaction in *R. pipiens* oocytes, no details were given. Masui and Markert's results may be summarized as follows:

1. Progesterone-treated oocytes develop a cytoplasmic activity that causes GVBD when injected into recipient oocytes some time before GVBD (12 hr after progesterone stimulation at 20°C).
2. Recipient oocytes induced to undergo GVBD by cytoplasmic transfer also mature completely during culture, i.e., complete the first meiosis, and acquire all of the developmental capacities possessed by normal mature eggs. Therefore, the putative cytoplasmic factor responsible for this effect has been called maturation-promoting factor (MPF).
3. The effect of MPF is dose-dependent, i.e., the frequency with which GVBD is induced in oocytes that receive injection of MPF-containing cytoplasm gradually increases with volumes of injected cytoplasm. This dose-dependent relationship is expressed as a sigmoidal curve with the threshold at 5 nl per oocyte. The dose-dependent effect of MPF has enabled assay of its activity by referring to the percentage of oocytes in which GVBD can be induced by MPF injection.
4. When relative levels of MPF in the cytoplasm of oocytes at different stages of maturation are measured, MPF activity first appears in the cytoplasm several hours before GVBD, rises rapidly as GVBD approaches, and reaches its maximum at the onset of GVBD.
5. High MPF activity is detected in metaphase II oocytes matured *in vitro* as well as in unfertilized eggs in the ovisac, but little activity is found in artificially activated oocytes and fertilized eggs. However, low MPF activities are still detectable in the cytoplasm of cleaving embryos.

2. MPF in Other Animals

Using a similar approach, MPF was also found in *X. laevis* and *Ambystoma mexicanum* oocytes by Schorderet-Slatkine and Drury (1973) and Reynhout and Smith (1974), respectively. In starfish, Kishimoto and Kanatani (1976) demonstrated the development of MPF in 1-MA-treated oocytes by employing a technique devised by Hiramoto (1962) that allowed them to microinject several picoliters of cytoplasm from maturing donor oocytes into immature recipient oocytes. MPF was shown to develop in mouse oocytes cultured *in vitro* by Bałakier and Czolowska (1977). These authors fused anucleate oocyte fragments cultured for a few hours with small immature oocytes that were unable to undergo spontaneous GVBD. They found that the cytoplasmic fragments developed the ability to cause GVBD in the cell hybrids.

Maturation-promoting factor is not a species-specific cytoplasmic factor. Its transfer from oocytes of a given species into immature oocytes of another species causes maturation in the recipient oocytes. This was first shown by Reynhout and Smith (1974) using the amphibian species, *X. laevis, R. pipiens*, and *A. mexicanum*. Later, similar interspecific transfers of oocyte cytoplasm were carried out by Kishimoto and Kanatani (1977) using starfish; by Dettlaff and Felgengauer (1980) using the fish, *Acipenser stellata*, and toad, *X. laevis;* and finally by Kishimoto *et al.* (1982) using animals of different phyla, starfish and amphibians. More recently, Kishimoto *et al.* (1984) reported that successful cytoplasmic transfers from mouse as well as surf clam (*Spisula*) oocytes into starfish oocytes induce GVBD in the latter. Further, Sorensen *et al.* (1985) were able to induce GVBD in immature *Xenopus* oocytes by injection of an extract from mature mouse oocytes. This proved for the first time that mammalian oocytes could develop MPF equivalent to amphibian MPF.

B. Molecular Characteristics of MPF

1. Extraction of MPF

Characterization of MPF has been hampered by its instability. Masui and Markert (1971) studied MPF activities in homogenates of mature *R. pipiens* oocytes made with a minimum volume of external medium. However, all extracts made from homogenized mature oocytes invariably showed little MPF activity. Therefore, to determine which cell components possess MPF activity, Masui (1972) stratified the cytoplasm of intact progesterone-treated oocytes of *R. pipiens* by applying a moderate centrifugal force to the oocytes placed in the interface between Ringer's solution and 40% Ficoll. In this way, the cytoplasm was separated into five layers: (from top to bottom) lipid, fluid hyaline, gel hyaline, pigment, and yolk. The most MPF activity was found in the hyaline layers. Reynhout and Smith (1974) obtained a cytosol containing MPF from mature oocytes of *X. laevis* which were gently broken in mineral oil and centrifuged at 13,500 *g* for 30 min. However, in *R. pipiens* it was noted that MPF activity in the hyaline layer of stratified oocytes remained more stable than the cytosol extracted from manually broken oocytes. Thus, a new method of extracting oocyte cytosol was devised. This method, described by Masui (1974), is as follows: "Dejellied eggs immersed in the extraction medium were packed in a tube by centrifugation (80 *g*, 1 min), and the excess medium above the eggs was removed. By further centrifugation (25,000 *g*, 1 min) the packed eggs were crushed and the coarse sediments discarded. The supernatant was again centrifuged (150,000 *g*, 60 min) to remove fine particles." This method has since been used with slight modifications for all the studies of cytoplasmic factors in our laboratory.

2. Ion Sensitivity and Dependency

Wasserman and Masui (1976) were successful in extracting a large quantity of cytosols containing MPF. They found that MPF could be stabilized to a certain extent by the presence of Mg ions and the Ca^{2+}-chelating agent, ethylene glycol bis (β-aminoethyl ether)N,N'-tetraacetic acid) (EGTA), although its activity lasted for only 3 days at 0°C. According to these authors, MPF is Ca^{2+}-sensitive and Mg^{2+}-dependent, since an addition of Ca ions or ethylenediaminetetraacetic acid (EDTA), which chelates both Ca and Mg ions, causes a rapid loss of MPF activity from the cytosol. Also MPF is protease sensitive, but not RNase sensitive. When cytosols are fractionated by centrifugal sedimentation through a sucrose density gradient, MPF activity can be recovered in the 4, 13, and 30 S fractions as discrete peaks. Based on these results, Wasserman and Masui (1976) suggested that MPF is a cytosolic protein or a protein complex, which may exist in multiple molecular forms as a result of being polymerized in different orders, or being associated with other molecules of different sizes.

3. Phosphorylation of MPF

In *X. laevis,* further stabilization of MPF in oocyte extracts was achieved by Drury (1978) who discovered that an inhibition of phosphatase activity with NaF as well as an enhancement of protein phosphorylation with ATP could support MPF activity. Similarly, Wu and Gerhart (1980) succeeded in stabilizing MPF in *Xenopus* oocyte extracts by using extraction medium containing high levels of β-glycerophosphate. These authors further characterized MPF after its partial purification, which involved protein precipitation with $(NH_4)_2SO_4$ and column chromatography with pentyl agarose followed by arginine agarose. This was a fundamental achievement in the course of MPF study. In this work the authors defined a unit of MPF activity as "that amount of activity which, in an injected volume, 20 nl, causes 50% of the recipient oocytes to mature." They also found that MPF samples purified 30- to 50-fold became less Ca^{2+}-sensitive, and contained polypeptides of 0.7 to 25×10^5 daltons. In addition, proteins in these samples were constantly phosphorylated and dephosphorylated in the presence of ATP which, in fact, doubles MPF activity in the samples. Recently, Hermann *et al.* (1983) reported a similar line of work on stabilization and purification of *Xenopus* MPF. These authors found that MPF could be precipitated with 5% polyethylene glycol or 20% ethanol and that its activity was stabilized and became less Ca^{2+}-sensitive in the presence of γ-thio-ATP. Taken together, the foregoing results strongly suggest that MPF is a phosphoprotein complex whose activity depends on the degree of phosphorylation.

C. Mode of MPF Action

1. Autocatalytic Amplification

According to Wasserman and Masui (1975b), the time required for 50% of oocytes to develop detectable MPF activity following progesterone stimulation, designated MPF_{50}, is variable among batches of oocytes derived from different animals. Similarly variable is the time required for 50% of the oocytes to begin GVBD, designated $GVBD_{50}$. However, the ratio between these two variables, i.e., $MPF_{50}/GVBD_{50}$ for a given batch of oocytes remains fairly constant from batch to batch. In *X. laevis,* this ratio is 0.65 ± 0.11 or $MPF_{50} = 0.65\ GVBD_{50}$. The implication of this observation may be that the processes involved in MPF production and GVBD are somehow closely correlated in such a way that the process leading to GVBD may be triggered when MPF accumulated in the cytoplasm reaches a threshold concentration.

The process of MPF production involves a reaction that resembles autocatalysis. It appears self-evident that MPF injection should give rise to more MPF in the cytoplasm of recipient oocytes if the recipient is able to reproduce the whole process of maturation, including the production of MPF. The evidence for this conjecture was provided by Masui and Markert (1971), using *R. pipiens* oocytes. In their experiment, cytoplasm was transferred first from progesterone-treated oocytes into untreated oocytes, then from these first recipients to the second, and so forth. Thus, 3–5% of the oocyte cytoplasm was transferred three times every 24 hr. In this serial transfer of oocyte cytoplasm, GVBD was induced in at least two-thirds of the recipient oocytes at every transfer. The same result was obtained when more extensive experiments were carried out by Reynhout and Smith (1974) with *X. laevis,* by Kishimoto and Kanatani (1976) with starfish, and by Dettlaff *et al.* (1977) with sturgeon oocytes. The important point of all these observations is that high MPF levels in the cytoplasm of recipient oocytes are detected throughout extensive serial dilutions of the original MPF. This means that MPF must have been amplified in the cytoplasm of the recipient oocytes between successive transfers, possibly by a process similar to autocatalysis.

2. Propagation of MPF

MPF amplification mechanisms may play an essential role in conducting the maturation-inducing signal received by the oocyte surface to the internal cytoplasm. It is not difficult to imagine that a small quantity of MPF introduced into a local area of the cytoplasm acts as a seed to induce MPF production in neighboring areas, thus propagating in space with time.

It appears that the initial maturation-inducing signal given by progesterone is localized mainly in the animal half surface of the oocyte as shown by Cloud and Schuetz (1977) in experiments in which progesterone was locally applied, either to the animal or vegetal hemisphere of *R. pipiens* oocytes. It seems that MPF appears first in the cytoplasm just beneath the cortex of the animal half, which in turn induces MPF production in the adjacent cytoplasm by an autocatalytic process, thus triggering its propagation into the rest of the cytoplasm. This model for MPF propagation was examined by Masui (1972), using *R. pipiens* oocytes as follows. First, the cytoplasm was removed from the animal and vegetal half separately at various times after progesterone stimulation of the oocytes and assayed for MPF activity. It was found that MPF became detectable in the cytoplasm of the animal half at earlier times and remained at higher levels than in the cytoplasm of the vegetal half. Second, the GV was displaced either to the animal or vegetal pole by a moderate centrifugal force and confined to the respective location by constricting the equatorial region with thread. The oocytes were then stimulated by progesterone. The GVs in the animal half underwent GVBD sooner than those in the vegetal half. Third, in the same experiment, it was found that the rate of GVBD which occurred in the vegetal half varied with the diameter of equatorial constriction, i.e., the narrower the cytoplasmic passage connecting the animal and vegetal halves, the lower the GVBD rate. However, the rate of GVBD that occurred in the animal half was not affected by the size of the equatorial constriction. These results suggest a propagation of MPF activity from the animal to vegetal half through cytoplasmic continuity.

Apparently, in the initial step of oocyte maturation, local physiological changes in subcortical cytoplasm brought about by a maturation-inducing agent persist until the local concentration of MPF reaches a threshold necessary to initiate its autocatalytic amplification. Once this happens, self-sustaining propagation leads to the MPF development in the entire cytoplasm of the oocyte.

3. Threshold of MPF

The existence of such a threshold, characteristic of oocyte maturation as an "all or none" process, may be explained by the capacity for autocatalytic amplification and the cooperativity of MPF. A clue to the explanation may be found in the equation proposed by Wu and Gerhart (1980), which relates V_{50}, the volume of an MPF sample required to be injected into an individual oocyte in order to induce GVBD in 50% of recipient oocytes, and C, the concentration of MPF:

$$V_{50} = k \, C^{-n}$$

where k and n are constant. The implication of this equation was interpreted, rather arbitrarily, by Masui (1982) as suggesting that in order for MPF to have an effect, possibly autocatalytic production, n MPF molecules must react with each other. Although n is variable, depending on the degree of purity of MPF samples, it is approximately 2 for MPF in crude cytosols. Thus, the rate of MPF production induced by cytosolic MPF may be proportional to the square of its concentration (X^2). However, MPF is an unstable molecule that decays with a certain probability, being lost at a rate proportional to its concentration (X). The rate of net increase in MPF concentration (dx/dt) is, therefore, given by:

$$\frac{dx}{dt} = k_1 X^2 - k_2 X$$

where k_1 and k_2 are rate constants for the autocatalytic synthesis and decay of MPF molecules, respectively. It is easy to see that MPF can continuously increase with time only when the initial concentration exceeds k_2/k_1, the threshold.

In summary, a maturation-inducing agent acting on the cortex of an oocyte changes physiological conditions in the subcortical cytoplasm to induce the production of MPF. This initial action of the maturation-inducing agent must be continued until the MPF in the peripheral cytoplasm reaches a threshold to trigger an autocatalytic reaction that leads to the propagation of MPF throughout the rest of the oocyte cytoplasm.

D. Factors Affecting MPF Activity

1. The GV Factor

The development of MPF activity requires certain physiological activities as well as cellular components of the oocyte. In *R. pipiens* oocytes, MPF can be developed in enucleated oocytes following progesterone stimulation, as shown by Masui and Markert (1971). Similar results were obtained in *X. laevis* by Schorderet-Slatkine and Drury (1973). Furthermore, Reynhout and Smith (1974) showed that MPF amplification also could occur in enucleated oocytes of *X. laevis*. Similar tests to determine the dispensability of the GV for MPF production were carried out in nonamphibian species. In the mouse, Bałakier and Czolowska (1977) found that anucleate fragments of oocytes produced before GVBD exhibited MPF activity after being cultured a few hours (see Section III,A,2). However, recent experiments have shown that if cultured for a longer period of 15 hr, these fragments lose MPF activity, unlike their nucleated counterparts (Bałakier and Masui, 1986). Thus, it appears that some factors from the GV are necessary for the maintenance of MPF activity, rather than its initial appearance. Kishimoto *et al.* (1981) found that after removal of the GV, starfish oocytes could develop MPF following 1-MA treatment, although the level reached was significantly lower than that in nucleated fragments. However, the amplification of MPF and its maintenance at a high level both required the presence of the GV.

Recent work by Picard and Dorée (1984) with European starfish yielded rather intriguing results. These authors observed that MPF could appear in enucleated fragments only if the oocytes had been collected late in the breeding season and also had been exposed to an extraordinarily high dose of hormone. Thus, it was concluded that the presence of the GV factor was essential for the appearance of MPF in starfish oocytes under ordinary conditions. The discrepancy between the observations of these two groups of researchers has not yet been reconciled.

2. Ca Ions

Various chemical treatments are known to inhibit oocyte maturation (see Masui and Clarke, 1979, for review). However, in many cases maturation can be induced in these inhibited oocytes by injection of MPF. This indicates that the development of MPF requires physiological activities or cellular components that have been perturbed by chemical inhibition, but the process of maturation induced by MPF is not prevented.

In *X. laevis,* if EGTA is injected into progesterone-treated oocytes before MPF appearance, GVBD can be blocked. However, after MPF appears in the cytoplasm, EGTA injections fail to prevent GVBD (Masui *et al.,* 1977). This suggests that a Ca^{2+}-dependent process is a prerequisite for the initial appearance of MPF, but that MPF amplification and its action in causing GVBD do not require Ca ions. Indeed, Ca ions have an inhibitory effect on these latter processes considering the Ca^{2+}-sensitivity of MPF. Wasserman and Masui (1975a) observed that the maturation-inducing effect of ionophore A23187 was suppressed at a very high external Ca^{2+} concentration (> 20 mM). According to Moreau *et al.* (1980), the Ca^{2+} level in oocytes decreases after MPF appearance.

3. pH

Utilizing various methods, Lee and Steinhardt (1981), Houle and Wasserman (1983), Cicirelli *et al.* (1983), and Morrill *et al.* (1984) all observed a significant increase in intracellular pH values (pH_i) following progesterone stimulation of *Rana* and *Xenopus* oocytes. Further, Lee and Steinhardt (1981) showed that an alkalinization of oocyte cytoplasm by triethylamine was effective in inducing GVBD. Conversely, Houle and Wasserman (1983) inhibited progesterone-induced GVBD in oocytes by using acetate to acidify the cytoplasm. Therefore, as suggested by Cicirelli *et al.* (1983), a rise in pH_i may be necessary, though not sufficient, for the development of MPF. The reason for this conjecture may be that immature oocytes in the ovary often exhibit high pH_i values and that the positive effect of cytoplasmic alkalinization can be suppressed by protein synthesis inhibition (Cicirelli *et al.,* 1983; Wasserman *et al.,* 1984).

4. cAMP

Since the discovery that oocyte maturation can be inhibited in the mouse by dbcAMP, an undegradable analog of cAMP (Cho *et al.,* 1974), and in *X. laevis* by a phosphodiesterase inhibitor, theophyllin (O'Connor and Smith, 1976), many studies have been carried out to investigate roles of cAMP and protein phosphorylation in oocyte maturation (Morrill *et al.*, 1981; Baulieu and Schorderet-Slatkine, 1983; Maller, 1983; see this volume for review). Results of these studies may be summarized as follows: To produce MPF, the cAMP levels in oocytes must be lowered to reduce the activity of cAMP-dependent protein kinase, because this enzyme exerts an inhibitory effect on oocyte maturation by phosphorylating a protein inhibitor that is active when phosphorylated. Therefore, suppression of cAMP-dependent protein kinase activity and dephosphorylation of the phosphoprotein inhibitor are necessary and sufficient for the oocyte to initiate maturation. However, when MPF first appears in the oocyte, cAMP-independent protein phosphorylation takes place. Therefore, in the oocyte, a shift in the pattern of protein phosphorylation from cAMP-dependence to cAMP-independence is essential for MPF development. More information concerning this inhibitory effect in amphibian and rodent oocytes can be found in Masui (1985).

5. *Protein Synthesis*

Progesterone-induced maturation of amphibian oocytes is always inhibited by protein synthesis inhibitors, as found by Smith *et al.* (1966), Dettlaff (1966), Brachet (1967), and Schuetz (1967a,b). Wasserman and Masui (1975b) and Drury and Schorderet-Slatkine (1975) found that although cycloheximide-treated *Xenopus* oocytes could not develop MPF following progesterone treatment, these oocytes were able to undergo GVBD in response to injected MPF, suggesting that MPF action does not require protein synthesis to cause GVBD. These authors also investigated whether or not autocatalytic amplification of MPF requires protein synthesis. In these experiments, cytoplasm was transferred from progesterone-treated oocytes into cycloheximide-treated oocytes, after which these recipients were used as donors. Wasserman and Masui (1975b) found that the cytoplasm taken from cycloheximide-treated oocytes 7 hr after receiving an MPF injection had the same effect as that taken from progesterone-treated oocytes undergoing GVBD. Since the volume of cytoplasm transferred was only 4% of the entire oocyte volume, this result implies that MPF can be amplified in the cycloheximide-treated oocyte. On the other hand, Drury and Schorderet-Slatkine (1975) observed that the cytoplasm taken from cycloheximide-treated oocytes 2 hr after receiving an MPF injection had a marginal effect when injected into cycloheximide-treated oocytes, which in turn developed little MPF during the next 2 hr. This result indicates that MPF amplification requires protein synthesis. Masui and Clarke (1979) ascribed the discrepancy between these results to the difference in the intervals between cytoplasmic transfers adopted by the two groups. Using *Xenopus,* Dettlaff *et al.* (1977) reexamined the problem and found that MPF activity never diminished during five successive cytoplasmic transfers through cycloheximide-treated oocytes. Recent studies by Gerhart *et al.* (1984) have shown that a 15-fold amplification of MPF occurs in both cycloheximide-treated oocytes and control oocytes within 90 min of injection. All in all, these results strongly suggest that in *Xenopus* oocytes neither MPF amplification nor the subsequent processes of maturation requires protein synthesis. Therefore, protein synthesis inhibition cannot prevent an oocyte from maturing once the MPF level in its cytoplasm reaches the threshold. However, if an oocyte's protein synthesis is inhibited before reaching this point, the oocyte cannot produce MPF. Therefore, it appears that under the influence of maturation-inducing agents, some protein that initiates maturation must be synthesized. Apparently, the synthesis of this initiator protein mediates the connection between the action of all maturation-inducing agents, from progesterone to cAMP-dependent protein kinase inhibitor, and the first appearance of MPF in the oocyte cytoplasm (see Maller, 1983, for review).

However, situations vary in different groups of animals. Dettlaff *et al.* (1977) and Schuetz and Samson (1979a) found that in the sturgeon (*A. stellata*) and the frog (*R. pipiens*), respectively, MPF injection, as well as progesterone treatment, failed to induce GVBD in cycloheximide-treated oocytes. It appears that the oocytes of these animals lack the protein factors required for MPF amplification, possibly MPF precursor. On the other hand, Stern *et al.* (1972) and Houk and Epel (1974) showed that in mouse and starfish oocytes, respectively, protein synthesis inhibition did not prevent GVBD, although meiosis was arrested near metaphase I, suggesting that MPF development is independent of protein synthesis. Evidently, in the oocytes of these animals, MPF precursor as well as the initiator protein have already been provided.

E. Biochemical Effects of MPF

1. Changes in Biochemical Parameters

As seen in the preceding section, MPF development and GVBD in amphibian oocytes are inhibited by conditions that prevent changes in some biochemical factors during early phases of maturation following progesterone stimulation. The biochemical factors involved are Ca ion level, pH_i, cAMP-dependent and independent protein phosphorylation, and protein synthesis activities. Apparently, changes in these factors are a prerequisite for MPF to appear in the oocyte. However, it is important to note that little change in these factors or in the oocyte's dependency on these factors can be observed until the first appearance of MPF. The time at which the maturation process becomes independent of protein synthesis activities coincides exactly with the appearance of MPF (Wasserman and Masui, 1975b). Coincidently, at this time the maturation process also becomes Ca^{2+}-independent (Masui *et al.*, 1977). In addition, the maturation process remains susceptible to an elevated level of cAMP until MPF appears in the oocyte (Schorderet-Slatkine *et al.*, 1978, 1982) and, simultaneously, cAMP-independent protein kinase activity rapidly increases (Maller *et al.*, 1977; Maller and Krebs, 1977). Lee and Steinhardt (1981) also showed that a rise in pH_i occurs at almost the same time as MPF appears, i.e., 0.6 to 0.7 $GVBD_{50}$.

2. Biochemical Changes Induced by MPF

As previously discussed, the changes in the biochemical factors and the appearance of MPF are so closely related temporally that it seems rather difficult to discuss the causal relationship between these two events. It is true that the former is a prerequisite for the latter and that some artificially produced changes, such as raising pH_i and inhibiting of cAMP-dependent protein kinase activity, cause the development of MPF, as mentioned previously. However, it is also true that an injection of MPF into oocytes not only causes changes in these biochemical factors that lead to MPF amplification, but it can induce maturation by overcoming the inhibitory conditions that have prevented these biochemical changes.

The effect of MPF in changing the biochemical factors necessary for the development of MPF was first described by Maller *et al.* (1977). Using *Xenopus* oocytes, these authors demonstrated that MPF injection almost immediately induced a burst of cAMP-independent protein phosphorylation. Further, Maller and Krebs (1977) found that MPF injection could overcome the inhibitory effect of previously injected catalytic subunits of cAMP-dependent protein kinase. Recently, Schorderet-Slatkine *et al.* (1982) reported that MPF injection also causes a decrease in cAMP level in the oocyte. A similar line of evidence was provided by Clarke and Masui (1985) using mouse oocytes. These authors found that the inhibition of GVBD in the presence of dbcAMP could be overcome by introducing cytoplasm from metaphase cells by cell fusion. MPF also stimulates protein synthesis activity (Wasserman *et al.*, 1982) and elevates pH_i to levels required for the development of MPF (Wasserman *et al.*, 1984). Clearly, MPF has a remarkable ability to change some oocyte biochemical factors leading to the recruitment of MPF. Apparently, it is this characteristic positive feedback that enables MPF to amplify itself.

F. Oscillation of MPF Activity

1. MPF Cycles in Mitosis and Meiosis

Gurdon (1968) observed that brain nuclei injected into maturing *X. laevis* oocytes were induced to undergo nuclear envelope breakdown and chromosome condensation to a metaphase state. This suggested that MPF had a similar effect on both somatic and oocyte nuclei. On the other hand, Masui and Markert (1971) found that in *R. pipiens* "cytoplasm from early embryos retains some capacity to induce oocyte maturation." Using both *R. pipiens* and *X. laevis,* Wasserman and Smith (1978) assayed MPF activity in the cytoplasm of early embryos at various times after fertilization and discovered that MPF completely disappeared shortly after fertilization, reappeared before mitosis, and disappeared again after mitosis, exhibiting cyclical behavior, and that following activation enucleated eggs also underwent the same MPF cycles as normal embryos. In the mouse, Bałakier (1978) demonstrated that the cytoplasm of colchicine-arrested blastomeres also had MPF activity. When these blastomeres were fused to small immature oocytes using Sendai virus, GVBD occurred.

Recently, Gerhart *et al.* (1984) confirmed the cyclical behavior of MPF in *Xenopus* embryos. However, these authors found that microtubule poisons did not affect MPF cycles, whereas Wasserman and Smith (1978) showed that colchicine treatment could hold MPF activity at a high level in *R. pipiens* embryos. The discrepancies between these results in amphibians may possibly reflect a species-specific difference in the susceptibility of embryonic cells to microtubule poisons, similar to that found by Yoneda and Schroeder (1984) in sea urchin embryos. These authors found that colchicine arrested cell divisions of zygotes in both *Dendrogaster excentricus* and *Strongylocentrotus purpuratus,* but it could arrest the nuclear cycle at metaphase only in the latter species. Therefore, a cytological examination of the chromosome cycles in blastomeres arrested by microtubule poisons in both amphibian species would help in reconciling the discrepancy.

MPF also cycles during meiosis. Dorée *et al.* (1983) found in starfish and Gerhart *et al.* (1984) in *X. laevis* that during both the first and the second meiosis, MPF activity rose shortly before metaphase, but it fell when the polar body was given off. These authors also found that the reappearance of MPF at metaphase II, after the first polar body expulsion, was completely suppressed by protein synthesis inhibitors.

2. *Autonomy of MPF Cycles*

As shown by Wasserman and Smith (1978), MPF cycles are entirely cytoplasmic activities independent of chromosome cycles. Therefore, it is conceivable that the cyclical behavior of MPF may be governed by the oscillatory activities of some cytoplasmic biochemical factors that change in a certain direction required for the development of MPF. As discussed in the preceding section, these biochemical factors may include the Ca ion level, pH_i, cAMP-dependent and independent protein phosphorylation, and protein synthesis activity. However, in view of the fact that MPF changes these biochemical factors in a way favorable for its amplification, as mentioned previously, MPF may be a self-oscillating system in the cell, i.e., the control of MPF cycles is autonomous.

This unique situation may be caused by the autocatalytic properties of MPF that are part of the MPF-generating system in the cell. The existence of such an autonomous oscillation of MPF activity is suggested by the recent observation of MPF cycles *in vitro* by Masui (1982). This was a rather fortuitous finding made in the course of testing the effects of EGTA on the stability of MPF during storage. Dejellied *R. pipiens* eggs were crushed by centrifugation in extraction medium at pH 6.5, and containing 5 m*M* EGTA. Clear extracts were prepared by further centrifugation at 150,000 *g* for 2 hr. The extracts were then kept at 0°C for 2 to 3 weeks. At 24-hr intervals during the storage period, the extracts were assayed for MPF activity. It was found that "MPF activity disappeared from the extract on day 3 or 4, but reappeared on the following days and persisted for 2 or 3 days at high levels before disappearing again. The cycle of MPF activity in extracts was repeated fairly regularly a few times during the storage of the extracts. The average period of the oscillation was 5.05 ± 1.25 days." It is unlikely that MPF cycles *in vitro* depend on oscillation of other cellular activities such as protein synthesis and respiratory functions since the extracts were completely devoid of both ribosomes and mitochondria. To explain the cyclical behavior of MPF activity, Masui (1982) adopted a chemical oscillator model similar to that proposed by Prigogine and Lefever (1968) for the Belousov-Zhabotinsky reaction (see Glansdorff and Prigogine, 1971, for review). In this model, it is assumed that

1. Inactive MPF (*Y*) is supplied at a certain rate from a large pool of cellular components (*A*):

$$A \rightarrow Y$$

2. This inactive MPF precursor (Y) is then activated by reaction with two active MPF molecules $(2X)$, as explained previously in relation to the concentration dependency of MPF (see Section III,C,3).

$$2X + Y \rightarrow 3X$$

3. However, active MPF (X) is reversibly inactivated to return to its precursor from (Y) at a certain rate:

$$X \rightarrow Y$$

or

4. irreversibly decayed to become a cellular component (B) at a certain rate:

$$Y \rightarrow B$$

Now, assuming that a relevant rate constant can be found for each of the preceding reactions and that all the reactions follow the law of mass action, a system of differential equations may be constructed which describes rates of changes in the concentrations of active and inactive MPF, X and Y. The mathematical inferences drawn from this system may be first, that MPF activity is stabilized at a high level if the rate constants for the recruitment and activation of MPF precursor (Y) are sufficiently large in relation to those for the decay of active MPF (X). Second, if the rate constants for MPF decay are increased to a certain limit, the system becomes unstable and a small perturbation may cause the so-called limit cycle, resulting in periodic oscillation of MPF level in the system. In the first situation, the persistence of MPF at a high level may lead to arrest of the chromosome cycle at metaphase; and in the second situation, the chromosome cycle may begin, alternating chromosome condensation of a metaphase state with chromosome decondensation to an interphase state.

IV. CYTOSTATIC FACTOR (CSF)

A. Metaphase Arrest and Activation

1. Egg Activation

Oocytes arrested at metaphase I or II resume meiosis following activation either normally by fertilization or artificially by physical and chemical agents. In amphibians, oocytes arrested at metaphase II are easily activated by both electrical shock and mechanical injury such as pricking with a glass needle. *Xenopus* eggs complete meiosis, expelling the second polar body within 10 min of activation, and *R. pipiens* eggs within 20 min of activation. Gerhart *et al.* (1984) observed such a rapid loss of MPF activity in the cytoplasm of *Xenopus* eggs that in less than 5 min after activation, activity was not detectable. This loss of MPF activity leads to decondensation of chromosomes with formation of a pronucleus. Further, activated egg cytoplasm develops the capability to induce enlargement of nuclei transplanted from other cells and promote DNA synthesis, as first noted by Gurdon (1967). Activation of egg cytoplasm not only releases the chromosome cycle from metaphase arrest, but also brings about cytoplasmic events such as cortical granule breakdown, surface contraction, and the formation of cleavage furrows.

2. Factors Responsible for Metaphase Arrest

The mechanism responsible for the arrest of meiosis at metaphase has been investigated since the early work by Bataillon and Tchou-Su (1930) (see Masui *et al.*, 1980, for review). Brachet (1951), summarizing classical theories of fertilization, states that "in a number of cases, the unfertilized egg appears as a poisoned cell and fertilization brings it out of its inertia by a purifying reaction." In the case of amphibian eggs, Bataillon and Tchou-Su (1930) hypothesized that the egg is anesthetized by CO_2 during passage through the female genital tract. Although this was confirmed by Brachet in 1951, he also remarked that "if the elimination of toxic substance is a necessary condition of fertilization, it is not sufficient to ensure development. It is evident that one does not achieve parthenogenesis in the frog egg simply freeing the eggs of their CO_2." Heilbrunn (1952) proposed that an antimitotic substance similar to heparin may exist in unfertilized eggs and its removal by fertilization causes development. Similarly, Monroy and Tyler (1967) hypothesized that the appearance of inhibitory factors in unfertilized eggs was a result of oocyte maturation. These factors may possibly include "the formation of specific enzyme inhibitors, allosteric effectors, the upsetting of the equilibrium of key metabolic reactions either as a result of the inhibition of an enzymatic intermediate, and the alteration of the steric configuration of some proteins, either enzymatic or nonenzymatic."

More recently, Chulitskaia (1970) examined effects of unfertilized egg cytoplasm on embryonic development of the sturgeon, *Acipenser güldenstädti*, and the frog, *R. temporaria*, by injecting the cytoplasm "into the animal region of developing embryos of the same female, in the region between furrows at the stage of the 4th cleavage." Although, she did not find any significant alteration in the recipient's cell division, a delay in the timing of cleavage desynchronization was observed. On the other hand, while investigating the effects of MPF on the mitotic process of *R. pipiens* embryos, Masui and Markert (1971) found, unexpectedly, that cytoplasm from maturing oocytes, when injected into one blastomere of a two-cell zygote, arrests cleavage as well as mitosis at metaphase, whereas injection of cytoplasm from cleaving blastomeres had no effect on development. Thus, the arrest of mitosis and cleavage can be attributed to a specific CSF in the cytoplasm of the maturing oocyte. However, Masui *et al.* (1980) and later Masui (1985) proposed that any inhibitory factor assumed to be responsible for the developmental arrest of unfertilized eggs, must meet the following four criteria, regardless of its chemical nature, in order to be recognized as authentic: (1) The inhibitory factor must be absent in fertilized eggs; (2) it must be inactivated under the conditions that cause activation of egg cytoplasm; (3) the zygote treated with this factor must be arrested, showing the same characteristics as the unfertilized egg; (4) the inhibition caused by the factor must be reversible. Therefore, we will examine CSF with respect to these criteria in the following sections.

B. Effects of Egg Cytoplasm on Blastomeres

1. Perfect and Imperfect Cleavage Arrests

As mentioned previously, when one blastomere of a two-cell *R. pipiens* embryo is injected with cytoplasm taken from an oocyte matured *in vitro* by progesterone, the recipient blastomere often stops cleaving for at least 24 hr without any sign of deterioration; the uninjected blastomere, however, continues to cleave to form a hemiblastula (see Fig. 6, Masui and Markert, 1971). This type of cleavage arrest only occurs when the cytoplasm is taken from donor oocytes that have competed GVBD. The cytoplasm taken from oocytes before GVBD or after activation has little inhibitory effect on the recipient blastomeres.

However, Meyerhof and Masui (1979a) noted that eggs squeezed out of the ovisac (40–50 hr after ovulation in *R. pipiens* and 9–19 hr after ovulation in *X. laevis*) were easier to activate by pricking with a glass needle, and their cytoplasm showed less cleavage-inhibiting activity than oocytes matured *in vitro* for the same length of time. In other words, the eggs that had matured in the ovisac appeared to be less resistant to activation stimuli than those matured *in vitro*, their CSF activity being easily lost. In order to prevent a loss of CSF activity, Meyerhof and Masui (1979a) injected a small dose of EGTA into donor eggs prior to withdrawal of cytoplasm. This treatment was found to be effective not only in preserving CSF activity but also in preventing donor eggs from activation, while the same treatment, when applied to donor two-cell blastomeres, did not alter their cytoplasmic activity. However, Ryabova (1983), who recently carried out a similar experiment using *R. temporaria, X. laevis,* and *A. stellata,* obtained different results. She reported that in these animals CSF activity could be detected in the egg cytoplasm only if donor eggs were given a prior EGTA injection. This activity increased in proportion to the EGTA doses, and even the cytoplasm of blastomeres gave rise to CSF activity if they were preinjected with EGTA. Therefore, the author concluded that "the cytoplasm of mature non-activated eggs of *R. temporaria* and *A. stellata,* unlike that of *R. pipiens,* exerts no cytostatic effect on the nuclei of the cleaving embryo, but acquires such a capacity after being treated with EGTA."

However, an essential difference exists between the features of arrested blastomeres described in these two papers. In the description by Masui and Markert (1971) and Meyerhof and Masui (1979a), CSF-arrested blastomeres did not show any sign of deteriorating changes in surface structure, which remained completely smooth and glossy. In these blastomeres there was neither irregular accumulation of the pigment in the cortex nor in the endoplasm, and only a single mitotic apparatus arrested at metaphase was found. This is clearly seen in Fig. 3 in the paper by Meyerhof and Masui (1979a). On the other hand, the comparable case of an arrested blastomere presented by Ryabova (1983) in Fig. 1 of her paper shows a surface contraction as well as irregular pigment accumulation in the cortex and subcortical cytoplasm. This blastomere also contains several interphase nuclei. Apparently, some of the blastomeres she described as "arrested" were, in fact, those that failed to continue normal cleavage, but underwent abortive cleavage processes, including surface contraction and nuclear multiplication.

Actually, similar blastomeres were also found in our earlier experiments, but they were discarded when the results were scored because these blastomeres, unlike those perfectly arrested by CSF, do not immediately cease cortical activity following cytoplasmic injection. Instead, although they repeatedly undergo abortive surface contractions similar to those seen in parthenogenetically activated eggs, they fail to form true cleavage furrows. Eventually these blastomeres show irregular distribution of pigment, a sign of degeneration, and are often multinucleate, as described previously. Therefore, it may be considered that in these blastomeres, cleavage activities were not completely suppressed. It is important to note that this imperfect inhibition of cleavage is not a specific effect of CSF, since it could occur in blastomeres injected with either unfertilized egg cytoplasm or fertilized egg cytoplasm, whereas perfectly arrested cleavage was observed only in the blastomeres injected with the cytoplasm of unfertilized eggs. A similar cleavage arrest without arrest of nuclear cycling was found by Baker and Warner (1972) after injecting one blastomere of a two-cell *Xenopus* embryo with Ca^{2+}-EGTA buffers. These authors were able to prevent further cleavage in the recipient blastomere. However, cytological examination revealed that these blastomeres contained numerous nuclei, in contrast to the singular metaphase spindle found in CSF-arrested blastomeres.

2. Procedures for CSF Assay

The frequencies with which the nonspecific imperfect cleavage inhibition occur are variable, depending on egg quality and the conditions of cytoplasmic transfer. We have succeeded in eliminating this nonspecific inhibition by using the following procedures: For experiments with *R. pipiens*, frogs are usually injected with a primary dose of 1/6 to 1/8 pituitary each, and kept at 18°C for 24 hr before receiving injections of 1 to 2 pituitaries and 1 mg progesterone each to induce ovulation. For both *R. pipiens* and *X. laevis*, frogs with eggs in the ovisac, as well as eggs to be used as cytoplasmic donor or recipients, should not be exposed to temperatures below 10°C. During cytoplasmic transfer, both donors and recipients are kept in a medium that prevents donor eggs from activation due to penetration by the micropipette. For *R. pipiens*, 25–50 mM NaH_2PO_4 has been used, and for *X. laevis*, Ringer's solution gassed with CO_2. Finally, it is important to note that the bore size of the micropipette tip used for cytoplasmic transfer has a significant effect on the result. Apparently, the shearing force that the cytoplasm experiences during passage through a pipette tip with a narrow bore has the same effect as homogenization, resulting in a loss of cytoplasmic activities and increasing the changes for imperfect arrest of the recipient blastomeres. To avoid this undesirable effect, micropipettes with tip sizes suitable for nuclear transplantation have been used.

As shown by Masui and Markert (1971), in *R. pipiens*, the volume of the cytoplasm injection into each two-cell blastomere altered not only the frequency but also the timing of cleavage arrest. The larger the volume of injected cytoplasm, the more frequent the cleavage arrest. Conversely, if the volume of injected cytoplasm is reduced, cleavage arrest becomes less frequent and is delayed, resulting in smaller arrested blastomeres. Thus, CSF activity has been assayed by referring to both the percentages of blastomere arrested by injection of a certain volume of cytoplasm (usually

30 nl for *Xenopus* and 60 nl for *R. pipiens*) and the size of arrested blastomeres.

C. Changes in CSF Activity

1. CSF Activity during Maturation and Fertilization

Cytostatic factor activity was assayed in the cytoplasm of oocytes maturing *in vitro* at various times after progesterone stimulation and after activation by pricking with a glass needle. As shown by Masui and Markert (1971) with *R. pipiens,* and Meyerhof and Masui (1979a) with *X. laevis,* the CSF activity first appears shortly after GVBD, and increases as meiosis progresses, reaching its highest level when oocytes become able to be activated. During activation, CSF activity in the egg cytoplasm quickly disappears, however.

To determine if the appearance of CSF was a consequence of GVBD, its activity was assayed in enucleated oocytes of *R. pipiens* induced to mature *in vitro.* Masui and Markert (1971) found that CSF could develop in the cytoplasm of these oocytes, and, in addition, Meyerhof and Masui (1979a) did not find any CSF activity before GVBD in the nucleoplasm of progesterone-treated *Xenopus* oocytes. Also, in both *R. pipiens* and *X. laevis,* CSF activity disappeared from the cytoplasm of mature oocytes lacking GV material after activation by pricking with a needle. Clearly, both the development of CSF as well as its inactivation are processes that do not require contributions from the GV, indicating the nonnuclear origin of CSF. To determine which cellular components possess CSF activity, Masui (1974) stratified the cytoplasm of unfertilized eggs of *R. pipiens* into five zones, as previously described (Section III,B,1) and found that activity was localized in the two hyaline zones. However, if activated eggs were stratified, CSF activity was not detectable in any of the zones. Evidently, CSF activity is purely cytoplasmic, being associated with cytosolic components.

2. Changes in CSF Activity in Vitro

To further investigate the process of CSF inactivation by fertilization, CSF has been extracted from unfertilized eggs of *R. pipiens* in the same way as MPF. Meyerhof and Masui (1977) found that CSF could remain active in cytosols extracted from eggs in medium containing EGTA and Mg ions, whereas, activity was lost rapidly in the presence of Ca ions or EDTA. These results strongly suggest that CSF, like MPF, is Ca^{2+}-sensitive and Mg^{2+}-dependent. The observation, mentioned earlier, that an injection of EGTA into eggs stabilizes CSF activity in the cytoplasm withdrawn from these eggs also indicates the high sensitivity of CSF to Ca ions. Thus, the quick disappearance of CSF activity from the cytoplasm of activated eggs can be explained from the point of view that egg activation causes a surge of free Ca ions released from intracellular compartments. This has recently been substantiated in amphibian eggs by Busa and Nuccitelli (1985). Their measurements with a Ca^{2+}-sensitive electrode indicated an increase in the intracellular free Ca ion level from 0.4 to 1.2 μM following activation of *Xenopus* eggs. The development of the ability to neutralize CSF in activated eggs was demonstrated in *R. pipiens* by Meyerhof and Masui (1977). They observed that if zygotes were injected with unfertilized egg cytoplasm or its extract within 45 min of insemination they continued to cleave without being arrested, whereas those injected later than 60 min were arrested at metaphase of the first mitosis. This suggests that in *R. pipiens* eggs the ability to neutralize both endogenous and exogenous CSF develops and remains during, at least, the first 30 min of activation.

However, the notion of the Ca^{2+} sensitivity of CSF activity appears to contradict the earlier observations of Masui (1974). He found that while CSF activity of the extract from unfertilized eggs obtained using the Ca^{2+}-free extraction medium was diminished within 24 hr during cold storage, no decrease in CSF activity of the extract was observed if the extraction medium contained Ca ions. To reconcile this contradiction, Meyerhof and Masui (1977) and Masui *et al.* (1977) compared changes in CSF activities in Ca^{2+}-free and Ca^{2+}-containing egg extracts. It was found that Ca ions could evoke CSF activity when added to extracts after loss of the original CSF activity which persists only for a short while after extraction. The CSF activity appearing *after* Ca^{2+} addition in the extracts is highly stable and Ca^{2+}-resistant. Therefore, the original unstable and Ca^{2+}-sensitive CSF found only in fresh extracts made with Ca^{2+}-free medium has been designated "primary CSF," to distinguish it from the stable and Ca^{2+}-resistant CSF referred to as "secondary CSF." However, the effects of these two CSFs on cleaving blastomeres are, at least, indistinguishable in morphological aspects, the blastomeres being arrested in the same fashion at metaphase. The only difference between the two factors was found when they were injected into zygotes shortly after insemination, i.e., while the zygotes injected with primary CSF continued to cleave, those injected with secondary CSF were arrested, reflecting the difference in Ca^{2+}-sensitivity between these CSFs (see Masui *et al.*, 1980, 1984, for review).

3. Molecular Characterization of CSF

Mainly due to its instability, primary CSF has not been biochemically characterized. However, since it can be sedimented from *R. pipiens* eggs extracts by centrifugation at 150,000 *g* for 6 hr, apparently the factor may, itself, be a large molecule or associated with a large molecule. To date, primary CSF activity has not been recovered from chemically fractionated egg extracts. On the other hand, secondary CSF could be precipitated by $(NH_4)_2SO_4$ at concentrations between 20 and 30% of saturation, as reported by Masui (1974). Preliminary experiments by Masui *et al.* (1980) showed that it was eluted into the void volume fractions from a BioGel 15-*M* (Bio-Rad) column. This indicates that secondary CSF itself is a large molecule or is associated with a large molecule (over 15×10^6 daltons). This has been corroborated by the recent findings of Shibuya (unpublished) that secondary CSF activity is developed in the supernatant after the primary CSF had been sedimented. In addition, the fraction precipitated with $(NH_4)_2SO_4$ at concentrations between 60 and 80% of saturation developed secondary CSF within a few days of incubation in the cold after addition of Ca ions. Therefore, it is highly probable that secondary CSF is a product of polymerization of some smaller molecules in the cytosol. Masui (1974) previously reported that the CSF precipitated with $(NH_4)_2SO_4$ was sensitive to heat treatment at 55°C, but not sensitive to proteolytic enzymes. It is intriguing that this CSF appeared to be sensitive to RNAse treatment. However, reexamination of these results using more purified samples is necessary.

D. Properties of CSF-Arrested Blastomeres

1. Similarities to Unactivated Eggs

Shortly after CSF-injected blastomeres cease cleavage, their surface becomes completely smooth and glossy and develops numerous long microvilli resembling those on unactivated mature eggs, unlike uninjected cleaving control blastomeres that form few microvilli (Masui *et al.*, 1980). In the cytoplasm of CSF-arrested blastomeres, generally a single mitotic spindle with associated metaphase chromosomes is found, though occasionally two metaphase spindles are found side by side. According to Meyerhof (1978), the size of the CSF-arrested spindle is often larger and more densely fibrous than that found in control blastomeres undergoing mitosis. The chromosomes on the spindle are highly condensed and shorter than those found in actively cleaving cells. Asters at the poles of the spindle are absent. They were only found in blastomeres that had been fixed within 5 hr of CSF injection. These observations clearly indicate a transformation of the CSF-arrested metaphase spindle from a mitotic spindle with asters into one resembling the metaphase-arrested meiotic spindle lacking asters.

In *R. pipiens,* Meyerhof and Masui (1979b) and Shibuya and Masui (1982) found that CSF-arrested blastomeres developed the cytoplasmic ability to induce the formation of metaphase chromosomes from injected brain or sperm nuclei. More recently, Karsenti *et al.* (1984) examined effects of CSF on the cytoplasmic ability to form asters by injecting centrosomes with or without nuclei (both isolated from mammalian cells) into CSF-inhibited zygotes of *X. laevis.* They found that CSF-inhibited cell cytoplasm was much less capable of supporting centrosome-induced aster formation than that of the control, uninhibited zygotes. Instead these centrosomes had a strong tendency to form spindles in the presence of chromosomes.

Effects of CSF on MPF activity in zygotes and blastomeres have also been investigated. Using eggs ovulated from *R. pipiens,* which had been primed with a dose of pituitary hormone, Shibuya and Masui (1982) found MPF activity in the cytoplasm of CSF-arrested blastomeres. This result differs from that previously obtained by Masui *et al.* (1980) with eggs ovulated from unprimed frogs. The persistence of MPF activity in CSF-inhibited cells was also reported for *Xenopus* by Gerhart *et al.* (1984) and Newport and Kirschner (1984). These authors suggest that CSF could maintain high MPF activity in zygotes, thus blocking their cell cycles. All in all, the observations cited above strongly indicate that CSF can convert the cytoplasmic state of zygote cells into that similar to the unactivated egg arrested at metaphase II.

2. Reversibility of CSF Arrest

The functional equivalence between the CSF-arrested zygote and unactivated egg, as well as the reversibility of the inhibition caused by CSF, have been further examined in the following experiments. Shibuya and Masui (1982) injected demembranated sperm into CSF-arrested blastomeres of two-cell embryos of *R. pipiens.* When fewer than 100 sperm were injected into a blastomere, they were induced to form metaphase chromosomes immediately. However, as the number of the sperm injected was increased, the nuclei were more frequently decondensed to form pronuclei. If blastomeres were given more than 300 sperm each, these blastomeres underwent surface contraction to form incomplete cleavage furrows, and the sperm nuclei began to synthesize DNA and later underwent chromosome condensation to a metaphase state. At the same time, MPF activity in the blastomeres fell to a low level and then oscillated. More recently, Newport and Kirschner (1984) observed that in *Xenopus* zygotes arrested with CSF, injected nuclei were induced to condense chromosomes to a metaphase state. However, these authors also found that an injection of Ca ions into these blastomeres could release the chromosomes from metaphase arrest, allowing DNA synthesis.

E. Meiotic Arrest at Metaphase in the Oocytes of Nonamphibian Species

1. Role of Protein Synthesis

To date, no concrete evidence has been presented for the existence of a cytoplasmic factor that is responsible for meiotic arrest in the secondary oocytes of nonamphibian species. However, to conclude this chapter, possible mechanisms of meiotic arrest in these animals will be considered (see Masui, 1985, for more detailed discussion).

The possibility that CSF exists in mouse oocytes arrested at metaphase II was suggested by Masui *et al.* (1977, 1980). This conjecture was based on the fact that blastomeres of two-cell embryos could be arrested at metaphase when fused with unactivated metaphase II oocytes, as reported by Bałakier and Czolowska (1977). Moreover, Ca ions have a promoting effect but Mg ions have an inhibitory effect on the ionophore A23187-induced activation of mouse oocytes (Masui *et al.*, 1977).

On the other hand, because the eggs of some animals, including the mouse, are readily activated by an inhibition of protein synthesis, it may be speculated that the continuous synthesis of certain proteins by these eggs may be responsible for their meiotic arrest. As found in the brine shrimp, *Artemia salina,* by Fautrez and Fautrez-Firlefyn (1961), and in the marine polychaete, *Chaetoptorus,* by Zampetti-Bosseler *et al.* (1973), eggs arrested at metaphase I can be activated parthenogenetically by cycloheximide. In the mouse, it was found by Siracusa *et al.* (1978) and confirmed by Clarke and Masui (1983) that eggs arrested at metaphase II could be activated parthenogenetically by protein synthesis inhibitors. However, the latter authors found that oocyte chromosomes at metaphase I could undergo a transient decondensation to an interphase state after inhibition of protein synthesis. When inhibitors were removed, chromosomes returned to a metaphase I-like state. Although parthenogenetic activation by protein synthesis inhibitors has not been demonstrated in amphibian eggs arrested at metaphase II, Ziegler and Masui (1976) observed that a loss of chromosome condensation activity could be induced by protein synthesis inhibitors in *R. pipiens* oocytes at metaphase I. Schuetz and Samson (1979b) also reported that *R. pipiens* oocytes treated with cycloheximide after GVBD underwent changes, such as surface contraction, similar to those observed after activation. Gerhart *et al.* (1984) found that MPF activity could not reappear in *Xenopus* oocytes after protein synthesis inhibition before metaphase II. The results cited here suggest that chromosome condensation activity at metaphase I depends on the continuous synthesis of protein in oocytes of most animals, including amphibians. Thus, constant synthesis of a certain protein that causes chromosome condensation to a metaphase state may be responsible for the meiotic arrest at metaphase I. The same mechanism may also hold true for the meiotic arrest in mouse oocytes at metaphase II. However, in amphibian oocytes, CSF develops after completing meiosis I, thus stabilizing MPF. This situation renders chromosome condensation activity as well as the meiotic arrest at metaphase independent of protein synthesis. Thus, the inactivation of CSF is a prerequisite for the activation of these oocytes.

2. Role of Sperm Factor

Normally, activation of eggs by sperm penetration causes the resumption of arrested meiosis. However, in some cases, the sperm enters oocytes precociously without activating them, then disengages meiosis from the arrest at metaphase. In the marine polychaetes, *Pectinaria* and *Sabellaria,* Hylander *et al.* (1981) and Peaucellier *et al.* (1982) observed that oocytes inseminated at the GV stage could mature without being arrested at metaphase I, which would normally occur in unfertilized oocytes. Recently Clarke and Masui (1986) observed that mouse oocytes inseminated at metaphase I were not activated and became polyspermic with varying numbers of sperm nuclei incorporated in the cytoplasm. If the number of sperm nuclei was less than four per oocyte, then all the nuclei were induced to form metaphase chromosomes. However, if more than four nuclei were incorporated into a single oocyte, then all of them were decondensed to varying degrees. Thus, in this situation the nucleocytoplasmic relationship may be analogous to that observed when variable numbers of sperm are injected into CSF-arrested blastomeres of *R. pipiens* (see Section IV,D,2). In both cases, an excess of the sperm factor introduced into metaphase cells could deprive the cytoplasm of the ability to maintain chromosome condensation activity which is required for metaphase arrest. It is plausible that the sperm factor titrates chromosome condensation factor in the cytoplasm.

V. CONCLUDING REMARKS

In this chapter, we have discussed the area of oocyte maturation with particular attention to research in amphibians during the past two decades. It appears that the cytoplasm of maturing oocytes undergoes changes that are independent of any contribution or activity of the oocyte nucleus. The cytoplasm also develops the ability to induce the entire maturation process when transferred into immature oocytes. Moreover, when isolated somatic cell nuclei are injected into maturing oocytes, the cytoplasm induces changes in the chromosomal morphology, mimicking those reflected in the host nucleus. The cytoplasmic factors responsible for maturation and arrest of meiosis at metaphase prior to activation have been designated MPF and CSF, respectively. Furthermore, because these factors are also effective during the mitotic cell cycle, their activities may be of more general importance. Future study of these cytoplasmic factors is necessary to resolve the various events involved in both the meiotic and mitotic cell cycles.

ACKNOWLEDGMENTS

The authors thank Rossana Soo for her assistance in preparing the manuscript. The studies carried out in our laboratory and cited in this chapter were supported by a grant (A5855) from the Natural Sciences and Engineering Council (NSERC), Canada. One of the authors (E.K.S.) has been supported by an Ontario Graduate Scholarship, Predoctoral Fellowship from NSERC, Canada, and University of Toronto Open Fellowship.

REFERENCES

Ayalon, D., Tsafriri, A., Lindner, H. R., Cordova, T., and Harell, A. (1972). Serum gonadotrophin levels in pro-estrous rats in relation to the resumption of meiosis by the oocytes. *J. Reprod. Fertil.* **31,** 51–58.

Baker, P. F., and Warner, A. E. (1972). Intracellular calcium and cell cleavage in early embryos of *Xenopus laevis. J. Cell Biol.* **53,** 579–581.

Bałakier, H. (1978). Induction of maturation in small oocytes from sexually immature mice by fusion with meiotic and mitotic cells. *Exp. Cell Res.* **112,** 137–141.

Bałakier, H., and Czolowska, R. (1977). Cytoplasmic control of nuclear maturation in mouse oocytes. *Exp. Cell Res.* **110**, 466–469.

Bałakier, H., and Tarkowski, A. K. (1980). The role of germinal vesicle karyoplasm in the development of male pronucleus in the mouse. *Exp. Cell Res.* **128**, 79–85.

Bałakier, H., and Masui, Y. (1986). Chromosome condensation activity in the cytoplasm of anucleate and nucleate fragments of mouse oocytes. *Dev. Biol.* **113**, 155–159.

Bataillon, E., and Tchou-Su. (1930). Etude analytiques et expérimentales sur les rhythms cinétique dans l'ouef (*Hyla arborea, Paracentrotus lividus, Bombyx mori*). *Arch. Biol.* **40**, 439–553.

Baulieu, E. E., and Schorderet-Slatkine, S. (1983). Steroid and peptide control mechanisms in membrane of *Xenopus laevis* oocyte resuming meiotic division. *Ciba Found. Symp.* [N. S.] **98**, 137–158.

Baulieu, E. E., Godeau, F., and Schorderet-Slatkine, S. (1978). Steroid-induced meiotic division in *Xenopus laevis* oocytes: Surface and calcium. *Nature (London)* **257**, 593–598.

Brachet, J. (1951). "Chemical Embryology" (L. L. Barth, transl.) Wiley (Interscience), New York.

Brachet, J. (1967). Effects of actinomycin, puromycin, and cycloheximide upon the maturation of amphibian oocytes. *Exp. Cell Res.* **48**, 233–236.

Busa, W. B., and Nuccitelli, R. (1985). An elevated free cytosolic Ca^{2+} wave follows fertilization in eggs of the frog, *Xenopus laevis*. *J. Cell Biol.* **100**, 1325–1329.

Cho, W. K., Stern, S., and Biggers, J. D. (1974). Inhibitory effect of dibutyryl cAMP on mouse oocyte maturation *in vitro*. *J. Exp. Zool.* **187**, 383–386.

Chulitskaia, E. V. (1970). Desynchronization of cell divisions in the course of egg cleavage and an attempt at experimental shift of its onset. *J. Embryol. Exp. Morphol.* **23**, 359–374.

Cicirelli, M. F., Robinson, K. R., and Smith, L. D. (1983). Internal pH of *Xenopus laevis* oocytes: A study of the mechanism and role of pH changes during meiotic maturation. *Dev. Biol.* **100**, 133–146.

Clarke, H. J., and Masui, Y. (1983). The induction of reversible and irreversible chromosome decondensation by protein synthesis inhibition during meiotic maturation in mouse oocytes. *Dev. Biol.* **97**, 291–301.

Clarke, H. J., and Masui, Y. (1985). Inhibition by dibutyryl cyclic AMP of the transition to metaphase of mouse oocyte nuclei and its reversal by cell fusion to metaphase oocytes. *Dev. Biol.* **108**, 32–37.

Clarke, H. J., and Masui, Y. (1986). Transformation of mouse sperm nuclei to metaphase chromosomes in the cytoplasm of maturing oocytes. *J. Cell Biol.* **102**, 1039–1046.

Cloud, J. G., and Schuetz, A. W. (1977). Interaction of progesterone with all or isolated portions of the amphibian (*Rana pipiens*) oocyte surface. *Dev. Biol.* **60**, 359–370.

Delage, Y. (1901). Etudes expérimentales sur la maturation cytoplasmique et sur la parthenogénèse artificielle chez les Echinodermes. *Arch. Zool. Exp. Gen.* [3] **9**, 284–336.

Dettlaff, T. A. (1966). Action of actinomycin and puromycin upon frog oocyte maturation. *J. Embryol. Exp. Morphol.* **16**, 183–195.

Dettlaff, T. A., and Felgengauer, P. E. (1980). Maturation of the stellate sturgeon oocytes following the injection of cytoplasm of the mature stellate sturgeon eggs and the maturing *Xenopus laevis* oocytes and the effect of cycloheximide on this process. *Ontogenez* **11**, 229–233.

Dettlaff, T. A., and Skoblina, M. N. (1969). The role of the germinal vesicle in the process of oocyte maturation in Anura and Acipenseridae. *Ann. Embryol. Morphog., Suppl.* **1,** 133–151.

Dettlaff, T. A., Nikitina, L. A., and Stroeva, O. G. (1964). The role of the germinal vesicle in oocyte maturation in Anurans as revealed by removal and transplantation of nuclei. *J. Embryol. Exp. Morphol.* **12.** 851–873.

Dettlaff, T. A., Felengauer, P. E., and Chulitzkaya, E. V. (1977). Effect of cycloheximide on disintegration of the germinal vesicle membrane and changes of the cortical layer in the oocytes of *Xenopus laevis* and *Acipenser stellatus* during their maturation under the influence of active cytoplasm. *Ontogenez* **8,** 478–486.

Dorée, M., Peaucellier, G., and Picard, A. (1983). Activity of the maturation-promoting factor and the extent of protein phosphorylation oscillate simultaneously during meiotic maturation of starfish oocytes. *Dev. Biol.* **99,** 489–501.

Drury, K. C. (1978). Methods for the preparation of active maturation promoting factor (MPF) from *in vitro* matured oocytes of *Xenopus laevis*. *Differentiation* **10,** 181–186.

Drury, K. C., and Schorderet-Slatkine, S. (1975). Effects of cycloheximide on the "autocatalytic" nature of the maturation promoting factor (MPF) in oocytes of *Xenopus laevis*. *Cell (Cambridge, Mass.)* **4,** 268–274.

Eisen, A., and Reynolds, G. T. (1984). Calcium transients during early development in single starfish (*Asterias forbesi*) oocytes. *J. Cell Biol.* **99,** 1878–1882.

Elinson, R. P. (1977). Fertilization of immature frog eggs: Cleavage and development following subsequent activation. *J. Embryol. Exp. Morphol.* **37,** 187–201.

Fautrez, J., and Fautrez-Firlefyn, N. (1961). Activation expérimentale des oeufs d'une race amphigonique d'*Artemia salina*. *Arch. Biol.* **72,** 611–626.

Fortune, J. E., Concannon, P. W., and Hansel, W. (1975). Ovarian progesterone levels in *in vitro* oocyte maturation and ovulation in *Xenopus laevis*. *Biol. Reprod.* **13,** 561–567.

Fostier, A., Jalabert, B., and Terqui, M. (1973). Action predominante d'un dérive hydroxyle de la progesterone sur la maturation *in vitro* des ovocytes de la Truite arc-en-ciel, *Salmo gairdnerii*. *C. R. Hebd. Seances Acad. Sci.* **277,** 421–424.

Gerhart, J. C., Wu, M., and Kirschner, M. W. (1984). Cell cycle dynamics of an M-phase specific cytoplasmic factor in *Xenopus laevis* oocytes and eggs. *J. Cell Biol.* **98,** 1247–1255.

Glansdorff, P., and Prigogine, I. (1971). "Thermodynamic Theory of Structure, Stability and Fluctuations." Wiley (Interscience). New York.

Godeau, J. F., Schorderet-Slatkine, S., Hubert, P., and Baulieu, E. E. (1978). Induction of maturation in *Xenopus laevis* oocytes by a steroid linked to a polymer. *Proc. Natl. Acad. Sci. U.S.A.* **75,** 2353–2357.

Gurdon, J. B. (1967). On the origin and persistence of a cytoplasmic state inducing DNA synthesis in frog eggs. *Proc. Natl. Acad. Sci. U.S.A.* **58,** 545–552.

Gurdon, J. B. (1968). Changes in somatic cell nuclei inserted into growing and maturing amphibian oocytes. *J. Embryol. Exp. Morphol.* **20,** 401–414.

Heidemann, S. R., and Kirschner, M. W. (1975). Aster formation in eggs of *Xenopus laevis:* Induction by isolated basal bodies. *J. Cell Biol.* **67,** 105–117.

Heilbrunn, L. V. (1952). "An Outline of General Physiology," 3rd ed. Saunders, Philadelphia, Pennsylvania.

Heilbrunn, L. V., Daugherty, K., and Wilbur, K. M. (1939). Initiation of maturation in the frog egg. *Physiol. Zool* **12,** 97–100.

Hermann, J., Bellé, R., Tso, J., and Ozon, R. (1983). Stabilization of the maturation-promoting factor (MPF) from *Xenopus laevis* oocytes. Protection against calcium ions. *Cell Differ.* **13,** 143–148.

Hiramoto, Y. (1962). Microinjection of live spermatozoa into sea urchin eggs. *Exp. Cell Res.* **27**, 416–426.

Houk, M. S., and Epel, D. (1974). Protein synthesis during hormonally induced meiotic maturation and fertilization in starfish oocytes. *Dev. Biol.* **40**, 298-310.

Houle, J. G., and Wasserman, W. J. (1983). Intracellular pH plays a role in regulating protein synthesis in *Xenopus* oocytes. *Dev. Biol.* **97**, 302–312.

Hylander, B. L., Anstrom, J., and Summers, R. G. (1981). Premature sperm incorporation into the primary oocyte of the polychaete *Pectinaria:* Male pronuclear formation and oocyte maturation. *Dev. Biol.* **82**, 382–387.

Ishikawa, K., Hanaoka, Y., Kondo, Y., and Imai, K. (1977). Primary action of steroid hormone at the surface of amphibian oocyte in the induction of germinal vesicle breakdown. *Mol. Cell. Endocrinol.* **9**, 91–100.

Kanatani, H., and Hiramoto, Y. (1970). Site of action of 1-methyladenine in inducing oocyte maturation in starfish. *Exp. Cell Res.* **61**, 280–284.

Kanatani, H., and Shirai, H. (1967). *In vitro* production of meiosis inducing substance by nerve extract in ovary of starfish. *Nature (London)* **216**, 284–286.

Kanatani, H., Shirai, H., Nakanishi, K., and Kurokawa, T. (1969). Isolation and identification of meiosis-inducing substances in starfish, *Asterias amurensis. Nature (London)* **221**, 273–274.

Karsenti, E., Newport, J., Hubble, R., and Kirschner, M. W. (1984). Interconversion of metaphase and interphase microtubule arrays, as studied by the injection of centrosomes and nuclei into *Xenopus* eggs. *J. Cell Biol.* **98**, 1730–1745.

Katagiri, C., and Moriya, M. (1976). Spermatozoa response to the toad egg matured after removal of the germinal vesicle. *Dev. Biol.* **81**, 177–181.

Kishimoto, K., and Kanatani, H. (1976). Cytoplasmic factor responsible for germinal vesicle breakdown and meiotic maturation in starfish oocytes. *Nature (London)* **221**, 273–274.

Kishimoto, K. and Kanatani, H. (1977). Lack of species specificity of starfish maturation-promoting factor. *Gen. Comp. Endocrinol.* **33**, 41–44.

Kishimoto, K., Hirai, S., and Kanatani, H. (1981). Role of germinal vesicle material in producing maturation-promoting factor in starfish oocytes. *Dev. Biol.* **81**, 177–181.

Kishimoto, T., Kuriyama, R., Kondo, H. K., Shirai, H., and Kanatani, H. (1982). Generality of the action of various maturation promoting factors. *Exp. Cell Res.* **137**, 121–126.

Kishimoto, T., Yamazaki, Y., Kato, S., Koide, S., and Kanatani, H. (1984). Induction of starfish oocyte maturation by maturation-promoting factor of mouse and surf clam oocytes. *J. Exp. Zool.* **231**, 293–295.

Lee, S. C., and Steinhardt, R. A. (1981). pH changes associated with meiotic maturation of oocytes in *Xenopus laevis. Dev. Biol.* **85**, 358–369.

Lohka, M. J., and Masui, Y. (1983). The germinal vesicle material required for sperm pronuclear formation is located in the soluble fraction of egg cytoplasm. *Exp. Cell Res.* **148**, 481–491.

Maller, J. L. (1983). Interaction of steroids with the cyclic nucleotide synthesis in Amphibian oocytes. *Adv. Cyclic Nucleotide Res.* **15**, 295–336.

Maller, J. L., and Krebs, E. G. (1977). Progesterone-stimulated meiotic cell division in *Xenopus* oocytes. *J. Biol. Chem.* **252**, 1712–1718.

Maller, J. L., Wu, M., and Gerhart, J. C. (1977). Changes in protein phosphorylation accompanying maturation of *Xenopus laevis* oocytes. *Dev. Biol.* **58**, 298–312.

Masui, Y. (1967). Relative roles of the pituitary, follicle cells and progesterone in the induction of oocyte maturation in *Rana pipiens. J. Exp. Zool.* **166**, 365–376.

Masui, Y. (1972). Distribution of the cytoplasmic activity inducing germinal vesicle breakdown in frog oocytes. *J. Exp. Zool.* **179**, 365–378.

Masui, Y. (1973a). Effects of ionizing radiation on meiotic maturation of frog oocytes. I. *In vivo* studies. *J. Embryol. Exp. Morphol.* **29**, 87–104.

Masui, Y. (1973b). Effects of ionizing radiation on meiotic maturation of frog oocytes. II. *In vitro* studies. *J. Embryol. Exp. Morphol.* **29**, 105–116.

Masui, Y. (1974). A cytostatic factor in amphibian oocytes: Its extraction and partial characterization. *J. Exp. Zool.* **187**, 141–147.

Masui, Y. (1982). Oscillatory activity of maturation promoting factor (MPF) in extracts of *Rana pipiens* eggs. *J. Exp. Zool.* **224**, 389–399.

Masui, Y. (1985). Meiotic arrest in animal oocytes. *In* "Biology of Fertilization" (C. B. Metz and A. Monroy, eds.), Vol. 1, pp. 189–219. Academic Press, New York.

Masui, Y., and Clarke, H. J. (1979). Oocyte maturation. *Int. Rev. Cytol.* **57**, 185–282.

Masui, Y., and Markert, C. L. (1971). Cytoplasmic control of nuclear behavior during meiotic maturation of frog oocytes. *J. Exp. Zool.* **177**, 129–145.

Masui, Y., Meyerhof, P. G., Miller, M. A., and Wasserman, W. J. (1977). Roles of divalent cations in maturation and activation in vertebrate oocytes. *Differentiation* **9**, 49–57.

Masui, Y., Meyerhof, P. G., and Miller, M. A. (1980). Cytostatic factor and chromosome behavior in early development. *Symp. Soc. Dev. Biol.* **38**, 235–256.

Masui, Y., Lohka, M. J., and Shibuya, E. K. (1984). Roles of Ca ions and ooplasmic factors in the resumption of metaphase-arrested meiosis in *Rana pipiens* oocytes. *Symp. Soc. Exp. Biol.* **38**, 45–66.

Merriam, R. W. (1972). On the mechanism of action in gonadotropin stimulation of oocyte maturation in *Xenopus laevis*. *J. Exp. Zool.* **180**, 421–426.

Meyerhof, P. G. (1978). Studies on cytoplasmic factors from amphibian eggs which cause metaphase and cleavage arrest. Doctoral Thesis, University of Toronto.

Meyerhof, P. G., and Masui, Y. (1977). Ca and Mg control of cytoplasmic factors from *Rana pipiens* oocytes which cause metaphase and cleavage arrest. *Dev. Biol.* **61**, 214–229.

Meyerhof, P. G., and Masui, Y. (1979a). Properties of a cytostatic factor from *Xenopus laevis* eggs. *Dev. Biol.* **72**, 182–187.

Meyerhof, P. G., and Masui, Y. (1979b). Chromosome condensation activity in *Rana pipiens* eggs matured *in vivo* and in blastomeres arrested by cytostatic factor (CSF). *Exp. Cell Res.* **123**, 345–353.

Monroy, A., and Tyler, A. (1967). The activation of the egg. *In* "Fertilization" (C. B. Metz and A. Monroy, eds.), Vol. 1, pp. 369–412. Academic Press, New York.

Moreau, M., Dorée, M., and Guerrier, P. (1976). Electrophoretic introduction of calcium ions into the cortex of *Xenopus laevis* oocytes triggers meiosis reinitiation. *J. Exp. Zool.* **197**, 443–449.

Moreau, M., Vilain, J. P., and Guerrier, P. (1980). Free calcium changes associated with hormone action in amphibian oocytes. *Dev. Biol.* **78**, 201–214.

Moreau, M., Guerrier, P., and Vilain, J. P. (1985). Ionic regulation of oocyte maturation. *In* "Biology of Fertilization" (C. B. Metz and A. Monroy, eds.), Vol. 1, pp. 299–345. Academic Press, New York.

Morrill, G. A., Ziegler, D., and Kostellow, A. B. (1981). The role of Ca^{2+} and cyclic nucleotides in progesterone initiation of the meiotic divisions in amphibian oocytes. *Life Sci.* **29**, 1821–1835.

Morrill, G. A., Kostellow, A. B., Mahajan, S., and Gupta, R. K. (1984). Role of calcium in regulating intracellular pH following the stepwise release of the metabolic blocks at first-meiotic prophase and second-meiotic metaphase in Amphibian oocytes. *Biochim. Biophys. Acta* **804**, 107–117.

Nadamitsu, S. (1953). Ovulation *in vitro* in several species of amphibia. *J. Sci. Hiroshima Univ., Ser. B, Div. 1* **14**, 1–7.

Newport, J. W., and Kirschner, M. W. (1984). Regulation of the cell cycle during early *Xenopus* development. *Cell (Cambridge, Mass.)* **37**, 731–742.

O'Connor, C. M., and Smith, L. D. (1976). Inhibition of oocyte maturation by theophylline: Possible mechanism of action. *Dev. Biol.* **52**, 318–322.

Peaucellier, G., Dorée, M., and Demaille, J. G. (1982). Stimulation of endogenous protein phosphorylation in oocytes of *Sabellaria alveolata* at meiosis reinitiation induced by protease, fertilization or ionophore A23187. *Gamete Res.* **5**, 115–123.

Picard, A., and Dorée, M. (1984). The role of the germinal vesicle in producing maturation-promoting factor (MPF) as revealed by the removal and transportation of nuclear material in starfish oocytes. *Dev. Biol.* **104**, 357–365.

Pincus, G., and Enzman, E. V. (1935). The comparative behaviour of mammalian eggs *in vivo* and *in vitro*. *J. Exp. Med.* **62**, 655–675.

Prigogine, I., and Lefever, R. (1968). Symmetry breaking instabilities in dissipative systems. II. *J. Chem. Phys.* **48**, 1695–1700.

Reynhout, J. K., and Smith, L. D. (1974). Studies on the appearance and nature of a maturation-inducing factor in the cytoplasm of amphibian oocytes exposed to progesterone. *Dev. Biol.* **38**, 394–400.

Robinson, K. R. (1985). Maturation of *Xenopus laevis* oocytes is not accompanied by electrode-detectable calcium changes. *J. Cell Biol.* **109**, 504–508.

Rugh, R. (1934). Induced ovulation and artificial fertilization in the frog. *Biol. Bull. (Woods Hole, Mass.)* **66**, 22–29.

Ryabova, L. V. (1983). The cytostatic effect of the cytoplasm of mature, non-activated and cleaving eggs of *Rana temporaria, Acipenser stellata* and *Xenopus laevis*. *Cell Differ.* **13**, 171–175.

Schatz, F., and Ziegler, D. H. (1979). The role of follicle cells in *Rana pipiens* oocyte maturation induced by 5-pregnenolone. *Dev. Biol.* **73**, 59–67.

Schorderet-Slatkine, S., and Drury, K. C. (1973). Progesterone-induced maturation in oocytes of *Xenopus laevis*. Appearance of a "maturation-promoting factor" in enucleated oocytes. *Cell Differ.* **2**, 247–254.

Schorderet-Slatkine, S., Schorderet, M., and Baulieu, E. E. (1977). Progesterone-induced meiotic reinitiation *in vitro* in *Xenopus laevis* oocytes. A role for displacement of membrane-bound calcium. *Differentiation* **9**, 67–76.

Schorderet-Slatkine, S., Schorderet, M., Boquet, P., Godeau, F., and Baulieu, E. E. (1978). Progesterone-induced meiosis in *Xenopus laevis* oocytes: A role for cAMP at the maturation-promoting factor level. *Cell (Cambridge, Mass.)* **15**, 1269–1275.

Schorderet-Slatkine, S., Schorderet, M., and Baulieu, E. E. (1982). Cyclic AMP-mediated control of meiosis: Effects of progesterone, choleratoxin, and membrane active drugs in *Xenopus laevis* oocytes. *Proc. Natl. Acad. Sci. U.S.A.* **79**, 850–854.

Schuetz, A. W. (1967a). Effect of steroids on germinal vesicle of oocytes of the frog (*Rana pipiens*) *in vitro*. *Proc. Soc. Exp. Biol. Med.* **124**, 1307–1310.

Schuetz, A. W. (1967b). Action of hormones on germinal vesicle breakdown in frog oocytes (*Rana pipiens*). *J. Exp. Zool.* **166**, 347–354.

Schuetz, A. W. (1972). Hormones and follicular functions. *In* "Oogenesis" (J. D. Biggers and A. W. Schuetz, eds.), pp. 479–511. University Park Press, Baltimore, Maryland.

Schuetz, A. W., and Biggers, J. D. (1967). Regulation of germinal vesicle breakdown in starfish oocytes. *Exp. Cell Res.* **46**, 624–628.

Schuetz, A. W., and Samson, D. (1979a). Protein synthesis requirement for Maturation Promoting Factor (MPF) initiation of meiotic maturation in *Rana* oocytes. *Dev. Biol.* **68**, 636–642.

Schuetz, A. W., and Samson, D. (1979b). Nuclear requirement for postmaturational cortical differentiation of amphibian oocytes: Effects of cycloheximide. *J. Exp. Zool.* **210,** 307–319.

Shapiro, H. A. (1936). Induction of ovulation by testosterone and certain related compounds. *J. Soc. Chem. Ind., London* **55,** 1031.

Shibuya, E. K., and Masui, Y. (1982). Sperm-induced cell cycle activities in blastomeres arrested by the cytostatic factor of unfertilized eggs in *Rana pipiens. J. Exp. Zool.* **220,** 381–385.

Siracusa, G., Whittingham, D. G., Molinaro, M., and Vivarelli, E. (1978). Parthenogenetic activation of mouse oocytes induced by inhibitors of protein synthesis. *J. Embryol. Exp. Morphol.* **43,** 157–166.

Skoblina, M. N. (1969). Independence of the cortex maturation from germinal vesicle material during the maturation of amphibian and sturgeon oocytes. *Exp. Cell Res.* **55,** 142–144.

Skoblina, M. N. (1974). Behavior of sperm nuclei injected into intact ripening and ripe toad oocytes and into oocytes ripening after removal of the germinal vesicle. *Ontongenez* **5,** 334–340.

Smith, L. D., and Ecker, R. E. (1969). Role of the oocyte nucleus in physiological maturation in *Rana pipiens. Dev. Biol.* **19,** 281–309.

Smith, L. D., and Ecker, R. E. (1971). The interaction of steroids with *Rana pipiens* oocytes in the induction of maturation. *Dev. Biol.* **25,** 233–247.

Smith, L. D., Ecker, R. E., and Subtelney, S. (1966). The initiation of proetin synthesis in eggs of *Rana pipiens. Proc. Natl. Acad. Sci. U.S.A.* **56,** 1724–1728.

Smith, L. D., Ecker, R. E., and Subtelney, S. (1968). *In vitro* induction of physiological maturation in *Rana pipiens* oocytes from ovarian follicles. *Dev. Biol.* **17,** 627–643.

Sorensen, R. A., Cyert, M. S., and Pedersen, R. A. (1985). Active maturation-promotion factor is present in mature mouse oocytes. *J. Cell Biol.* **100,** 1637–1640.

Stern, S., Rayyis, A., and Kennedy, J. F. (1972). Incorporation of amino acids during maturation *in vitro* by the mouse oocyte: Effect of puromycin on protein synthesis. *Biol. Reprod.* **7,** 341–346.

Sundararaj, B. I., and Goswami, S. V. (1977). Hormonal regulation of *in vitro* oocyte maturation in the catfish, *Heteropneutes fossilis. Gen. Comp. Endocrinol.* **32,** 17–28.

Tchou-Su, and Wang Yu-Lan. (1958). Etudes comparatives sur l'ovulation et la maturation *in vivo* et *in vitro* chez le crapaud asiatique (*Bufo bufo asiaticus*). *Acta Biol. Exp. Sin.* **6,** 129–180.

Tso, J., Thibier, C., Mulner, O., and Ozon, R. (1982). Microinjected progesterone reinitiates meiotic maturation of *Xenopus laevis* oocytes. *Proc. Natl. Acad. Sci. U.S.A.* **79,** 5552–5556.

Tyler, A. (1941). Artificial parthenogenesis. *Biol. Rev. Cambridge Philos. Soc.* **16,** 291–336.

Van der Hurk, R., and Richter, C. J. J. (1980). Histochemical evidence for granulosa steroids in follicle maturation in the African catfish, *Clarias lazera. Cell Tissue Res.* **211,** 345–348.

Wasserman, W. J., and Masui, Y. (1974). A study of gonadotropin action in the induction of oocyte maturation in *Xenopus laevis. Biol. Reprod.* **11,** 133–144.

Wasserman, W. J., and Masui, Y. (1975a). Initiation of meiotic maturation in *Xenopus laevis* oocytes by the combination of divalent cations and ionophore A23187. *J. Exp. Zool.* **193,** 369–375.

Wasserman, W. J., and Masui, Y. (1975b). Effects of cycloheximide on a cytoplasmic factor initiating meiotic maturation in *Xenopus laevis* oocytes. *Exp. Cell Res.* **91,** 381–388.

Wasserman, W. J., and Masui, Y. (1976). A cytoplasmic factor promoting oocyte maturation: Its extraction and preliminary characterization. *Science* **191,** 1266–1268.

Wasserman, W. J., and Smith L. D. (1978). The cyclic behavior of a cytoplasmic factor controlling nuclear membrane breakdown. *J. Cell Biol.* **78**, R15–R22.

Wasserman, W. J., Pinto, L. H., O'Connor, C. M., and Smith, L. D. (1980). Progesterone induces a rapid increase in Ca^{2+} in *Xenopus laevis* oocytes. *Proc. Natl. Acad. Sci. U.S.A.* **77**, 1534–1536.

Wasserman, W. J., Richter, J. D., and Smith, L. D. (1982). Protein synthesis during maturation-promoting factor- and progesterone-induced maturation in *Xenopus* oocyte. *Dev. Biol.* **89**, 152–158.

Wasserman, W. J., Houle, J. G., and Samuel D. (1984). The maturation response of stage IV, V, and VI *Xenopus* oocytes to progesterone *in vitro*. *Dev. Biol.* **105**, 315–324.

Wilson, E. B. (1903). Experiments on cleavage and localization in the nemertine-egg. *Wilhelm Roux' Arch. Entwicklungsmech. Org.* **16**, 411–461.

Wilson, E. B. (1925). "The Cell in Development and Heredity," 3rd ed. Macmillan, New York.

Wright, P. A. (1945). Factors affecting *in vitro* ovulation in the frog. *J. Exp. Zool.* **100**, 565–575.

Wright, P. A. (1961). Induction of ovulation *in vitro* in *Rana pipiens* with steroids. *Gen. Comp. Endocrinol.* **1**, 20–23.

Wu, M., and Gerhart, J. C., (1980). Partial purification and characterization of the maturation-promoting factor from eggs of *Xenopus laevis*. *Dev. Biol.* **79**, 465–477.

Yatsu, N. (1905). The formation of centrosomes in enucleated egg-fragments. *J. Exp. Zool.* **2**, 287–312.

Yoneda, M., and Schroeder, T. E. (1984). Cell cycle timing in colchicine-treated sea urchin eggs: Presistent coordination between the nuclear cycles and the rhythm of cortical stiffness. *J. Exp. Zool.* **231**, 367–378.

Zampetti-Bosseler, F., Huez, G., and Brachet, J. (1973). Effects of several inhibitors of macromolecular synthesis upon maturation of marine invertebrate oocytes. *Exp. Cell Res.* **78**, 383–393.

Ziegler, D. H., and Masui, Y. (1976). Control of chromosome behavior in amphibian oocytes. II. The effect of inhibitors on RNA and protein synthesis on the induction of chromosome condensation in transplanted brain nuclei by oocyte cytoplasm. *J. Cell Biol.* **68**, 620–628.

Zwarenstein, H. (1937). Experimental induction of ovulation with progesterone. *Nature (London)* **139**, 112.

2

Dynamics of the Nuclear Lamina during Mitosis and Meiosis

REIMER STICK

Max-Planck-Institut für Entwicklungsbiologie
Abteilung für Zellbiologie
D-7400 Tübingen
Federal Republic of Germany

I. INTRODUCTION

Nuclei of eukaryotic cells are surrounded by an envelope consisting of the inner and outer nuclear membrane, the pore complexes, and the nuclear lamina. The lamina is a protein shell which appears in electron micrographs as a morphologically distinct layer between the inner nuclear membrane and the peripheral chromatin (Fig. 1). At the positions of pore complexes, the lamina layer is fenestrated, allowing access to the pores. Different cell types may characteristically differ by the thickness of their lamina (in the range between 15 and 100 nm). A nuclear lamina has been demonstrated in organisms as phylogenetically diverse as protozoans, molluscs, insects, and vertebrates (Fawcett, 1966, 1981). It is present in all somatic cell nuclei thus far studied and can be visualized by immunofluorescence techniques using lamin-specific antibodies (Fig. 2) as well as by electron microscopy (EM). Numerous electron micrographs showing the lamina are found in the textbook "The Cell" by D. W. Fawcett (1981). In addition, differential EM staining techniques, which leave lipid membranes unstained, allow the visualization of very thin lamina layers that might hardly be detectable by conventional staining techniques (Schellens et al., 1979; Stick and Schwarz, 1982, 1983). It can now be assumed that

43

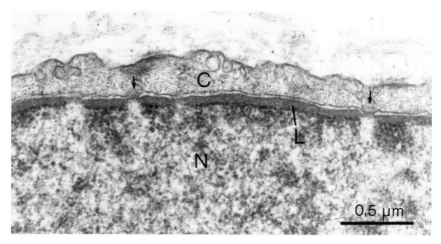

Fig. 1. Ultrathin section of the nuclear envelope of a human epithelial cell. A tissue sample of tongue was fixed and embedded according to standard procedures. The section was stained with uranyl acetate and lead citrate. C = cytoplasm; N = nucleoplasm; L = nuclear lamina. Arrows point to nuclear pores. (The micrograph was kindly provided by Dr. Heinz Schwarz.)

the nuclear lamina is a ubiquitous component of the nuclear envelope of somatic cells.

Nuclear lamina structures have been isolated as "pore complex-lamina fractions" from several cell types (Aaronson and Blobel, 1975; Dwyer and Blobel, 1976; Cobbs and Shelton, 1978; Shelton *et al.,* 1980; Stick and Hausen, 1980; Krohne *et al.,* 1981; Fisher *et al.,* 1982; Baglia and Maul, 1983) by exploiting the fact that the lamina is stable in high- and low-ionic strength buffers and retains its integrity in the absence of nuclear membranes and chromatin. The isolated lamina consists of only a few polypeptides, named lamins (Gerace *et al.,* 1978), which range in molecular weight from 60,000 to 75,000. Immunological crossreactivity between individual polypeptides from different tissues and species has been demonstrated with both monoclonal antibodies (Burke *et al.,* 1983; Krohne *et al.,* 1984; Stick and Hausen, 1985) and lamin-specific polyclonal antisera (Krohne *et al.,* 1978; Stick and Hausen, 1980; Gerace and Blobel, 1981; Högner *et al.,* 1984). Peptide maps of particular lamins show striking similarities (Lam and Kasper, 1979; Shelton *et al.,* 1980; Gerace and Blobel, 1981; Kaufmann *et al.,* 1983). It is therefore assumed that lamins form a family of evolutionarily related proteins. In mammals and birds, three major lamin polypeptides have been characterized thus far (lamins

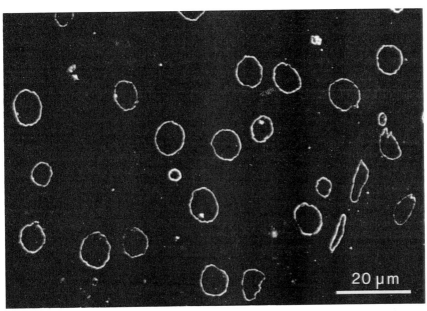

Fig. 2. Thin section of chicken liver stained with lamin-specific antibodies and viewed by indirect immunofluorescence. The rabbit anti-chicken lamin serum used has been previously described by Stick and Hausen (1980). Fluorescein-conjugated second antibody was used. Note that antibodies react with nuclei of all cell types.

A, B, and C) (Gerace *et al.,* 1978; Shelton *et al.,* 1980). In amphibians, cell type-specific differences in the lamin composition have been observed (Krohne *et al.,* 1981). While two major lamin components, L_I and L_{II}, exist in nuclei of amphibian somatic cells (probably the homologues of lamins A and B in other vertebrates), the lamina of oocyte and early cleavage nuclei is formed by a single polypeptide, L_{III} (Stick and Hausen, 1985). L_{III} is also found in later developmental stages and in certain adult tissues together with L_I and L_{II} (Benavente *et al.,* 1985). Lamin polypeptides have also been described in molluscs and insects (Fisher *et al.,* 1982; Baglia and Maul, 1983). The lamins of *Drosophila* seem to differ in the solubility characteristic of the vertebrate lamins (for details, see Fisher *et al.,* 1982).

It has been suggested that the nuclear lamina provides a framework for nuclear envelope organization (Gerace *et al.,* 1978). It furthermore seems to be involved in DNA-loop organization in interphase nuclei (Hancock and Hughes, 1982; Lebkowsky and Laemmli, 1982), and it may also contribute to the spatial arrangement of pore complexes and the maintenance of nuclear shape (Stick and Schwarz, 1982, 1983).

During mitotic and meiotic divisions, the nuclear lamina disintegrates and is reassembled during re-formation of telophase nuclei. In this chapter, the dynamics of the nuclear lamina during mitotic and meiotic divisions as well as the changes in lamina structure during meiotic prophase are described. Sections II and IV discuss the reversible lamin disassembly during mitosis and meiosis, and Section III deals with the disappearance and reappearance of the lamina structure during meiotic prophase.

II. DYNAMICS OF THE NUCLEAR LAMINA IN MITOSIS

The structural changes of the nuclear envelope during mitosis have been studied by electron microscopy (for references, see Roos, 1973). Disassembly of the nuclear envelope begins at prophase when the pore complexes disappear and the nuclear membranes are fragmented, forming small vesicles that disperse throughout the cytoplasm and become indistinguishable from endoplasmic membranes. Between late anaphase and early telophase, nuclear membranes begin to assemble on the surface of the still condensed chromosomes. This process probably occurs by fusion of small membrane vesicles. After the nuclear membranes have reformed, pore complexes reappear at telophase.

In vertebrates, the nuclear envelope disassembles completely (open mitosis). In other organisms (for example, *Drosophila*), fragmentation of the nuclear envelope occurs at the spindle poles only (Stafstrom and Staehelin, 1984) (semiopen mitosis). In some organisms, such as yeast and many protozoans, nuclear divisions take place inside the intact nuclear envelopes (closed mitosis). To date, the dynamics of the nuclear lamina has been studied only in organisms with open and semiopen mitosis.

A. Immunocytochemical Observations

Numerous investigators have followed the fate of the lamina proteins during mitosis using immunohistochemical methods with lamin-specific polyclonal or monoclonal antibodies (Ely *et al.*, 1978; Gerace *et al.*, 1978; Krohne *et al.*, 1978, 1984; Stick and Hausen, 1980; Jost and Johnson, 1981; Burke *et al.*, 1983; McKeon *et al.*, 1983, 1984; Stick and Schwarz, 1983). Most of these investigators analyzed cells in culture. (For mitotic lamin distribution in tissues, see Stick and Schwarz, 1983). In mitotic vertebrate cells the lamina is completely disassembled, the lamins being homogeneously dispersed in metaphase cells leaving the chromosomes devoid of antigen. (For a nonhomogeneous distribution of lamin fluorescence in metaphase obtained by a monoclonal IgM antibody, see Burke, *et al.*, 1983.)

A series of different mitotic stages of chicken fibroblast cells is shown in Fig. 3. The cells were double-stained with lamin-specific antibodies and with a DNA-specific fluorescent dye (DAPI) for identification of the mitotic stages. During initial chromosome condensation at early prophase, the lamina appears structurally unchanged (Fig. 3c,d; see also in McKeon *et al.*, 1984, Fig. 2b), and the cytoplasm is still devoid of antigen, as it is in interphase (Fig. 3b). At late prophase, disassembly of the lamina begins, and lamin-specific fluorescence is detected in the cytoplasm (Fig. 3f). The appearance of the lamina strikingly changes at prometaphase when chromosomes arrange in the metaphase plate (Fig. 3g); it becomes highly irregular and begins to fragment (Fig. 3h). At metaphase and early anaphase, the lamin antigens are diffusely distributed throughout the whole cell but are excluded from the chromosomes (Figs. 3k,m). Re-formation of the lamina in telophase and early G_1 phase is shown in Fig. 3o and 3q. Significant cytoplasmic fluorescence is still observed in early G_1, at a time when nucleoli have already formed (Fig. 3q). Formation of the nucleoli was detected by a nucleolus-specific antibody (results not shown).

A pattern differing from the one described here for vertebrate cells has been observed in *Drosophila* by Fuchs *et al.* (1983). In *Drosophila*, lamina disassembly is incomplete, with the lamin antigens remaining in a particulate state throughout mitosis. This mitotic lamin distribution corresponds well to the electron microscopic observation that the nuclear envelope in *Drosophila* disassembles only near the spindle poles, characteristic of semiopen mitosis (Stafstrom and Staehelin, 1984).

B. Change of Lamin Epitope Accessibility for a Monoclonal Antibody during Mitosis

In the course of lamin disassembly during mitosis, a new antigenic determinant becomes accessible to a particular monoclonal antibody, L7-4A2, directed against *Xenopus* lamin L_I (Klein, 1984). In fixed *Xenopus* tissue culture cells, this antibody recognizes lamin polypeptides exclusively in mitotic cells but does not bind to interphase lamins (Fig. 4d). A modification of the lamin protein occurring during mitotic disassembly would not explain this observation because in immunoblotting experiments the antibody recognizes SDS-denatured lamin L_I isolated from interphase nuclei (Fig. 4b). The reaction of this antibody with mitotic cells

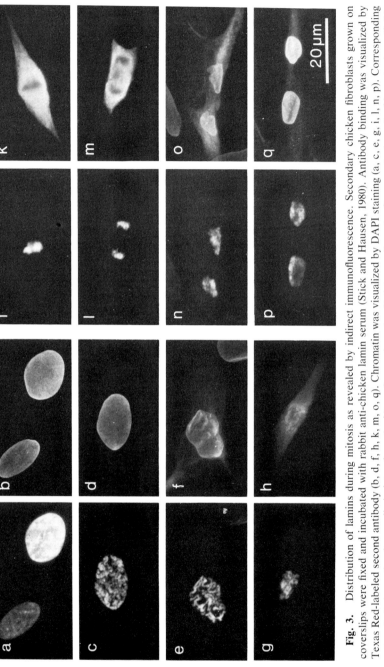

Fig. 3. Distribution of lamins during mitosis as revealed by indirect immunofluorescence. Secondary chicken fibroblasts grown on coverslips were fixed and incubated with rabbit anti-chicken lamin serum (Stick and Hausen, 1980). Antibody binding was visualized by Texas Red-labeled second antibody (b, d, f, h, k, m, o, q). Chromatin was visualized by DAPI staining (a, c, e, g, i, l, n, p). Corresponding pairs of photographs show cells in interphase (a, b); early prophase (c, d); late prophase (e, f); prometaphase (g, h); metaphase (i, k); anaphase (l, m); telophase (n, o); and early G_1 phase (p, q). In early G_1 phase (p, q), nucleoli had re-formed, as revealed by double fluorescence using a nucleolus-specific antibody (not shown). (The micrographs were kindly provided by B. Bühler.)

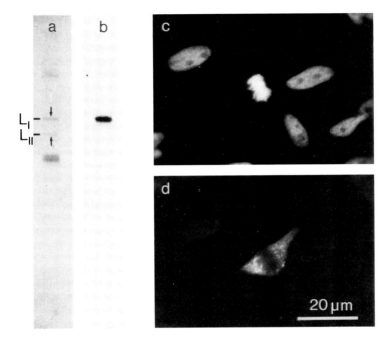

Fig. 4. A lamin L_I-specific monoclonal antibody reacts with an epitope, which is accessible only during disaggregation of the lamina in mitosis. Pore complex-lamina fraction of *Xenopus laevis* erythrocytes was separated by SDS–PAGE and proteins were either stained with Coomassie blue (a) or blotted to a nitrocellulose filter and probed with antibody L7-4A2 using radioiodinated second antibody (b). Arrows in (a) mark positions of lamins L_I and L_{II}. Reaction of antibody L7-4A2 with *Xenopus* culture cells was visualized by indirect immunofluorescence using fluorescein-conjugated second antibody (d). (c) Same cells as in (d) stained with DAPI to visualize nuclei and chromosomes. (These data were kindly provided by G. Klein.)

exclusively must be due to the exposure on disassembled lamins of an antigenic site hidden in the intact lamina.

C. Reversible Disassembly of Lamin Polypeptides during Mitosis

Immunocytological observations suggest that the lamin polypeptides are reused during telophase for the formation of the nuclear envelopes of the daughter cells. Several lines of evidence support this view. Gerace and Blobel (1980) showed that soluble lamin polypeptides can be isolated from mitotic cells. When cells are pulse-labeled in G_2 phase with radiolabeled amino acids, similar quantities of each of the isotopically labeled lamins can be immunoprecipitated from cells in the following M and G_1 phase (Gerace and Blobel, 1980). Finally, inhibition of protein synthesis during cell division and G_1 phase does not inhibit lamin re-formation in cultured cells (Jost and Johnson, 1981). Thus, there is no indication of a significant turnover of the lamina polypeptides during mitosis.

Lamin A is synthesized as a larger precursor molecule (Gerace et al., 1984; R. Stick, unpublished). The precursor is processed after its integration into the lamina structure. The reutilization of lamin A after mitotic disassembly demonstrates, however, that a precursor form is not a precondition for the assembly of lamin A into the lamina. The functional significance of this precursor form is still unknown.

D. State of Lamin Polypeptides during Metaphase

During interphase, lamin polypeptides form a supramolecular aggregate that can be isolated intact and that is only disrupted by treatment with strong protein denaturants. In a detailed study, Gerace and Blobel (1980) showed that during mitosis all three lamins of Chinese hamster ovary (CHO) cells become monomeric. Lamins A and C occur in a soluble form, not bound to membranes. They can be immunoprecipitated from supernatants obtained after homogenization of mitotic cells and high-speed centrifugation of cytoplasmic extracts. They sediment as monomers during sucrose gradient centrifugation. In contrast, lamin B may be membrane-associated because it can be sedimented by low-speed centrifugation and is immunoprecipitated from mitotic extracts only when membranes are solubilized during extraction by treatment with nonionic detergents. After such treatment, lamin B also sediments as a monomer during sucrose gradient centrifugation. Analysis of a chicken erythroid cell line produced similar results (R. Stick, unpublished).

On the basis of these results, models have been proposed for nuclear envelope breakdown and re-formation (Alberts et al., 1983; Gerace et al., 1984). According to these models, lamin B remains associated with the membrane vesicles that form during the fragmentation of the nuclear membranes, while lamins A and C are released in soluble form. Affinity of lamin B for chromosome constituents may allow reassociation of the nuclear membrane vesicles with the decondensing chromosomes at telophase. Fusion of the vesicles may then initiate lamina reaggregation and re-formation of the nuclear envelope.

E. Possible Mechanisms of Lamina Disassembly

It has been suggested that phosphorylation may be important for disassembly of the nuclear lamina structure during cell division (Gerace and Blobel, 1980, 1981; Ottaviano and Gerace, 1985). Lamin phosphorylation occurs at a low level throughout interphase. When the lamina is disassembled during cell division, the level of phosphorylation increases four- to sevenfold. Each of the mitotic lamins carries approximately one to two phosphates per molecule (Gerace and Blobel, 1980; Ottaviano and Gerace, 1985). The sites of lamin phosphorylation have been mapped by Ottaviano and Gerace (1985). Phosphorylation occurs predominantly at serine residues distributed over numerous tryptic peptides. Many sites are phosphorylated during both mitosis and interphase, whereas others seem to be mitosis specific.

More recently, Miake-Lye and Kirschner (1985) showed that nuclear envelope breakdown can be induced by maturation-promoting factor (MPF) in somatic interphase nuclei incubated in a cell-free extract of *Xenopus* eggs. Using this cell-free system, these investigators showed that the lamins are hyperphosphorylated within 15 min after addition of MPF, followed by depolymerization of the nuclear lamina and breakdown of the whole nuclear envelope 30 min later. This observation is consistent with the previously stated role of phosphorylation in lamina breakdown postulated by Gerace and Blobel (1980).

Concomitant with the re-formation of the lamina during telophase, however, lamin phosphorylation decreases to a level intermediate between that of mitotic and exponentially growing interphase cells. This intermediate level of phosphorylation might indicate either a slightly altered structure of the newly formed lamina at early G_1 phase or might be due to the persistence of soluble lamins in the cytoplasm of G_1 cells (see Fig. 3q). A causal relationship between lamin phosphorylation and modulation of lamin organization is very suggestive, but still remains to be proved.

F. Control of Lamina Disassembly and Reassembly

Little is known about the control of lamina dynamics. Jost and Johnson (1981) concluded from cell fusion experiments that lamin disassembly is under the positive control of diffusible factors. When interphase and mitotic cells were fused, breakdown of the lamina in interphase nuclei could be observed soon after fusion, while at later times the lamina was deposited around both the original interphase and metaphase nuclei, which suggests the inactivation of these control factors.

Miake-Lye *et al.* (1983) were able to induce lamina breakdown by microinjection of purified MPF into cycloheximide-arrested cleavage-stage cells of *Xenopus laevis*. However, as the intracellular action of MPF is not understood, it is not known whether MPF is acting directly or indirectly on the lamina.

Reassembly of the envelope-lamina occurs at the outer surface of the telophase nuclei. The site-specificity of this process can be perturbed by prolonged treatment of mitotic cells with Colcemid or by exposing them to low temperature (Jost and Johnson, 1981). In the absence of microtubules, individual mitotic chromosomes are enveloped by a lamina and membranes at telophase, leading to formation of mininuclei (Fig. 5; Jost and Johnson, 1981). Since perturbation of the actin filament system by treatment of cells with cytochalasin B does not affect the localized lamina formation (Bühler, 1984), microtubules may specifically influence the spatial organization of lamina reassembly (for discussion, see Jost and Johnson, 1981).

Such formation of mininuclei occurs naturally during formation of cleavage-stage nuclei in sea urchin and amphibians (for references, see Forbes *et al.*, 1983). These mininuclei, however, fuse together to form a single interphase nucleus. Therefore, the lamina has to be remodeled during this fusion. Formation of mininuclei in early development may be related to the very short cell cycle times (10–30 min) in cleavage embryos (for the lamin polypeptide composition of amphibian cleavage nuclei, see Section IV).

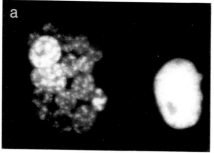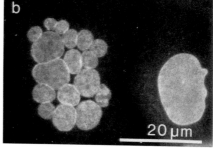

Fig. 5.   Formation of mininuclei after treatment of cells with Colcemid. *Xenopus* tissue culture cells were treated for 20 hr with 0.06 μg/ml of Colcemid. Cells were fixed and processed for indirect immunofluorescence using anti-chicken lamin antibodies (see legend to Fig. 3). DAPI staining (a); anti-lamin staining (b).

III. DYNAMICS OF THE NUCLEAR LAMINA DURING MEIOTIC PROPHASE

In mitosis, lamina disassembly is accompanied by partial or complete breakdown of the whole nuclear envelope concomitant with chromosome condensation and movement. During meiotic prophase, on the other hand, condensation and rearrangement of paired meiotic chromosomes occurs inside the nucleus with the nuclear membranes remaining intact. The dynamics of the lamina during meiotic prophase in oocytes has been studied using indirect immunofluorescence techniques and electron microscopy (Stick and Schwarz, 1983).

A. Meiotic Prophase in Oocytes: Immunocytological Observations

When sections of chicken ovaries were stained with anti-chicken lamin antibodies, all somatic cells reacted, but meiotic cells in certain stages of prophase remained unstained (Fig. 6; Stick and Schwarz, 1983). Analysis of squash preparations of these ovaries revealed that lamin antigen is present in oogonia and in leptotene stage oocytes, although the fluorescence intensity is reduced in the latter. During zygotene, it disappears and it remains absent in pachytene, when chromosomes are maximally condensed forming the synaptonemal complex. Lamin antigen reappears in diplotene when chromosomes decondense again. From these observations it can be assumed that a lamina structure disappears during meiotic prophase and reappears in diplotene oocytes. (For independent electron microscopic evidence, see Section III,E.)

B. Probable Presence of the Same Lamins in Oogonia and Diplotene Oocytes of the Chicken

To answer the question of whether the same lamins are present in oogonia, leptotene oocytes and, after the reappearance of the lamina, diplotene oocytes, we have produced monoclonal antibodies that specifically recognize chicken lamins A plus C or lamin B. The specificity of the antibodies was demonstrated by immunoblotting SDS-denatured lamin polypeptides and by immunoprecipitation of nondenatured newly synthesized lamins extracted from the cytoplasm of radiolabeled chicken erythroblast cells (R. Stick, unpublished).

Both lamin A plus C- and lamin B-specific antibodies recognize lamina of leptotene as well as diplotene oocytes. It may be assumed that both

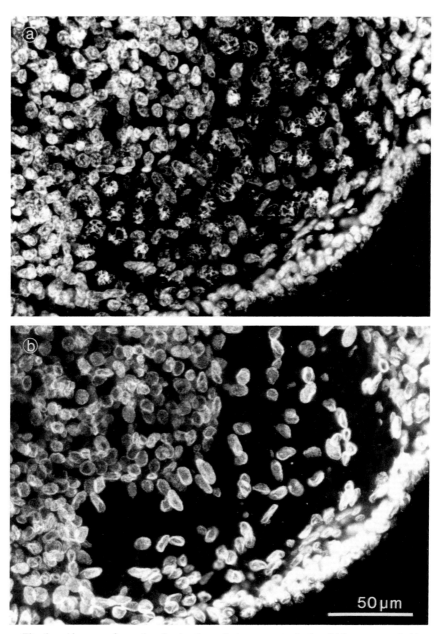

Fig. 6. Absence of a nuclear lamina in pachytene oocytes in the chicken as revealed by indirect immunofluorescence. An ovary from a 2-day-old chicken was fixed in 2% trichloroacetic acid, wax embedded, and sectioned (for details of wax embedding, see Stick and Schwarz, 1983). The section was stained with DAPI (a) and with anti-chicken lamin antibodies by indirect immunofluorescence using fluorescein-conjugated second antibody (b). Note the absence of lamin-specific immunofluorescence in both the nuclei and the cytoplasm of pachytene oocytes.

immunological types of lamins (A plus C and B) are present in diplotene oocytes in chicken. This conclusion, however, must be made with reservation until independent biochemical evidence unambiguously clarifies this question. Due to the difficulties in isolating sufficient quantities of chicken diplotene oocyte nuclei, these data are not yet available.

C. Appearance of the Lamina in Diplotene Accompanied by a Change in Lamin Composition in Xenopus

Results analogous to those reported in Section III,A for chickens were also obtained when oocyte development in *Xenopus* was investigated (Stick and Schwarz, 1983). However, in *Xenopus* the lamina re-formed at early diplotene has an altered polypeptide composition (Benavente *et al.*, 1985) when examined by two monoclonal antibodies which specifically recognize either lamins L_I and L_{II} (antibody PKB8) or lamin L_{III} (antibody L_{III}46F7). (For nomenclature of the *Xenopus* lamins, see Section I.) While antibody PKB8 reacted with all somatic cells by indirect immunofluorescence, it failed to bind to diplotene oocyte lamina. The opposite staining pattern was found with L_{III} 46F7. It reacted strongly with diplotene nuclei but not with oogonia. In addition, immunological and biochemical evaluation of the lamin polypeptide composition of the *Xenopus* oocyte nucleus (germinal vesicle) lead to the conclusion that lamin L_{III} is the only lamin component in *Xenopus* germinal vesicles (see Section I; Stick and Hausen, 1980; Krohne *et al.*, 1981; Stick and Krohne, 1982).

The monoclonal antibody PKB8 cross-reacts with lamins of other species, including mammals and birds. Contrasting with the results in *Xenopus*, this antibody recognizes diplotene oocyte nuclei in rats and chickens (Krohne *et al.*, 1984). In support of the data described in Section III,B, these results indicate that a change in lamina composition during meiosis is not necessarily a universal feature of oocyte development.

D. Dynamics of the Nuclear Lamina in Male Germ Cells

Disappearance of the nuclear lamina is not restricted to the female germ line but is also seen in male meiotic cells (Stick and Schwarz, 1982). Figure 7 shows a section of a seminiferous tubule of chicken which has been stained with lamin-specific antiserum (Fig. 7b) and with DAPI to visualize positions of nuclei in the section (Fig. 7a). Successive stages of germ cell development are spatially arranged in the seminiferous tubules of chicken in such a way that spermatogonia are located close to the

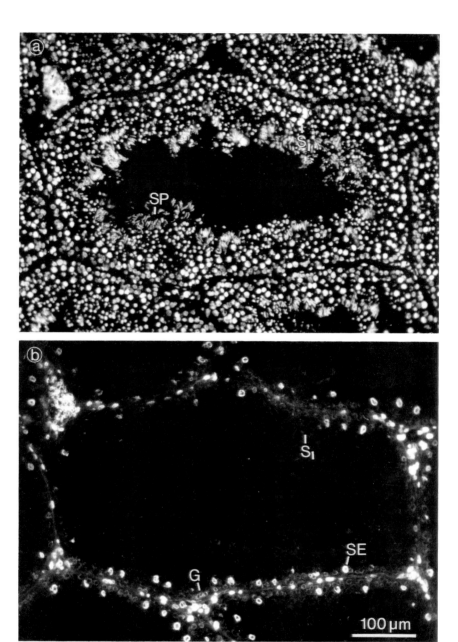

Fig. 7. Absence of a nuclear lamina in male meiotic cells as revealed by indirect immunofluorescence. A cryostat section of seminiferous tubules of an adult chicken was stained with DAPI (a) and sequentially with anti-chicken lamin antibodies and fluorescein-conjugated second antibody (b). In the seminiferous epithelium only nuclei of sertoli cells (SE) and, to a lesser extent, nuclei of spermatogonia (G) react with the antibodies. Primary (SI) and secondary spermatocytes and sperm nuclei (SP) are not stained.

circumference of the tubules and mature sperm are located next to the lumen of the tubule. Only somatic cell nuclei and, to a lesser extent, nuclei of spermatogonia show lamin-specific immunofluorescence, while primary and secondary spermatocytes and sperm are negative, indicating that these cells lack a lamina (see Section III,E). Absence of lamina in male meiotic germ cells has been observed using indirect immunofluorescence in other animals, including amphibians (Stick and Schwarz, 1982) and several mammals (Stick and Schwarz, 1982; Högner et al., 1984; Krohne et al., 1984). These results suggest that absence of the lamina from certain meiotic stages may be a general phenomenon. Absence of a lamina is also evident from immunoblotting experiments in which DNase I-digested, high salt-extracted *Xenopus* sperm are probed with lamin-specific antiserum after denaturation by SDS (Stick and Hausen, 1985).

More recently, Benavente and Krohne (1985) reported the presence of a sperm-specific laminlike polypeptide (L_{IV}) which is recognized by an antilamin monoclonal antibody. So far, this polypeptide has been detected only in spermatids and spermatozoa, but in contrast to other lamins it is located in small patches at the periphery of sperm nuclei rather than in a typical lamina structure. A continuous lamina shell surrounding the whole nucleus is not detectable in demembranated sperm nuclei (Stick and Schwarz, 1982).

E. Evidence for the Absence of a Lamina in Specific Stages of Germ Cell Differentiation from Electron Microscopy

The absence of a lamina in pachytene oocytes and in male meiotic cells has been demonstrated by electron microscopic techniques using either tissue preparations, in combination with selective staining, or preparations of isolated demembranated nuclei (Stick and Schwarz, 1982, 1983). These investigations rule out the possibility that the immunofluorescence observations were due to stage-specific masking of lamina antigens, since in pathytene oocytes, male meiotic cells, and sperm, absence of the lamina structure can be demonstrated. These studies indicate that pore complexes and nuclear membranes can exist without a lamina in certain cell stages. Furthermore, the terminal attachment plates of the synaptonemal complex do not contain lamin proteins.

F. State of Lamins during the Absence of a Lamina in Oocytes

This is an important difference between the reversible disaggregation of the lamina during mitosis and its disappearance and reappearance in meiotic prophase. During mitotic metaphase, lamin polypeptides are present in the cytoplasm as detected by immunofluorescence and are reused during nuclear formation at telophase. However, in meiotic prophase, lamins are not detectable by immunofluorescence in pachytene oocytes (Fig. 6). Whether this is due to a masking of their antigenic sites by extensive modifications or to a complete absence of lamin polypeptides cannot be decided at present.

G. Possible Significance of the Absence of the Lamina in Meiotic Cells

Disappearance of the lamina during meiosis is paralleled by the condensation and movements of meiotic chromosomes. This correlation might indicate that interactions between lamina and DNA, which are thought to be functionally important in interphase nuclei (Hancock and Hughes, 1982; Lebkowsky and Laemmli, 1982), must be interrupted to allow chromosome condensation and/or movement, and that this disruption may only be achieved by complete lamina disassembly. (For further discussion, see Stick and Schwarz, 1983.)

IV. DYNAMICS OF THE NUCLEAR LAMINA DURING EGG MATURATION AND EARLY DEVELOPMENT

Because oocytes and embryos of *Xenopus* are easily manipulated and their lamin polypeptides are well characterized, they were chosen for studies of the fate of the lamina during egg maturation and early development. During maturation, the germinal vesicle disintegrates and nuclear proteins are shed into the cytoplasm. Meiotic divisions proceed up to the second metaphase, where mature eggs arrest.

Isolated oocytes, induced to mature *in vitro* by incubation with progesterone, were fixed at different times after hormone induction, embedded in wax, and sections stained with a lamin-specific monoclonal antibody (L6-8A9). (For further characterization of this antibody, see Stick and Hausen, 1985.) Disassembly of the lamina began at the basal side of the nucleus even before the maturation spot occurred at the animal pole of the oocyte. Patches of lamina became distributed throughout the cytoplasm in the manner illustrated in Fig. 8b. At later stages of maturation, lamins were distributed more evenly and finally became undetectable by immunofluorescence techniques, probably due to their dispersal in the cytoplasm (Hausen *et al.*, 1985).

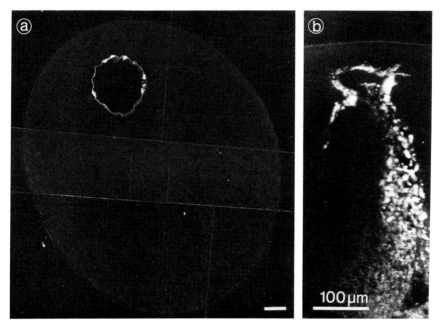

Fig. 8. Lamin distribution in oocytes and maturing eggs of *Xenopus*. Oocytes of *Xenopus laevis* were fixed before and after progesterone-induced egg maturation and embedded in wax (for details, see Hausen *et al.*, 1985). Sections were stained by indirect immunofluorescence techniques using a lamin-specific monoclonal antibody (L6-8A7), described previously by Stick and Hausen (1985). (a) Oocyte, lamin staining is found only in the periphery of the nucleus; (b) part of a maturing egg fixed shortly after the maturation spot had appeared at the animal pole. Germinal vesicle breakdown is most advanced at the basal part of the nucleus; patches of lamins are shed into the cytoplasm.

A. State of the Lamins during Meiotic Metaphase

Lamin polypeptides are detectable in extracts of unfertilized eggs by immunoblotting (Fig. 9) and ELISA (Benavente *et al.*, 1985). When egg extracts are fractionated by centrifugation, almost all of the lamin protein is found in a 100,000 g supernatant, irrespective of whether the extracts were made in the presence or absence of nonionic detergents. Solubility of these lamins, therefore, resembles that of mitotic lamins A and C, rather than lamin B, in mammals and birds (see Section II,D). The lamin of *Xenopus* eggs has been identified as lamin L_{III} either by use of an L_{III}-specific monoclonal antibody (Benavente *et al.*, 1985), two-dimensional gel analysis, or by tryptic peptide mapping. According to tryptic peptide

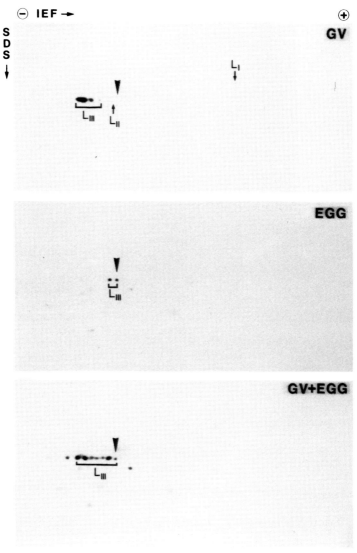

Fig. 9.   Two-dimensional gel electrophoresis and immunoblotting of whole germinal vesicles (GV), cytoplasmic egg extracts (EGG), and a mixture of GVs and egg extract (GV + EGG) of *Xenopus laevis*. Egg extract was prepared from unfertilized eggs. After removal of the jelly coat, eggs were homogenized in isolation buffer (83 mM NaCl, 17 mM KCl, 10 mM Tris-HCl, pH 7.2, 0.2 mM PMSF, 2 μg/ml leupeptin, and 50 U/ml trasylol). The homogenate was centrifuged for 15 min at 40,000 g and aliquots of the supernatant, equivalent to 20 embryos, were applied onto the gels. Twenty manually isolated germinal vesicles were applied each. Gel electrophoresis and immunoblotting was carried out essentially as described by Stick and Hausen (1985). Monoclonal antibody L6-8A7 was used to detect lamin

analysis, lamin L_{III} of eggs and early embryos is identical to lamin L_{III} found in the oocyte lamina (Stick and Hausen, 1985).

When egg extracts were fractionated by two-dimensional gel electrophoresis and lamin polypeptides were visualized by immunoblotting using a lamin-specific monoclonal antibody (L6-8A7), two to three isoelectric variants of L_{III} could be detected (Fig. 9). Comparison of the isoelectric points of L_{III} to an internal marker [bovine serum albumin (BSA), arrowheads in Fig. 9] and to the insoluble lamin L_{III} derived from whole germinal vesicles (Fig. 9) indicates a shift in isoelectric points of the soluble L_{III} toward the acidic side of the isoelectric focusing gradient. Preliminary data suggest that this shift might be due to phosphorylation, since it can be reversed by treatment of L_{III} with alkaline phosphatase (results not shown). Since the monoclonal antibody recognizes all three lamins, it is evident that neither lamin L_I nor L_{II} is present in a significant amount in mature eggs.

The sedimentation coefficient of the soluble lamin L_{III} was determined by sucrose gradient centrifugation of egg extracts (Benavente et al., 1985; R. Stick unpublished). Although s values vary between 9 S and 7 S in these determinations, both values are significantly larger than that expected for a monomeric L_{III}. Further experiments will be needed to clarify whether L_{III} forms homogeneous oligomers or whether it is associated with other components in the egg cytoplasm.

B. Two Different States of Lamin L_{III} Existing in the Same Cell during Early Development

We have previously shown that lamin L_{III} is the only lamin component in pronuclei and in cleavage nuclei up to the midblastula transition (MBT) in *Xenopus* (Stick and Hausen, 1985). The protein forming the oocyte nuclear lamina is stored in soluble form in the ooplasm after egg maturation and is reused in the assembly of lamina structures of the cleavage nuclei up to MBT. L_{III} is not synthesized at a significant rate at these early stages of development (R. Stick, unpublished). Synthesis of L_{III} begins only after the onset of the MBT when lamin L_I first appears. At this stage

---

polypeptides. Note that only one lamin (L_{III}) is present in germinal vesicles as well as in eggs. The positions where lamins L_I and L_{II} would be expected to be observed are indicated by arrows. The lamin found in eggs and embryos up to MBT has previously been shown by tryptic peptide map analysis to be identical to that in germinal vesicles, i.e., L_{III} (Stick and Hausen, 1985). The brackets designate the isoelectric variants of lamin L_{III}. The arrowheads mark the basic spot of BSA, which served as an internal reference protein.

of development, both L_I and L_{III} are translated from prexisting, maternal messengers (Stick and Hausen, 1985).

During cleavage, the embryonic cell cycle can be arrested in interphase by inhibition of protein synthesis (Miake-Lye *et al.*, 1983). This treatment allows the isolation of L_{III} in two different states from the same cells: assembled in the lamina of cleavage-stage nuclei or stored in the cytoplasm in the soluble form. The cytoplasmic L_{III} is indistinguishable from L_{III} of mature eggs by two-dimensional gel analysis (R. Stick, unpublished). Whether this is also true for the aggregated lamins is currently under investigation. These experiments may elucidate whether protein modification (e.g., phosphorylation) is an essential step in lamina assembly in cleavage-stage nuclei, as has been assumed for lamina formation during mitotic telophase in mammalian cells (see Section II,E; Ottaviano and Gerace, 1985).

C. *In Vitro* Assembly of Lamina in Amphibian Egg Extracts

Recently, Forbes *et al.* (1983) showed that nucleuslike structures are formed around DNA injected into *Xenopus* eggs. These "nuclei" are surrounded by a nuclear envelope consisting of a double membrane, pore complexes, and a lamina. Formation of nuclei can also proceed *in vitro*, using a cytoplasmic extract from activated eggs (Lohka and Masui, 1983). These nuclei respond to modulators of the cell cycle, e.g., MPF and cytostatic factor (CSF). Although formation of nuclei is governed by addition of DNA to these extracts, lamin assembly occurs independent of specific DNA sequence information. Therefore, this *in vitro* system might be very suitable for studying lamin disassembly and re-formation in molecular detail.

V. COMPARISON OF THE DYNAMICS OF THE NUCLEAR LAMINA DURING MITOSIS AND MEIOSIS

The molecular events that occur during mitotic and meiotic divisions have many features in common. Factors regulating nuclear envelope breakdown and chromosome condensation (MPF and CSF) are effective both in mitosis and in meiosis, indicating that the molecular mechanism of both of these events may be related. This may also hold for the mechanisms by which disassembly and reassembly of the lamins is achieved in these processes. In mitosis and meiosis, disassembly of the lamina is accompanied by modification of lamins (influencing the electric charge of the molecules), and in both processes, lamin polypeptides are released into the cytoplasm in soluble form and are reused in the subsequent nuclear assembly process. It remains to be clarified whether the observed differences (monomeric versus oligomeric forms and membrane-associated versus free lamins) are characteristic differences between mitosis and meiosis, or whether they reflect species-specific differences of the animals used as experimental systems in the studies reported here.

The dynamics of the lamina observed during meiotic prophase, described in Section III, differ from those observed during both mitotic and meiotic divisions. During meiotic prophase, lamina disassembly and formation occurs within the intact nuclear membranes. Lamin polypeptides are not detected in the cytoplasm of these cells during the absence of the lamina structure. In *Xenopus,* this might be due to the change in lamin composition which occurs after reappearance of a lamina in diplotene oocytes. In chickens, it might be related to the fact that meiotic prophase lasts much longer (days) than mitotic and meiotic divisions; lamins might be degraded during disappearance of the lamina and newly synthesized at diplotene stage.

Unfortunately, analysis of molecular details of lamin dynamics during meiotic prophase is hindered by the fact that sufficient quantities of cells in defined meiotic stages are difficult to obtain in the species thus far studied. Since the nuclear lamina is a major skeletal element of eukaryotic nuclei, analysis of the changes of the lamina during mitosis and meiosis described here may be an important step toward an understanding of these cellular processes and might also give further insights into the principles of the organization of interphase nuclei.

ACKNOWLEDGMENTS

I am grateful to Brigitte Bühler, Gerd Klein, and Heinz Schwarz for providing the data presented in Figs. 1, 3, 4, and 5. I would like to thank Karin Herrmann for excellent technical assistance, Barbara Breidenbend and Roswitha Grömke-Lutz for help with the photographic work, and Brigitte Hieber for typing the manuscript. For helpful discussions and critical reading of the manuscript I thank Christine Dreyer, Peter Hausen, David Russell, and Rudolf Winklbauer.

REFERENCES

Aaronson, R., and Blobel, G. (1975). On the attachment of the nuclear pore complex. *J. Cell Biol.* **62,** 746–754.

Alberts, B., Bray, D., Lewis, J., Raff, M., Roberts, K., and Watson, J. D. (1983). "Molecular Biology of the Cell." Garland Publishing, New York.

Baglia, F. A., and Maul, G. G. (1983). Nuclear ribonucleoprotein release and nucleoside triphosphatase activity are inhibited by antibodies directed against one nuclear matrix glycoprotein. *Proc. Natl. Acad. Sci. U.S.A.* **80,** 2285–2289.

Benavente, R., and Krohne, G. (1985). Change of karyoskeleton during spermatogenesis of *Xenopus:* Expression of lamin L$_{IV}$, a nuclear lamina protein specific for the male germ line. *Proc. Natl. Acad. Sci. U.S.A.* **82,** 6172–6180.

Benavente, R., Krohne, G., and Franke, W. W. (1985). Cell type-specific expression of nuclear lamina proteins during development of *Xenopus laevis. Cell (Cambridge, Mass.)* **41,** 177–190.

Bühler, B. (1984). Immunhistologische Untersuchungen über das Verhalten von Kernproteinen während der Mitose. Diploma Thesis, University of Tübingen.

Burke, B., Tooze, J., and Warren, G. (1983). A monoclonal antibody which recognises each of the nuclear lamin polypeptides in mammalian cells. *EMBO J.* **2,** 361–367.

Cobbs, C. S., Jr., and Shelton, K. R. (1978). Major oligomeric structural proteins of the HeLa nucleus. *Arch. Biochem. Biophys.* **189,** 323–335.

Dwyer, N., and Blobel, G. (1976). A modified procedure for isolation of a pore complex-lamina fraction from rat liver nuclei. *J. Cell Biol.* **70,** 581–591.

Ely, S., D'Arcy, A., and Jost, E. (1978). Interaction of antibodies against nuclear envelope-associated proteins from rat liver nuclei with rodent and human cells. *Exp. Cell Res.* **116,** 325–331.

Fawcett, D. W. (1966). On the occurrence of a fibrous lamina on the inner aspect of the nuclear envelope in certain cells of vertebrates. *Am. J. Anat.* **119,** 129–146.

Fawcett, D. W. (1981). "The Cell," pp. 281–291. Saunders, Philadelphia, Pennsylvania.

Fisher, P. A., Berrios, M., and Blobel, G. (1982). Isolation and characterization of a proteinaceous subnuclear fraction composed of nuclear matrix, peripheral lamina, and nuclear pore complexes from embryos of *Drosophila melanogaster. J. Cell Biol.* **92,** 674–686.

Forbes, D. J., Kirschner, M. W., and Newport, J. W. (1983). Spontaneous formation of nucleus-like structures around bacteriophage DNA microinjected into Xenopus eggs. *Cell (Cambridge, Mass.)* **34,** 13–23.

Fuchs, J.-P., Giloh, H., Kuo, C.-H., Saumweber, H., and Sedat, J. (1983). Nuclear structure: Determination of the fate of the nuclear envelope in *Drosophila* during mitosis using monoclonal antibodies. *J. Cell Sci.* **64,** 331–349.

Gerace, L., and Blobel, G. (1980). The nuclear envelope lamina is reversibly depolymerized during mitosis. *Cell (Cambridge, Mass.)* **19,** 277–287.

Gerace, L., and Blobel, G. (1981). Nuclear lamina and the structural organization of the nuclear envelope. *Cold Spring Harbor Symp. Quant. Biol.* **46,** 967–978.

Gerace, L., Blum, A., and Blobel, G. (1978). Immunocytochemical localization of the major polypeptides of the nuclear pore complex-lamina fraction. *J. Cell Biol.* **79,** 546–566.

Gerace, L., Comeau, C., and Benson, M. (1984). Organization and modulation of nuclear lamina structure. *J. Cell Sci., Suppl.* **1,** 137–160.

Hancock, R., and Hughes, M. E. (1982). Organization of DNA in the interphase nucleus. *Biol. Cell.* **44,** 201–212.

Hausen, P., Wang, Y. H., Dreyer, C., and Stick, R. (1985). Distribution of nuclear proteins during maturation of the *Xenopus* oocyte. *J. Embryol. Exp. Morphol. (Suppl.)* **89,** 17–34.

Högner, D., Telling, A., Lepper, K., and Jost, E. (1984). Patterns of nuclear lamins in diverse animal and plant cells and in germ cells as revealed by immunofluorescence microscopy with polyclonal and monoclonal antibodies. *Tissue Cell* **16,** 693–703.

Jost, E., and Johnson, R. T. (1981). Nuclear lamina assembly, synthesis and disaggregation during the cell cycle in synchronized HeLa cells. *J. Cell Sci.* **47,** 25–53.

Kaufmann, S. H., Gibson, W., and Shaper, J. H. (1983). Characterization of the major polypeptides of the rat liver nuclear envelope. *J. Biol. Chem.* **258,** 2710–2719.

Klein, G. (1984). Untersuchungen der Lamine des Krallenfrosches *Xenopus laevis* mit Hilfe monoklonaler Antikörper. Diploma Thesis, University of Tübingen.

Krohne, G., Franke, W. W., Ely, S., D'Arcy, A., and Jost, E. (1978). Localization of a nuclear envelope-associated protein by indirect immunofluorescence microscopy using antibodies against a major polypeptide from rat liver fractions enriched in nuclear envelope-associated material. *Cytobiologie* **18,** 22–38.

Krohne, G., Dabauvalle, M.-C., and Franke, W. W. (1981). Cell type-specific differences in protein composition of nuclear pore complex-lamina structures in oocytes and erythrocytes of *Xenopus laevis*. *J. Mol. Biol.* **151,** 121–141.

Krohne, G., Debus, E., Osborn, M., Weber, K., and Franke, W. W. (1984). A monoclonal antibody against nuclear lamina proteins reveals cell type-specificity in *Xenopus laevis*. *Exp. Cell Res.* **150,** 47–59.

Lam, K., and Kasper, C. (1979). Electrophoretic analysis of three major nuclear envelope polypeptides. *J. Biol. Chem.* **254,** 11713–11720.

Lebkowsky, J. S., and Laemmli, U. K. (1982). Non-histone proteins and long-range organization of HeLa interphase DNA. *J. Mol. Biol.* **156,** 325–344.

Lohka, M. J., and Masui, Y. (1983). Formation in vitro of sperm pronuclei and mitotic chromosomes induced by amphibian ooplasmic components. *Science* **220,** 719–721.

McKeon, F. D., Tuffanelli, D. L., Fukuyama, K., and Kirschner, M. W. (1983). Autoimmune response directed against conserved determinants of nuclear envelope proteins in a patient with linear scleroderma. *Proc. Natl. Acad. Sci. U.S.A.* **80,** 4374–4378.

McKeon, F. D., Tuffanelli, D. L., Kobayashi, S., and Kirschner, M. W. (1984). The redistribution of a conserved nuclear envelope protein during the cell cycle suggests a pathway for chromosome condensation. *Cell (Cambridge, Mass.)* **36,** 83–92.

Maike-Lye, R., and Kirschner, M. W. (1985). Induction of early mitotic events in a cell-free system. *Cell (Cambridge, Mass.)* **41,** 165–175.

Miake-Lye, R., Newport, J. W., and Kirschner, M. W. (1983). Maturation-promoting factor induces nuclear envelope breakdown in cycloheximide-arrested embryos of *Xenopus laevis*. *J. Cell Biol.* **97,** 81–91.

Ottaviano, Y., and Gerace, L. (1985). Phosphorylation of the nuclear lamins during interphase and mitosis. *J. Biol. Chem.* **260,** 624–632.

Roos, U.-P. (1973). Light and electron microscopy of rat kangaroo cells in mitosis. *Chromosoma* **40,** 43–82.

Schellens, J. P. M., James, J., and Hoeben, K. A. (1979). Some aspects of the fine structure of the sex chromatin body. *Biol. Cell.* **35,** 11–14.

Shelton, K. R., Higgins, L. L., Cochran, D. L., Ruffolo, J. J., Jr., and Egle, P. M. (1980). Nuclear lamins of erythrocyte and liver. *J. Biol. Chem.* **255,** 10978–10983.

Stafstrom, J. P., and Staehelin, L. A. (1984). Dynamics of the nuclear envelope and of nuclear pore complexes during mitosis in the *Drosophila* embryo. *Eur. J. Cell Biol.* **34,** 179–189.

Stick, R., and Hausen, P. (1980). Immunological analysis of nuclear lamina proteins. *Chromosoma* **80,** 219–236.

Stick, R., and Hausen, P. (1985). Changes in the nuclear lamina composition during early development of *Xenopus laevis*. *Cell (Cambridge, Mass.)* **41,** 191–200.

Stick, R., and Krohne, G. (1982). Immunological localization of the major architectural protein associated with the nuclear envelope of the *Xenopus laevis* oocyte. *Exp. Cell Res.* **138,** 319–330.

Stick, R., and Schwarz, H. (1982). The disappearance of the nuclear lamina during spermatogenesis: An electron microscopic and immunofluorescence study. *Cell Differ.* **11,**
235–243.

Stick, R., and Schwarz, H. (1983). Disappearance and reformation of the nuclear lamina
structure during specific stages of meiosis in oocytes. *Cell* (*Cambridge, Mass.*) **33,**
949–958.

3

Regulation of Nuclear Formation and Breakdown in Cell-Free Extracts of Amphibian Eggs

MANFRED J. LOHKA AND JAMES L. MALLER

Department of Pharmacology
University of Colorado School of Medicine
Denver, Colorado 80262

I. INTRODUCTION

Embryonic development in most animals is initiated by the fusion of the sperm with the egg at fertilization. Prior to fertilization, the DNA of the sperm nucleus is usually in a highly condensed state and is often associated with basic proteins specific to the sperm chromatin (Bloch, 1969). The sperm nuclei are inactive in both RNA and DNA synthesis. However, upon entering the egg cytoplasm at fertilization the sperm nucleus undergoes a dramatic transformation in its morphology and synthetic activity as the male pronucleus is formed. The sperm-specific nuclear proteins are replaced, and the sperm chromatin decondenses and is reorganized into an interphase pronucleus that replicates its DNA during the first mitotic cell cycle of the zygote. Since both the male pronucleus and the female pronucleus form the diploid nucleus of the zygote, the transformation of the sperm nucleus into the male pronucleus has long been recognized as being crucial for ensuring that the paternal genome participates in embryonic development (Wilson, 1896). Because of its importance to embryonic development, male pronuclear formation has been examined in a large variety of species, including those in which normal fertilization occurs at either the germinal vesicle (GV) stage, the first meiotic meta-

MOLECULAR REGULATION OF NUCLEAR EVENTS
IN MITOSIS AND MEIOSIS

phase, the second meiotic metaphase, or the pronuclear stage. These studies, which have dealt mostly with the cytological and morphological events of pronuclear formation, have been the subject of thorough reviews (Longo, 1973, 1983; Longo and Kunkle, 1978).

It has become clear from ultrastructural observation of male pronuclear formation in many different organisms that, although the details may vary between organisms, the morphological changes that occur during pronuclear formation are remarkably consistent, regardless of the type of specific nuclear proteins found in the sperm or the stage of meiosis at which the egg is normally fertilized. In general, the sperm nucleus undergoes the following changes during pronuclear formation (Longo, 1973; Longo and Kunkle, 1978). Initially, the chromatin is highly condensed and consequently very electron dense when examined at an ultrastructural level. Upon entering the egg cytoplasm, the nuclear envelope (NE) surrounding the sperm chromatin breaks down, a process in which the inner and outer membrane of the NE fuse at many sites to form numerous vesicles that become dispersed among the membranous elements of the egg cytoplasm. The sperm chromatin, which is now in direct contact with the egg cytoplasm, begins to decondense; the peripheral chromatin becomes less electron dense and more fibrous in nature, while a core of highly condensed chromatin remains. Direct contact of the sperm chromatin with the egg cytoplasm is transient, however, as an NE again forms around the chromatin from membrane elements of the egg cytoplasm. Since the surface area of the pronucleus is many times greater than that of the sperm head, much of the newly assembled NE must be formed from egg membrane vesicles (Longo, 1976). Vesicles first accumulate along the peripheral chromatin and then fuse together and flatten against the chromatin until its entire periphery is delimited by nascent NE. Pores form at numerous sites in the pronuclear envelope. The chromatin continues to decondense, both during and after the assembly of the pronuclear envelope, until all of the highly condensed electron-dense chromatin disappears. Although the morphological events of male pronuclear formation have been described extensively, relatively little is known about the nature of the egg components that regulate these changes.

It is important to note that the formation of the female pronucleus entails several of the morphological changes seen during male pronuclear formation (Longo, 1973). Following anaphase II, membranous vesicles aggregate around the periphery of individual egg chromosomes and progressively fuse together to form an NE as the chromosomes decondense. The NE enclosing each chromosome fuses to that surrounding other chromosomes until all chromosomes become incorporated into a single nucleus, the female pronucleus. Once formed, the female pronucleus continues to enlarge and synthesizes DNA (Simmel and Karnofsky, 1961; Hinegardner *et al.*, 1964; Zimmerman and Zimmerman, 1967). Both the assembly of an NE from cytoplasmic membranes of the egg and the enlargement of the female pronucleus appear to occur in a manner similar to that described for male pronuclear formation. Furthermore, formation of the female pronucleus closely resembles the formation of the nucleus during telophase in mitotically dividing cells (Chang and Gibley, 1968; Longo, 1972; Gulyas, 1972; Chai *et al.*, 1974; Itoh *et al.*, 1921). Thus, an understanding of the control of pronuclear formation may provide insight into the control of NE assembly and chromatin decondensation in many different types of cells.

Amphibian eggs, like those of most vertebrates, are physiologically arrested at metaphase of the second meiotic division until fertilization, at which time meiosis is completed and both the male and female pronuclei form rapidly. The large size of amphibian eggs has prevented a detailed ultrastructural study of pronuclear formation. However, given the similarities in pronuclear formation for species as diverse as sea urchin and mouse, it is unlikely that in amphibian eggs this process differs markedly from that described for other species. On the other hand, the large size of amphibian eggs, together with their ability to withstand a variety of experimental manipulations, such as microinjection and nuclear transplantation, has favored their use in the study of the cytoplasmic control of pronuclear formation and other types of nuclear behavior. In addition, following fertilization, amphibian eggs enter a period of extremely rapid cell division without growth of the embryo. Within 8 hr the fertilized *Xenopus* egg divides into about 4000 cells. The nuclear constituents required for the proliferation of the embryonic nuclei during this period of rapid cell division are stored in the egg cytoplasm in quantities that greatly exceed those found in single somatic cells. Stored cytoplasmic pools of histones (Woodland and Adamson, 1977; Kleinschmidt and Franke, 1982; Kleinschmidt *et al.*, 1985), DNA polymerases (Benbow *et al.*, 1975; Zierler *et al.*, 1985), and nuclear lamins (Benavente *et al.*, 1985; Stick and Hausen, 1985) are all thought to contribute to the formation of blastomere nuclei. Since egg proteins enter into transplanted nuclei (Arms, 1968; Merriam, 1969; Hoffner and DiBerardino, 1977), it is likely that these proteins are also incorporated into the pronuclei.

The cytoplasmic components involved in pronuclear formation, whether or not they are the same as those involved in formation of blastomere nuclei, are abundant in eggs. In amphibians, as in other species, many male pronuclei form following polyspermic fertilization, both in urodeles, where normal fertilization is often polyspermic (Fankhauser and Moore, 1941; Wakimoto, 1979), and in anurans, where normal fertilization is monospermic (Graham, 1966). Similarly many pronuclei form from sperm nuclei that have been transplanted into the cytoplasm of fertilized or parthenogenetically activated eggs (Katagiri and Moriya, 1976; Moriya and Katagiri, 1976; Lohka and Masui, 1983a). Somatic cell nuclei can also be induced to undergo changes similar to those seen during pronuclear formation. When they are transplanted into amphibian egg cytoplasm, their chromatin becomes more decondensed and the nuclei enlarge and synthesize DNA (Graham *et al.*, 1966; Merriam, 1969; de-Roeper *et al.*, 1977). In fact, the egg cytoplasm even appears to be capable of inducing similar changes in purified DNA. When DNA from bacteriophages or plasmids is injected into eggs, the DNA is assembled into chromatin that becomes enclosed within an NE to form nucleus-like structures (Forbes *et al.*, 1983) whose replication is coordinate with that of the blastomere nuclei (Harland and Laskey, 1980). It is tempting to speculate that the changes in the injected DNA result from the activity of ooplasmic components that act either when sperm-specific nuclear proteins are replaced by those stored in the egg cytoplasm or during the assembly of newly replicated pronuclear DNA into chromatin. An NE would form around the newly assembled chromatin just as it does around the sperm chromatin during pronuclear formation.

As described previously, sperm nuclei are transformed into male pronuclei through the interaction of egg cytoplasmic components with the sperm chromatin. This transformation is the culmination of several complex processes, including NE assembly, chromatin decondensation, and nuclear enlargement. Although pronuclear formation has been examined extensively at an ultrastructural level, neither the biochemical characterization of the egg components involved nor the elucidation of the mechanism whereby they induce sperm chromatin to undergo this process has progressed rapidly. These shortcomings are due, at least in part, to the difficulty with many eggs of isolating pronuclei during their formation (for an exception, see Chapter 6, this volume) and in the inability to manipulate ooplasmic conditions that may affect pronuclear formation. These problems may be overcome if pronuclear formation can be studied *in vitro*.

Previous studies have suggested that cytoplasmic egg extracts may be used to induce *in vitro* the changes in nuclear activity seen in the intact egg. Cytoplasmic preparations of *Xenopus* eggs have been shown to induce nuclear swelling and DNA synthesis in incubated somatic cell nuclei (Barry and Merriam, 1972; Benbow and Ford, 1975). Also, extracts of sea urchin eggs can induce sperm chromatin to decondense and accumulate proteins that normally associate with pronuclei in intact eggs (Kunkle *et al.*, 1978). In this chapter we describe a cell-free system, derived from amphibian egg cytoplasm, which can induce sperm nuclei to form pronuclei *in vitro*. The extracts we discuss were prepared from the eggs of the leopard frog, *Rana pipiens* (Lohka and Masui, 1983a; 1984a,b), or of the South African clawed toad, *Xenopus laevis* (Lohka and Maller, 1985), although similar extracts have been prepared from the Japanese toad, *Bufo bufo japonicus* (Iwao and Katagiri, 1984).

II. REGULATION OF PRONUCLEAR FORMATION

A. Preparation of Cytoplasmic Extracts

Cell-free cytoplasmic extracts capable of inducing pronuclear formation *in vitro* can be prepared either from unfertilized eggs or from parthenogenetically activated eggs. The procedure for preparing these extracts is similar for both *Rana* and *Xenopus* eggs, although the buffers utilized are somewhat different (Table I). The buffers were modified from those used by Masui (1982) to prepare extracts of *Rana* eggs and by Benbow and Ford (1975) to assay nuclear DNA synthesis in *Xenopus* egg extracts. Females were induced to ovulate by injection of either pituitary homogenates and progesterone, for *Rana* (see Shibuya and Masui, 1982), or human chorionic gonadotropin, for *Xenopus* (see Newport and Kirschner, 1982). Ovulated eggs were dejellied and washed well with 0.1 M NaCl and then with extraction buffer. Undamaged, dejellied eggs were transferred to centrifuge tubes containing ice-cold extraction buffer and allowed to

TABLE I

Composition of Buffers Used for the Preparation of Cytoplasmic Extracts from Amphibian Eggs

Component	Concentration (mM)	
	Rana	*Xenopus*
KCl	200	100
MgCl₂	1.5	5
Sucrose	250	—
2-Mercaptoethanol	2	2
Tris-HCl (pH 7.5)	10	—
HEPES (pH 7.5)	—	20
Phenylmethylsulfonyl fluoride	—	0.3
Leupeptin	—	3 μg/ml

settle. Usually about 5 ml of dejellied *Rana* eggs in Beckman Ultra-Clear centrifugation tubes (344057) or 3 ml of dejellied *Xenopus* eggs in Dupont centrifuge tubes (03100) were used. Once the eggs had settled, excess buffer was withdrawn and the tubes were centrifuged at 10,000 g for 10 min. During centrifugation, the eggs are crushed—the yolk, cortex, and much of the pigment are sedimented in the pellet, and other cytoplasmic constituents are released into the supernatant without homogenization. In *Rana* egg extracts, the supernatant consists of two layers: an upper layer, designated the light ooplasmic fraction, and a more viscous and heavily pigmented lower layer, designated the heavy ooplasmic fraction. Only the heavy ooplasmic fraction is capable of supporting pronuclear formation (Lohka and Masui, 1983b; 1984a). In contrast to the supernatants obtained from *Rana* eggs, those from *Xenopus* eggs did not have similar light and heavy ooplasmic fractions. In this case, the entire supernatant could be used for incubation with sperm nuclei. In addition, for *Xenopus* egg extracts, cytochalasin B was added to a concentration of 50 μg/ml to prevent the formation of actin gels, which can trap particulate cytoplasmic components and the incubated nuclei. The supernatant fractions from both *Rana* and *Xenopus* eggs were again centrifuged at 10,000 g for 10 min before they were mixed with sperm nuclei.

Mature sperm and late spermatids were isolated from *Xenopus* testes and permeabilized by treatment with lysolecithin (α-lysophosphatidyl-choline) in the presence of protease inhibitors to minimize damage to the chromatin by sperm-associated proteases. Both the sperm plasma membrane and nuclear envelope were removed by the lysolecithin treatment, while the chromatin remained in a highly condensed state characteristic of the intact sperm head (Lohka and Masui, 1983b). The demembranated sperm nuclei were mixed with 150–200 μl aliquots of the cytoplasmic extracts to give concentrations of 0.5–1 $\times$ 10^5 nuclei/ml for *Rana* and 1–2 $\times$ 10^6 nuclei/ml for *Xenopus*. Reaction mixtures were incubated at 18°C.

Although lysolecithin-treated sperm nuclei can be stored at -70°C for several months, we find that egg cytoplasmic extracts must be used soon after they are prepared, since in our hands they lose their ability to induce pronuclear formation if they are frozen at -70°C or kept at 4°C.

B. Pronuclear Formation in Cell-Free Extracts

Since lysolecithin treatment removes both the sperm plasma membrane and the nuclear envelope, the incubation of lysolecithin-treated sperm nuclei in egg cytoplasmic extracts may be equivalent to the condition that exists shortly after fertilization, when the sperm chromatin is directly exposed to the egg cytoplasm after the fragmentation of the NE. As in intact eggs, the sperm chromatin incubated in cytoplasmic extracts is transformed into a pronucleus—an NE is assembled around the sperm chromatin, the chromatin decondenses, and the pronucleus enlarges and synthesizes DNA (Lohka and Masui, 1983b, 1984a). These changes are shown in Fig. 1. When first mixed with the cytoplasmic extracts, the chromatin of the *Xenopus* sperm is highly condensed in a helically shaped nucleus that stains deeply with Giemsa (Fig. 1A). In *Rana* egg extracts, most of the sperm nuclei have become round or oval in shape by 30 min of incubation, yet their chromatin still stains deeply (Fig. 1B). By 60 min the nuclei have a core of condensed chromatin and a region of decondensed chromatin, usually at the periphery or at one end of the nucleus, which stains less intensely (Fig. 1C). The decondensation of the sperm chroma-

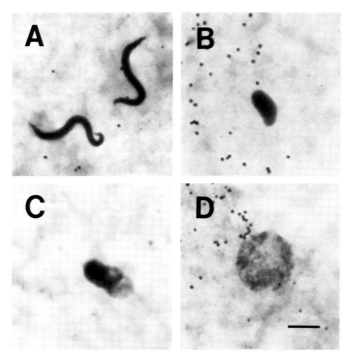

Fig. 1.   Changes in *X. laevis* sperm nuclei during incubation in cell-free extracts of *R. pipiens* eggs. (A) Condensed sperm nuclei, 0 min; (B) rounded sperm nucleus, 30 min; (C) partially decondensed sperm nucleus, 60 min; (D) completely decondensed sperm, 90 min. Scale, 10 μm.

tin continues until by 90 min the highly condensed core of chromatin is completely absent and pronuclei with fully decondensed chromatin have formed (Fig. 1D). The pronuclei continue to enlarge during further incubation.

Although the initial studies of pronuclear formation *in vitro* were with *Xenopus* sperm nuclei incubated in *Rana* egg extracts, *Rana* sperm nuclei also undergo a similar series of changes when incubated in *Rana* egg extracts, as do *Xenopus* sperm incubated in *Xenopus* egg extracts. However, in *Xenopus* extracts, fully decondensed pronuclei usually form more rapidly, within 45–60 min. Changes observed in the cell-free cytoplasmic extracts closely resemble those seen in fertilized amphibian eggs, but occur more slowly than in intact eggs. Pronuclear formation occurs within 45 min in *Rana* (Subtelny and Bradt, 1963) and within 20 min in *Xenopus* (Graham, 1966). Nevertheless, since pronuclear formation takes about twice as long in *Rana* extracts than in *Xenopus* extracts, the cell-free system appears to display the differences in the rate of pronuclear formation seen in intact eggs.

The ability of pronuclei formed *in vitro* to synthesize DNA was examined in *Rana* egg extracts (Lohka and Masui, 1983b). Tritium-labeled thymidine triphosphate ([³H]dTTP), a DNA precursor, was added to the cytoplasmic extracts and the incorporation of radiolabel by pronuclei was examined by autoradiography. As shown in Fig. 2, both partially decondensed (Fig. 2A) and fully decondensed pronuclei (Fig.2B) incorporated [³H]dTTP. However, in partially decondensed pronuclei, incorporation of radiolabel was not always restricted to the region of the nucleus that was decondensed (as shown in Fig. 2A), but often could be seen over the deeper staining condensed chromatin as well. The incorporation of [³H]dTTP by the pronuclei was inhibited by aphidicolin (Fig 2C), an inhibitor of DNA polymerase-α (Ikegami *et al.*, 1978), the major DNA polymerase involved in DNA replication (Weissbach, 1979). However, further experiments are necessary to determine whether the incorporation of radiolabel by pronuclei is due to replication rather than DNA repair by DNA polymerase-α.

The cell-free extracts of *Xenopus* eggs appear to be capable of inducing other nuclei to undergo changes in their morphology and activity similar to those seen in intact eggs. Just as somatic cell nuclei are induced to enlarge when transplanted into activated eggs (Graham *et al.*, 1966; Merriam, 1969; deRoeper *et al.*, 1977), *Xenopus* brain and liver nuclei also enlarge during incubation in cell-free extracts (Fig. 3). Furthermore, extracts of *Xenopus* eggs have been reported to induce human sperm nuclei to swell and initiate DNA synthesis, although it is not known if an NE is assembled around the swollen human sperm chromatin (Gordon *et al.*,

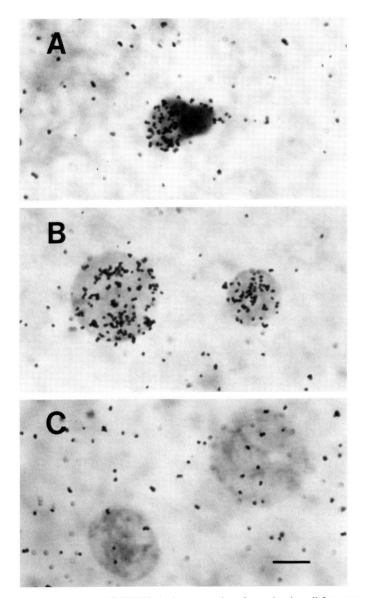

Fig. 2. Incorporation of [³H]dTTP during pronuclear formation in cell-free extracts of *R. pipiens* eggs. Sperm nuclei were incubated for 3 hr with egg extracts containing (A and B) 40 µCi/ml [³H]dTPP or (C) 40 µCi/ml [³H]dTPP and 5 µg/ml aphidicolin. Dimethyl sulfoxide, the vehicle for aphidicolin, had no effect on [³H]dTTP incorporation. Scale, 10 µm.

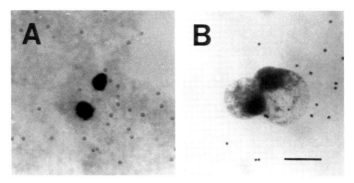

Fig. 3.   Enlargement of isolated *X. laevis* brain nuclei during incubation in activated *Xenopus* egg extracts. (A) 0 min; (B) 60 min. Scale, 10 μm.

1985). In this regard, it may be of interest to examine the behavior in the cell-free extracts of sperm nuclei from a variety of species, particularly those whose sperm-specific nuclear proteins differ greatly from *Rana* or *Xenopus*.

C. Nuclear Envelope Assembly *in Vitro*

During formation of the male pronucleus, membranous elements of the egg cytoplasm, including vesicles, contribute to the assembly of an NE around the decondensing sperm chromatin. Morphologically, NE assembly during male pronuclear formation is similar to that seen during female pronuclear formation (Longo, 1973) and during telophase, when mitotic chromosomes re-form a nucleus (Chang and Gibley, 1968; Longo, 1972; Gulyas, 1972; Chai *et al.*, 1974; Itoh *et al.*, 1981). Therefore, the molecular basis of NE assembly during pronuclear formation may be similar to that in many different dividing cells. However, in all cases, these mechanisms are, at best, only poorly understood.

In both *Rana* and *Xenopus* egg extracts, an NE is also assembled around the sperm chromatin during pronuclear formation *in vitro* (Fig. 4). The ultrastructure of NE assembly has been studied in detail in *Rana* egg extracts (Lohka and Masui, 1984a). The sperm chromatin, which is initially highly condensed and electron-dense when added to the cytoplasmic extracts, disperses within 5 min and becomes less electron-dense and more fibrous in nature. At this time the periphery of the chromatin is almost totally devoid of a NE. During further incubation, membrane vesicles aggregate at the periphery of the chromatin (Fig. 4A). These vesicles, many of which are about 180–200 nM in diameter and which contain electron-dense material, fuse to one another and flatten against the chromatin to form larger, flattened vesicles with both an outer membrane and an inner membrane adjacent to the chromatin (Fig. 4B). Electron-dense material accumulates at sites on the outer membrane (Fig. 4C) and pores form where the inner and outer membranes have joined (Fig. 4D). Although the mechanism of nuclear pore formation is not known, the material found on the outer membrane may be involved in this process. Nuclear envelope assembly is not synchronous over the entire periphery of the chromatin (Fig. 5); rather, short fragments of NE form first at many sites and then vesicles continue to fuse at the margins of these fragments (Fig. 4D) until a continuous envelope encloses the chromatin (Fig. 6). The time course of NE assembly in extracts is given in Table II. Only short fragments of NE, which together cover less than 50% of the periphery of the chromatin, are formed during the first 30 min of incubation. The amount of NE formed increases with continued incubation, but only after a 60- to 90-min incubation does the sperm chromatin become completely enclosed within an NE. Even after a continuous NE is formed, other vesicles continue to fuse with the NE as the pronucleus enlarges.

Stacks of NE resembling annulate lamellae were occasionally found in close apposition to the chromatin (Fig. 7), usually in sections where sev-

TABLE II

Time Course of NE Assembly on Sperm Chromatin Incubated in Activated *R. pipiens* Extracts[a]

Incubation time (min)	Number of nuclei	Percentage of nuclei based on percentage of the chromatin perimeter lined by NE			
		0	50	50	100
5	25	100	—	—	—
30	143	64	35	—	—
45	102	29	53	18	—
60	55	4	11	29	56
90	57	—	—	9	91

[a] *X. laevis* sperm nuclei were incubated in cytoplasmic extracts from activated eggs of *R. pipiens*. At various times, the mixture of nuclei and cytoplasmic extract was fixed and processed for ultrastructural examination. For each incubation time, the nuclei were classified according to proportion of the chromatin periphery that was covered by NE.

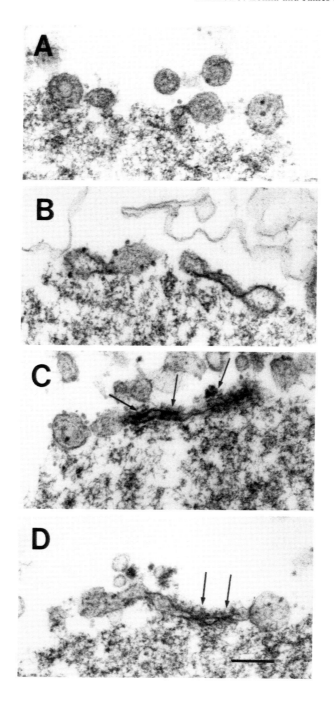

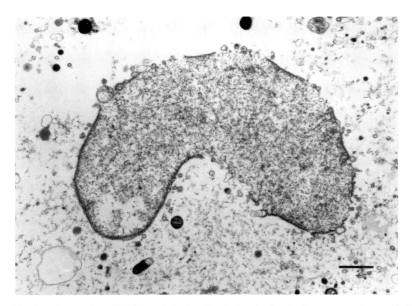

Fig. 5. Incomplete NE formed during 60-min incubation in *Rana* egg extracts. Scale, 1.0 μm.

eral nuclei were found near each other. It must be emphasized that NE were found only in close association with chromatin, and neither annulate lamellae nor membrane fragments resembling NE were found when cytoplasmic extracts were incubated in the absence of chromatin. Thus, it is likely that chromatin, cytoplasmic vesicles, and soluble cytoplasmic factors all interact to assemble an NE (Lohka and Masui, 1984a). However, the nature of these interactions awaits elucidation.

Although it is likely that the vesicles that form an NE *in vitro* also function in NE assembly during the period of rapid nuclear proliferation in the early embryo, nothing is known of their origin. Are they specialized vesicles whose sole function is to serve as precursors of NE assembly, or are they unspecialized membrane vesicles recruited into NE assembly by specific soluble factors? What is the role, if any, of their contents? What role does chromatin play in controlling the fusion of vesicles and their

Fig. 4. NE assembly in cell-free extracts of *R. pipiens* eggs. (A) Vesicles associated with the periphery of the sperm chromatin; (B) vesicles fused together and flattened against chromatin; (C) electron-dense material (arrows) accumulated on outer membrane; (D) pores (arrows) formed in short fragment of NE. Vesicles fused to lateral margins of fragment. Scale, 0.25 μm.

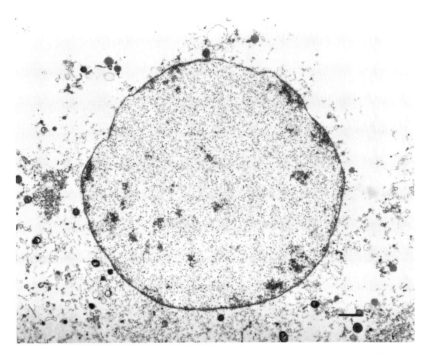

Fig. 6.   Complete NE formed during 90-min incubation in *Rana* egg extracts. Scale, 1.0 μm.

flattening at its periphery? How are nuclear pores formed and from where do the proteins of the nuclear pore complex originate? Are they soluble, and, if so, how do they recognize the membranous elements of the NE? Clearly, many interesting questions remain to be answered.

The membrane precursors of the NE assembly in sea urchin pronuclei are thought to be derived from the endoplasmic reticulum (Longo, 1976). In mitotically dividing cells, both the endoplasmic reticulum and remnants of the original NE, which breaks down prior to mitosis, are thought to contribute to NE assembly during telophase (Franke, 1974). The NE of the oocyte nucleus, the GV, may contribute many of the precursors of NE assembly required during formation of pronuclei and blastomere nuclei. Nuclear pore complexes occupy a large fraction of the total surface of the GV envelope (Franke and Scheer, 1970; Scheer, 1973). In addition, immunofluorescent staining has shown that the major protein of the nuclear lamina in *Xenopus* oocytes and early embryos, lamin L_{III}, is localized to the GV envelope (Stick and Krohne, 1982; Stick and Hausen, 1985; Benavente *et al.*, 1985; see also Chapter 2, this volume). During oocyte

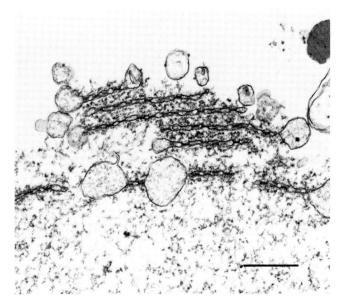

Fig. 7.   Stacks of NE formed at chromatin periphery during 60-min incubation in *Rana* egg extract. Scale, 0.5 μm.

maturation, the GV envelope breaks down before the oocyte completes its meiotic divisions and becomes capable of being fertilized (see Chapter 1, this volume). The NE fragments into vesicles devoid of any distinctive features and, therefore, cannot be distinguished from other cytoplasmic membranes once they have dispersed throughout the cytoplasm (Szollosi *et al.*, 1972a; Huchon *et al.*, 1981). At the time of NE breakdown, the nuclear lamina depolymerizes, and L_{III} becomes soluble (Stick and Hausen, 1985; Benavente *et al.*, 1985). Soluble L_{III} in the egg cytoplasm serves as a pool for the assembly of the nuclear lamina in pronuclei, both in fertilized eggs and in cell-free extracts (Stick and Hausen, 1985) and in the formation of blastomere nuclei. Just as L_{III} is utilized in NE assembly in pronuclei and blastomere nuclei, the membranes of the GV envelope and proteins of its nuclear pore complexes may also contribute to this process.

Circumstantial evidence suggests that components of the GV envelope are not the only source of precursors for NE assembly. In amphibians, as in many other species, it has been shown that sperm nuclei fail to form pronuclei when exposed to the cytoplasm of eggs from which the GV has been removed prior to oocyte maturation (Katagiri and Moriya, 1976;

Skoblina, 1976; Lohka and Masui, 1983a). Injection of only soluble GV contents, without the NE, is sufficient to restore the ability of these enucleated eggs to induce pronuclear formation. Therefore, if NE assembly is normal in pronuclei that form following the injection of soluble GV contents into enucleated eggs, the precursors of the nascent NE could not have come from the GV envelope. One must surmise that these precursors came from the soluble GV contents or a cytoplasmic store, or were synthesized *de novo* during oocyte maturation.

D. Chromatin Decondensation and Nuclear Enlargement

At least two different changes in chromatin morphology occur during the incubation of *Xenopus* sperm nuclei in cell-free cytplasmic extracts. The first is a rapid dispersion of the sperm chromatin. The second is the further decondensation of chromatin seen as the pronuclei enlarge, initially when the peripheral chromatin is no longer stained deeply by Giemsa and continuing until all of the chromatin is diffuse and uniformly lightly staining.

Within 5 min after sperm chromatin is mixed with the cytoplasmic extracts, the chromatin is no longer in a highly compact electron-dense state when examined ultrastructurally, but is less electron-dense and fibrous. Although this change is evident ultrastructurally, it cannot be readily detected at this time upon examination of Giemsa-stained nuclei with the light microscope. Since chromatin dispersion does not occur when sperm nuclei are incubated in extraction buffer alone (Lohka and Masui, 1984a), it appears to result from the action of egg cytoplasmic factors. These factors may be soluble and heat stable, since chromatin dispersion occurs in supernatants obtained following centrifugation of the cytoplasmic extracts at 150,000 g for 2 hr, and even after the supernatants have been heated to 100°C for 10 min.

Perhaps the most conspicuous change in the sperm nuclear morphology in the cell-free extracts is the formation of large, spherical, interphase pronuclei whose decondensed chromatin is more or less uniformly lightly staining (Fig. 1D). As mentioned earlier, this change is first evident in *Rana* egg extracts when the peripheral chromatin begins to enlarge after a 60-min incubation, and fully formed pronuclei with completely decondensed chromatin are seldom seen before 90 min. The first appearance of fully decondensed pronuclei between 60 and 90 min corresponds to the time when NE assembly around the entire periphery of the sperm chromatin is first completed. This correlation suggests that NE assembly is a prerequisite for the enlargement of the pronuclei. Support for this contention comes from the observation that pronuclei do not form when sperm nuclei are incubated either with the high-speed supernatants in the absence of the vesicles that contribute to NE assembly, or with the membrane vesicles under conditions in which NE assembly fails to occur (Lohka and Masui, 1984a).

In intact cells, many proteins are specifically localized to the nucleus and are found only in negligible amounts in the cytoplasm (Bonner, 1978). In amphibian oocytes, it has been clearly shown that nucleoplasmin (Mills *et al.*, 1980; Dingwall *et al.*, 1982), histones (Gurdon, 1970; Bonner, 1975a), and other nuclear proteins (Bonner 1975b; Feldherr, 1975; DeRobertis *et al.*, 1978; Feldherr *et al.*, 1983) selectively accumulate in the GV after they have been introduced into the oocyte cytoplasm by microinjection. This accumulation may result from selective entry through the pores of the NE (Feldherr *et al.*, 1984). Since the contents of the GV are known to be required for pronuclear formation, it is likely that GV proteins also accumulate in newly formed pronuclei in both intact eggs and their extracts, leading to the enlargement of the nuclei. One possible explanation of the requirement for an NE for nuclear enlargement may be that the assembly of an NE around the sperm chromatin establishes a nuclear compartment that can accumulate GV proteins which had been dispersed throughout the egg cytoplasm following GV breakdown. The NE may play a similar role in dividing cells during telophase when nuclear proteins that had been dispersed throughout the cytoplasm are again taken up by the re-forming nuclei.

Recently, nuclei assembled *in vitro* have been shown to accumulate nucleoplasmin (Newmeyer *et al.*, 1986). Whether or not the accumulation of proteins *in vitro* is as specific as *in vivo* has not yet been determined. For this, it would be necessary to demonstrate that the proteolytic fragments of nucleoplasmin (Dingwall *et al.*, 1982) or histone H1 (Dingwall and Allan, 1984) that accumulate in the GV of intact oocytes also accumulate in pronuclei, whereas those that do not accumulate in the nuclei of intact cells also fail to do so *in vitro*. Should the accumulation of proteins by the pronuclei *in vitro* prove to be as selective as in intact cells, the cell-free system described here may facilitate the study of the mechanism by which specific proteins accumulate in nuclei, not only during pronuclear formation and early development, but also throughout the cell cycle.

The presence of an NE may not be sufficient for the accumulation of proteins by nuclei. Isolated brain or liver nuclei enlarge when incubated in the cytoplasmic extracts (see Fig. 3) but not when incubated in the high-speed supernatants of these extracts. Therefore, it appears that the particulate components sedimented by centrifugation are also necessary for nuclear enlargement, either as components actively involved in transporting proteins into the nuclei or as precursors required for the increase in the surface area of the NE that must accompany enlargement.

In summary, cell-free extracts of amphibian eggs are able to transform sperm nuclei into pronuclei with structural and functional properties closely resembling those of pronuclei formed in intact eggs. The use of this cell-free system may advance our understanding not only of pronuclear formation but also of NE assembly and chromosome decondensation. The regulation of NE assembly–disassembly and chromosome condensation–decondensation may be particularly amenable to analysis since the cytoplasmic factors known to control these processes in intact cells are also active in cell-free extracts (see Section III).

III. REGULATION OF CHROMOSOME CONDENSATION

A. Nuclear Behavior in Unactivated Eggs

In fusing with the egg at fertilization, the sperm not only contributes the paternal genome for embryonic development, but also provides the stimulus required to initiate embryonic development. This stimulus is termed "activation." When embryonic development is activated by fertilization, sperm nuclei that enter the egg cytoplasm are quickly transformed into pronuclei. In contrast, pronuclei fail to form when sperm are exposed to egg cytoplasm under conditions in which activation has not occurred, such as when maturing amphibian oocytes are either inseminated precociously (Bataillon, 1928; Bataillon and Tchou-su, 1934; Elinson, 1977) or injected with sperm nuclei (Moriya and Katagiri, 1976). Rather, under these conditions, the sperm nuclei form metaphase chromosomes that become aligned on spindles. Similarly, when somatic cell nuclei are transplanted into maturing oocytes or unactivated eggs under conditions in which activation is suppressed, the NE of the transplanted nuclei breaks down and their chromatin condenses into metaphase chromosomes that become aligned on spindles (Gurdon, 1967, 1968; Ziegler and Masui, 1973, 1976). Since the unfertilized amphibian egg chromosomes are physiologically arrested at metaphase of the second meiotic division, the response of the transplanted nuclei is most likely due to the activity of the cytoplasmic factors that control the behavior of the egg chromosomes prior to fertilization. Although the exact nature of these cytoplasmic factors is not known, it is clear from these studies that the unfertilized amphibian egg cytoplasm can support NE breakdown, chromosome condensation, and spindle assembly in hundreds of transplanted nuclei. Thus, the cellular constituents contributing to these processes may be stored in the egg cytoplasm in much the same way that constituents of interphase blastomere nuclei are stored for the period of rapid proliferation following fertilization.

The behavior of egg chromosomes is controlled by cytoplasmic factors that appear during oocyte maturation (Masui and Clarke, 1979; Maller, 1985; see also Chapter 1, this volume). In the ovary, fully grown amphibian oocytes are arrested at prophase of the first meiotic division. Stimulation with a steriod hormone, such as progesterone, initiates oocyte maturation, during which the NE of the GV breaks down, the oocyte chromosomes condense and are incorporated into a spindle, and meiosis proceeds to metaphase II. GV breakdown results from the activity of a cytoplasmic factor(s) that develops after hormonal stimulation. Although injection of hormone alone into fully grown immature oocytes does not induce maturation, the cytoplasm taken from maturing oocytes several hours after hormonal stimulation induces GV breakdown and chromosome condensation in the recipients (Masui and Markert, 1971; Smith and Ecker, 1971). Since recipient oocytes undergo the entire process of maturation, the cytoplasmic factor that induces these changes is called maturation-promoting factor (MPF) (Masui and Markert, 1971). MPF activity can be detected not only in amphibian oocytes, but also in maturing oocytes from echinoderms (Kishimoto and Kanatani, 1976; Kishimoto *et al.*, 1982), molluscs (Kishimoto *et al.*, 1984), and mammals (Kishimoto *et al.*, 1984; Sorensen *et al.*, 1982). A similar activity is also present during mitosis in yeast (Weintraub *et al.*, 1982), cultured mammalian cells (Sunkara *et al.*, 1979; Nelkin *et al.*, 1980; Kishimoto *et al.*, 1982), and cleaving amphibian (Wasserman and Smith, 1978; Gerhart *et al.*, 1984) or echinoderm (Kishimoto *et al.*, 1982) embryos. Since MPF activity in these cells is very low or absent during interphase, but increases just prior to mitosis, MPF may be a universal regulator of NE breakdown, chromosome condensation, and spindle formation in both meiotically and mitotically dividing cells. In fact, the MPF activity from many different cell types is able to induce NE breakdown, chromosome condensation, and spindle assembly in both oocytes and cleavage-arrested embryos without any apparent species specificity (Kishimoto *et al.*, 1982, 1984; Halleck *et al.*, 1984). Therefore, characterization of MPF may be of general significance in understanding the control of cell division in all eukaryotic cells.

While the activity of MPF may be responsible for inducing NE breakdown and chromosome condensation during oocyte maturation, the arrest of meiosis at metaphase II may be due to the activity of a second cytoplasmic factor, cytostatic factor (CSF) (Masui and Markert, 1971; Meyerhof and Masui, 1977, 1979a). CSF can be detected in the cytoplasm of unfertilized eggs by its ability to arrest cell division after injection into cleaving blastomeres. In the arrested blastomere, the chromosomes are condensed at metaphase and found on a large anastral spindle, similar to that seen in unfertilized eggs (Masui and Markert, 1971; Meyerhof and Masui, 1977, 1979a,b). Therefore, the activity of at least two factors, MPF and CSF, can influence the behavior of nuclei that are exposed to unfertilized egg cytoplasm. Whereas MPF activity has been found in a variety of meiotic and mitotic cells, so far CSF has been found only in unfertilized eggs.

The stimulus that activates embryonic development appears to be an increase in the cytoplasmic free Ca^{2+} ion concentration ($[Ca^{2+}]_i$). In many species, including amphibians, agents that artificially elevate $[Ca^{2+}]_i$, such as the ionophore A23187, also trigger parthenogenetic activation of eggs (Steinhart et al., 1974; Belanger and Schuetz, 1975). Moreover, increases in ooplasmic $[Ca^{2+}]_i$ at fertilization have been directly demonstrated by microinjection of Ca^{2+}-sensitive indicators, such as aequorin (Ridgway et al., 1977; Gilkey et al., 1978; Cuthbertson et al., 1981) or fura-2 (Poenie et al., 1985), and by the use of Ca^{2+}-sensitive microelectrodes (Busa and Nuccitelli, 1985).

The MPF and CSF in ooplasmic extracts are both sensitive to Ca^{2+} ions, and both quickly disappear following fertilization, most likely in response to the increased $[Ca^{2+}]_i$. At the same time that MPF and CSF activity are lost, the cytoplasm also loses the ability to induce NE breakdown and chromosome condensation in transplanted nuclei, but develops, instead, the ability to form pronuclei. The increase in $[Ca^{2+}]_i$ following fertilization may be responsible for this change (for further discussion, see Masui et al., 1977, 1984).

B. Chromosome Condensation and Spindle Assembly in Cell-Free Extracts

Under the conditions described in Section II, cytoplasmic extracts from unactivated eggs, like those from parthenogenetically activated eggs, support the transformation of sperm nuclei into pronuclei in vitro. Therefore, the cytoplasmic activities responsible for NE breakdown and chromosome condensation in intact unactivated eggs are no longer functional in either of these extracts. The loss of these activities in extracts of unactivated eggs may be related to the loss of MPF and CSF activity. As mentioned previously, both MPF and CSF are sensitive to Ca^{2+} ions and their extraction from cells requires the presence of a Ca^{2+} chelating agent, such as ethylene glycol bis(β-aminoethyl ether)N,N'-tetraacetic acid (EGTA) (Wasserman and Masui, 1976; Meyerhof and Masui, 1977; Wu and

Gerhart, 1980; Sunkara *et al.*, 1982). Therefore, the effect of EGTA during the preparation of cytoplasmic extracts from unactivated eggs was examined. When sperm nuclei were incubated in extracts prepared in the presence of EGTA, pronuclei were not formed. Rather, in egg extracts from *Rana* (Lohka and Masui, 1984b) or *Xenopus* (Lohka and Maller, 1985), the sperm chromatin formed metaphase chromosomes and an NE was not assembled around the condensed chromatin (Lohka and Masui, 1984b). Furthermore, in *Xenopus* egg extracts, the metaphase chromosomes were incorporated into bipolar (Fig. 8) or multipolar spindles. Nuclei from *Xenopus* brain or liver also underwent similar changes when incubated in extracts from unactivated *Xenopus* eggs. The spindles that formed *in vitro* varied greatly in size and often incorporated chromosomes from more than one nucleus. Asters were not seen at the spindles poles, even though centrioles were in all likelihood present in the nuclear preparations. Therefore, the morphology of the spindles closely resembled that seen when nuclei are transplanted into the cytoplasm of intact unactivated eggs. Since the changes induced in incubated nuclei closely resemble those seen when either sperm or somatic cell nuclei are transplanted into unactivated eggs, we will designate the extracts that induce chromosome condensation and spindle assembly "unactivated" egg extracts to distinguish them from "activated" egg extracts, which induce pronuclear formation and nuclear enlargement.

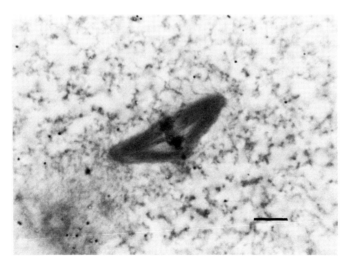

Fig. 8. Bipolar spindle formed during 90-min incubation of *Xenopus* sperm in unactivated *Xenopus* egg extract. Scale, 10 μm.

C. Cytoplasmic Components Involved in Chromosome Condensation and Spindle Assembly

Despite the fact that the process of spindle assembly has been described at an ultrastructural level in many cell types (Bajer and Molè-Bajer, 1969; London, 1972; Roos, 1973), our knowledge of the cytoplasmic components, other than microtubules, involved in this process is not complete. The use of unactivated egg extracts may help to provide a more detailed understanding of spindle assembly in amphibian eggs, as well as in other cells.

Neither chromosome condensation nor spindle formation occur when nuclei are incubated in the supernatant obtained following ultracentrifugation of the unactivated egg extracts, suggesting that particulate cytoplasmic components are required for these events. While neither the nature of these components nor their role in chromosome condensation and spindle assembly is known, the simplest explanation for this observation is that at least some of the factors involved in chromosome condensation and spindle assembly are bound to cytoplasmic particles, including membranes. MPF is not likely to be among these putative particulate factors, however, since its activity is detected in the soluble fraction of a variety of cell types (Wasserman and Masui, 1976; Sunkara et al., 1979; Wu and Gerhart, 1980). Similarly, CSF is also found in egg cytosol that is free of cytoplasmic particles (Masui, 1974; Meyerhof and Masui, 1977). Another possible requirement for cytoplasmic particles may be for structural elements of the spindle. Membrane vesicles are an abundant constituent of mitotic spindles in intact cells (Porter and Machado, 1960; Harris, 1961, 1962, 1975; also see Hepler and Wolniak, 1984). These vesicles are thought to play an important role in the regulation of Ca^{2+} ion concentration in the vicinity of the spindle during its assembly, as well as during mitosis (Harris, 1975, 1978; Petzelt and Auel, 1978; for review, see Hepler and Wolniak, 1984). Vesicles in the cell-free extracts may play a similar role during spindle assembly in vitro to provide the proper ionic environment for polymerization of tubulin into spindle microtubules and for the incorporation of other components into the spindle. If such vesicles are involved in spindle assembly it would be of interest to determine how they differ from those that form the NE in activated egg extracts.

In addition to the role of ooplasmic components in spindle assembly, the involvement of constituents found in the nuclear preparation also awaits investigation. For example, the role, if any, of centrioles or centrosomes, which are present in preparations of both sperm and somatic nuclei, is not clear. In mitotically dividing cells, centrosomes are found at the spindle poles and are thought to play a role in organizing microtubules in this region. However, the spindle poles of meiotic cells lack centrioles, but have only the electron-dense pericentriolar material found around the centrioles of mitotic cells (Calarco *et al.*, 1972; Szollosi *et al.*, 1972b; Calarco-Gillam *et al.*, 1983). As mentioned previously, nuclei injected into unactivated *Xenopus* eggs form chromosomes that are incorporated into anastral spindles. However, anastral arrays of microtubules are still assembled around the condensed chromosomes when the injected nuclear preparations are completely free of centrosomes (Karsenti *et al.*, 1984). Thus, the question of whether centrioles in the nuclear preparation or pericentriolar material from the egg cytoplasm are involved in organizing spindle assembly in the cell-free extracts merits further study.

Chromosome condensation in somatic cell nuclei that have been transplanted into the cytoplasm of unfertilized eggs is accompanied by the acquisition of ooplasmic proteins by the chromosomes (Masui *et al.*, 1979). Similarly, when chromosome condensation is induced by the fusion of mitotic cells with those in interphase, proteins from the mitotic cells accumulate on the condensing chromosomes (Rao and Johnson, 1974). These findings have led to the suggestion that the proteins that associate with the interphase chromatin are responsible for inducing chromosome condensation (Rao and Johnson, 1974; Masui *et al.*, 1979). It might be expected that similar proteins also associate with the sperm chromatin during the formation of metaphase chromosomes in the cell-free cytoplasmic extracts. Furthermore, since the proteins associated with the sperm chromatin differ from those of both interphase nuclei and metaphase chromosomes, the structural proteins of metaphase chromosomes are also likely to be acquired during chromosome condensation *in vitro*. Proteins of particular interest are those of the chromosome scaffold (Lewis and Laemmli, 1982; Lewis *et al.*, 1984), including DNA topoisomerase II (Earnshaw and Heck, 1985; Earnshaw *et al.*, 1985) and those of the kinetochore (Cox *et al.*, 1983; Earnshaw *et al.*, 1984).

D. Nuclear Envelope Breakdown in Cell-Free Extracts

Sperm nuclei that are incubated in cell-free cytoplasmic extracts of unactivated *Xenopus* eggs undergo changes similar to those seen following the injection of such nuclei into the cytoplasm of intact unactivated eggs. In both cases, chromosomes and spindles are formed. When somatic cell nuclei are transplanted into unactivated eggs, breakdown of the NE precedes chromosome condensation and spindle assembly. Since the NE had been removed from lysolecithin-treated sperm nuclei, these nuclei could not be used to characterize NE breakdown in unactivated egg

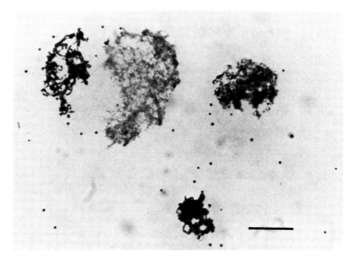

Fig. 9.   Early stages of chromosome condensation 15 min after mixing supernatant from unactivated egg extracts with sperm pronuclei. Scale, 10 μm.

extracts. Instead, sperm pronuclei or interphase nuclei isolated from *Xenopus* brain or liver were used. In these experiments, supernatants were obtained from unactivated egg extracts by ultracentrifugation and mixed with an equal volume of activated egg extracts in which pronuclei had formed during a previous 60-min incubation. Within 30–60 min the pronuclear envelope had broken down and the chromatin began to con-

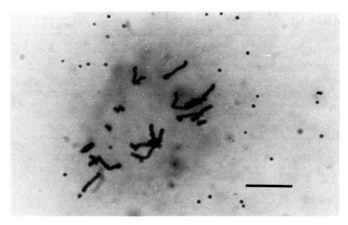

Fig. 10.   Condensed chromosomes formed 90 min after mixing supernatant from unactivated eggs with sperm pronuclei. Scale, 10 μm.

dense (Fig. 9). Fully condensed metaphase chromosomes and spindles had formed by 60–90 min. (Figs. 10 and 11). Similar results were obtained when liver (Lohka and Maller, 1985) or brain nuclei (Figs. 12 and 13) were used, although in these cases preincubation for 60 min in activated egg extracts was not necessary. However, the chromosomes formed following a 60-min preincubation were often better separated than those formed without preincubation, probably because the nuclei had enlarged during the preincubation. The ability to induce chromosome condensation in

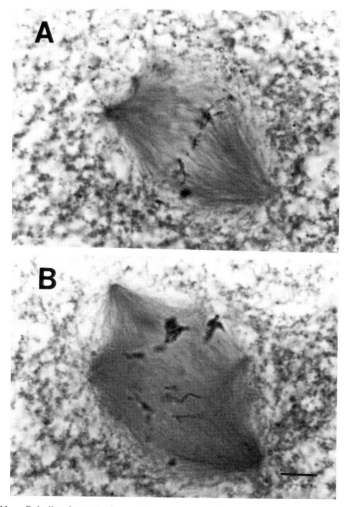

Fig. 11.   Spindles formed after mixing supernatant from unactivated egg extracts with sperm pronuclei. (A) Bipolar spindle; (B) multipolar spindle; Scale, 10 μm.

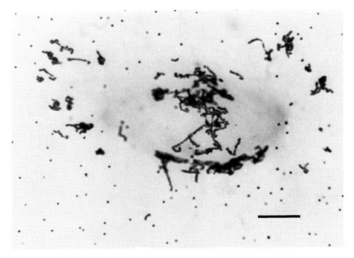

Fig. 12.   Condensed chromosomes formed from brain nuclei 90 min after the addition of supernatants from unactivated egg extracts. Scale, 10 μm.

certain types of nuclei may be improved by preincubation in activated egg extracts. Under these conditions we have recently found that even human sperm nuclei can be induced to form chromosomes in egg extracts (Fig. 14).

It is important to note that while supernatants from unactivated egg extracts could not by themselves induce chromosome condensation and

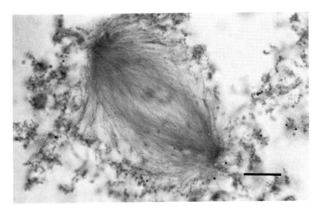

Fig. 13.   Spindle formed 90 min after mixing supernatants from unactivated egg extracts with extracts containing brain nuclei. Scale, 10 μm.

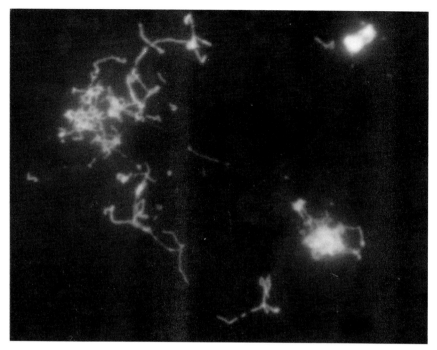

Fig. 14. Chromosomes formed from human sperm chromatin after mixing with supernatants from unactivated egg extracts.

spindle assembly in incubated nuclei, both of these changes were induced when the supernatants were added to extracts in which cytoplasmic particles were abundant, even though the recipient extracts could initially support only pronuclear formation. These results suggest that activated egg extracts, which normally induce pronuclear formation, contain all of the particulate cytoplasmic components required for chromosome condensation and spindle assembly. However, this particulate material can be utilized for this purpose only in the presence of soluble cytoplasmic factors extracted from metaphase-arrested unactivated eggs in the presence of EGTA. At the same time, the NE breaks down and the membrane components of the recipient extracts can no longer function in NE assembly. Thus, there appears to be a reversible control mechanism governing the activity of the particulate cytoplasmic material. An understanding of the mechanism whereby soluble factors control the ability of cytoplasmic particles, including membranes, to participate in either NE assembly or chromosome condensation and spindle assembly, may aid in determining how nuclear behavior is controlled during the cell cycle. MPF activity, Ca^{2+} ions, and protein phosphorylation may all be involved in regulating the type of nuclear behavior induced by the cytoplasmic extracts.

IV. THE ROLE OF MPF, Ca^{2+} IONS, AND PROTEIN PHOSPHORYLATION IN CONTROLLING NUCLEAR ENVELOPE BREAKDOWN, CHROMOSOME CONDENSATION, AND SPINDLE ASSEMBLY IN CELL-FREE EXTRACTS

A. MPF

The effect of adding supernatants from unactivated eggs to extracts containing sperm pronuclei or isolated somatic cell nuclei is similar to that seen when MPF is injected into either immature oocytes (Masui and Markert, 1971) or early embryos arrested in G_2 of the cell cycle (Miake-Lye et al., 1983; Halleck et al., 1984). To determine whether or not MPF is also responsible for inducing NE breakdown, chromosome condensation, and spindle assembly in the cell-free extracts, a partially-purified preparation of MPF (Wu and Gerhart, 1980; a gift from M. Wu and J. Gerhart, University of California, Berkeley) was mixed with extracts containing sperm pronuclei. The partially purified MPF rapidly induced NE breakdown, chromosome condensation (Fig. 15) and spindle assembly (Lohka and Maller, 1985). However, interphase nuclei often re-formed when the samples were incubated longer than 60 min after MPF addition. Similar results have been observed when partially purified preparations of MPF were added to nuclei from Chinese hamster ovary (CHO) cells or rat thymocytes that had been incubated in extracts from G_2-arrested embryos, although in these cases spindle assembly was not seen (Miake-Lye and Kirschner, 1985). From these results, it is clear that MPF is able to act both in intact cells and in vitro to induce NE breakdown, chromosome condensation, and spindle assembly. The ability to detect MPF-induced nuclear changes in vitro may provide an alternative method to oocyte microinjection for assaying this activity.

Although partially purified preparations of MPF induce NE breakdown and chromosome condensation when added to cytoplasmic extracts (Miake-Lye and Kirschner, 1985; Lohka and Maller, 1985) and when injected into G_2-arrested embryos (Miake-Lye et al., 1983; Newport and Kirschner, 1984), in both cases interphase nuclei re-form spontaneously within 1 hr. In contrast, condensed chromosomes and spindles often per-

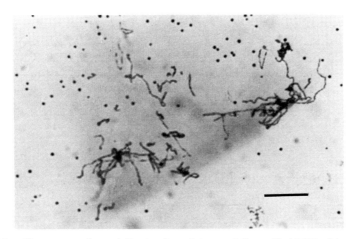

Fig. 15. Chromosome formed 60 min after mixing partially purified MPF with extracts containing sperm pronuclei. Scale, 10 μm.

sist for at least 3 hr when supernatants from unactivated egg extracts are added to extracts (Lohka and Maller, 1985), and for at least 18 hr when these supernatants are injected into cleaving blastomeres (Meyerhof and Masui, 1977). Therefore, chromosome condensation induced by supernatants seems to be more stable than that induced by partially purified MPF preparations. It is not likely that these results are due only to the presence of greater MPF activity in supernatants than in partially purified preparations; rather, they may be explained by the presence of CSF in supernatants and its absence from partially purified MPF preparations. This contention is supported by observations that CSF can stabilize not only the condensed chromosomes and spindles formed when dividing cells enter mitosis (Masui and Markert, 1971; Meyerhof and Masui, 1977) but also those induced by MPF injection (Newport and Kirscher, 1984). It is possible that the cell-free extracts such as those described here may also be suitable to analyze the action of CSF.

B. Ca^{2+} Ions

The increase in the concentration of cytoplasmic free Ca^{2+} ions at fertilization or during parthenogenetic activation is thought to play a crucial role in regulating the cytoplasmic factors that control nuclear behavior (Masui *et al.*, 1977, 1984). Prior to this increase, the amphibian egg cytoplasm is able to induce transplanted nuclei to undergo NE breakdown, chromosome condensation, and spindle assembly, whereas after the increase the cytoplasm induces NE assembly, chromosome decondensation, and nuclear enlargement. In a similar manner, Ca^{2+} ions may regulate the type of nuclear behavior induced by cytoplasmic extracts of unfertilized eggs. As described in Section II, extracts from unfertilized eggs made without EGTA are able to transform sperm nuclei into pronuclei, whereas extracts made with EGTA induce NE breakdown, chromosome condensation, and spindle assembly instead. Therefore, it appears that during the preparation of extracts Ca^{2+} ions are released, most likely when the eggs are crushed by centrifugation. Unless the Ca^{2+} ions are chelated by EGTA, the cytoplasmic factors responsible for inducing NE breakdown and the other changes are lost, and the ability to induce pronuclear formation develops. Such a role for Ca^{2+} ions can be demonstrated in both *Rana* and *Xenopus* by the addition of $CaCl_2$ solutions to unfertilized egg extracts made with EGTA. Prior to Ca^{2+} addition, these extracts induced chromosome condensation and, in the case of *Xenopus,* spindle assembly. However, following Ca^{2+} addition extracts were able to induce pronuclear formation (Lohka and Masui, 1984b; M. J. Lohka and J. L. Maller, unpublished results). In contrast, once the ability to induce chromosome condensation and spindle assembly is lost, it cannot be recovered simply by the addition of EGTA to chelate Ca^{2+}. Recovery of this ability requires the addition of MPF activity, either from supernatants of unactivated eggs or from partially purified preparations.

The concentration of the free Ca^{2+} ions required to convert an extract from one which induces chromosome condensation to one that induces pronuclear formation is not known, since neither the amount of EGTA in the final extracts nor the amount of Ca^{2+} released during centrifugation has been accurately determined. In addition, mitochondria and membrane vesicles have been shown to sequester Ca^{2+}, even in a cell-free system (for review, see Borle, 1981). In this regard, the Ca^{2+}-sequestering ability of vesicles of the mitotic spindle is perhaps the most impressive (Silver *et al.,* 1980; Kiehart, 1981; Izant, 1983). Injection of $CaCl_2$, at concentrations up to 5 mM, directly into the spindle of sea urchin embryos results in only a localized loss of spindle birefringence, which soon returns to normal as Ca^{2+} is sequestered (Kiehart, 1981). Since mitochondria and membrane vesicles are abundant in cell-free extracts, their capacity to either sequester or release Ca^{2+} must be considered in attempts to determine accurately the Ca^{2+} concentration necessary for development of the ability to transform sperm into pronuclei. The judicious use of Ca^{2+}–EGTA buffers or Ca^{2+}-sensitive indicators may help to resolve this question.

The addition of Ca^{2+} to cell-free extracts containing condensed chromosomes and spindles results in the re-formation of interphase nuclei

(Lohka and Maller, 1985). It is well known that microtubules are sensitive to Ca^{2+} ions (Weisenberg, 1972) and that exposure of mitotic spindles to Ca^{2+} disrupts their morphology (Salmon and Segall, 1980; Kiehart, 1981; Izant, 1983). However, in the cytoplasmic extracts, not only is the spindle disassembled following Ca^{2+} addition, but an NE is assembled around the chromosomes as they decondense to form an interphase nucleus. Therefore, in the cell-free system the addition of unactivated egg extracts to pronuclei induces chromosome condensation and spindle formation and, in turn, interphase nuclei are re-formed in the presence of Ca^{2+} ions. Since this series of nuclear changes resembles those seen when a cell divides, the ability to manipulate these events in such a manner in cell-free extracts may greatly facilitate analysis of nuclear behavior during cell division.

C. Protein Phosphorylation

The mechanism by which MPF controls NE breakdown, chromosome condensation, and spindle assembly is not clear. It is likely, however, that protein phosphorylation plays an important role in the regulation of these processes (see also other chapters in this volume). In maturing *Xenopus* oocytes, a three- to five-fold increase in total cell phosphoprotein occurs at about the time MPF activity is first detected in the cytoplasm, shortly before the GV breaks down (Maller *et al.,* 1977). A similar increase follows almost immediately after the injection of either crude or partially purified MPF activity into oocytes (Maller *et al.,* 1977; Wu and Gerhart, 1980). Examination of protein phosphorylation in individual oocytes demonstrated a strict correlation between the increase in protein phosphorylation and GV breakdown (Maller *et al.,* 1977). Analysis by two-dimensional polyacrylamide gel electrophoresis showed that many new phosphoproteins appear during the burst in phosphorylation, although some phosphoproteins appear even earlier (Maller and Smith, 1985). These results, together with findings that protein phosphorylation also increases during oocyte maturation in starfish (Guerrier *et al.,* 1977; Dorée *et al.,* 1983), *Urechis* (Meijer *et al.,* 1982), and annelids (Peaucellier *et al.,* 1982, 1984), have led to the hypothesis that MPF is either a protein kinase or a regulator of protein kinase activity. Although partially purified preparations of MPF have been shown to possess endogenous protein kinase activities (Wu and Gerhart, 1980), no specific protein kinase has been shown to cofractionate with MPF activity.

The idea that protein phosphorylation is important in the control of cell division is not new, and there are several examples of specific proteins phosphorylated in a cell-cycle dependent manner. Histones H1 and H3 are highly phosphorylated at metaphase, changes that are thought to play an important role in chromosome condensation (for reviews, see Gurley *et al.*, 1978 and Chapter 11, this volume). Both histones are dephosphorylated following metaphase, although this change may not be necessary for chromosome decondensation (Tanphaichitr *et al.*, 1976). In addition, the major components of the nuclear lamina, the lamins, are phosphorylated when the NE breaks down both in intact cells (Gerace and Blobel, 1980; Ottaviano and Gerace, 1985) and in cell-free extracts (Miake-Lye and Kirschner, 1985), and are dephosphorylated as the NE re-forms during telophase (Gerace and Blobel, 1980; Ottaviano and Gerace, 1985).

The possibility that protein phosphorylation plays a role in the processes described here has been examined by the addition of analogs of ATP to the cell-free cytoplasmic extracts and by two-dimensional electrophoresis followed by autoradiography of radiolabeled preparations (Lohka and Maller, 1987). The addition of App(NH)p, a nonhydrolyzable analog, did not affect NE assembly, but prevented NE breakdown and chromosome condensation. In contrast, γS-ATP, a hydrolyzable analog, prevented NE assembly but not NE breakdown and chromosome condensation. These results suggest that protein phosphorylation is necessary for NE breakdown and chromosome condensation. On the other hand, NE assembly may require protein dephosphorylation, since thiophosphorylated proteins are poor substrates for protein phosphatases. The effect of ATP analogs on NE assembly in egg extracts is similar to their effect on NE assembly in cell-free extracts of mitotic CHO cells (Burke and Gerace, 1986). In both cases, NE assembly is inhibited by γ-S-ATP, but not by App(NH)p.

The proteins phosphorylated in cytoplasmic extracts that induce NE assembly and chromosome condensation were compared by two-dimensional electrophoresis and autoradiography to those phosphorylated in extracts that induce NE breakdown and chromosome condensation. Although many of the proteins radiolabeled in the two extracts appeared to be the same, phosphoproteins with apparent molecular weights of 110K, 49K, 42K, 41K, 39K, and 35K were consistently found only in extracts that induced NE breakdown and chromosome condensation (Fig. 16). In some cases, proteins of similar molecular weight appeared as clusters on an autoradiogram, possibly representing multiply phosphorylated forms of the same protein. As yet, none of the specific phosphoproteins have been identified. However, their identification might prove useful in developing an assay for isolating protein kinase activities in MPF preparations. This approach has been used in the isolation of a protein kinase specific for ribosomal protein S6 (Erikson and Maller, 1985), one of the few substrates identified during the burst in protein phosphorylation in maturing oocytes. The phosphorylation of proteins specific to the unactivated egg extracts was either greatly reduced or absent when the ability to form pronuclei developed after the addition of Ca^{2+} ions. Instead, phosphoproteins similar to those radiolabeled in extracts that supported pronuclear formation were seen. Similar changes in protein phosphorylation also occur in intact eggs when the ability to form pronuclei develops following parthenogenetic activation (Lohka and Maller, 1987).

Taken together, the results suggest that protein phosphorylation may play an important role in controlling NE assembly–breakdown and chromosome condensation–decondensation in the cell-free extracts. While the role of any of the proteins phosphorylated in these experiments is not known, it would be of great interest to determine whether any are associated with the particulate fraction that is necessary for either NE assembly or chromosome condensation and spindle formation. It is an attractive hypothesis that phosphorylation and dephosphorylation may underlie the mechanism of reversible modification of the particulate elements necessary for the conversion between nuclear assembly and nuclear breakdown.

V. SUMMARY

Nuclei introduced into amphibian eggs by fertilization or by microinjection assume a morphology that resembles that of the resident egg nucleus. In unactivated eggs, whose chromosomes are arrested at metaphase II of meiosis, the transplanted nuclei undergo NE breakdown, chromosome condensation, and spindle formation, whereas in activated eggs NE assembly, chromatin decondensation, and nuclear enlargement are induced instead. Cell-free extracts of *R. pipiens* or *X. laevis* eggs, prepared under appropriate conditions, are able to induce *in vitro* the nuclear changes observed in intact eggs. Extracts prepared in the presence of EGTA from unactivated eggs can induce nuclei to form condensed chromosomes that, in *Xenopus* egg extracts, are incorporated into bipolar or multipolar spindles. In contrast, extracts made from unactivated eggs without EGTA or from pathenogenetically activated eggs induce somatic cell nuclei to enlarge and sperm nuclei to form pronuclei. During pronuclear formation in the cell-free extracts, cytoplasmic vesicles are assembled around the sperm chromatin to form an NE. The chromatin decondenses as the pronuclei enlarge and synthesize DNA.

Both chromosome condensation and pronuclear formation require par-

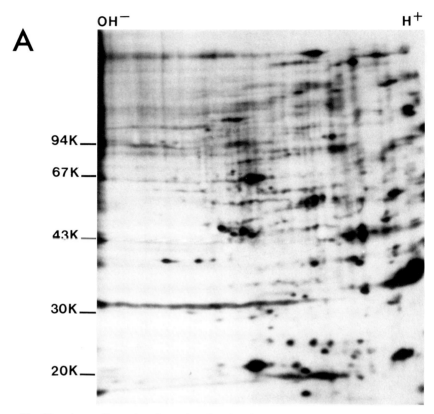

Fig. 16. Autoradiographs of proteins phosphorylated in egg extracts and separated by two-dimensional polyacrylamide gel electrophoresis. (A) Activated egg extracts; (B) unactivated egg extracts. Proteins were separated in the first dimension by isoelectric focusing in a mixture of ampholytes (1.4%, pH 5–7, and 0.6%, pH 3–10; SERVA). The second dimension was an SDS gel with a 10–17% acrylamide gradient having a low level of crosslinking, as previously described by Younglai *et al.* (1982). Arrows designate proteins that are consistently phosphorylated only in unactivated extracts made with EGTA.

ticulate cytoplasmic components, since neither process occurs when nuclei are incubated in only the soluble fraction obtained following the centrifugation of extracts at 150,000 *g* for 2 hr. Although the soluble fraction of unactivated egg extracts is unable by itself to support chromosome condensation and spindle assembly, mixing this fraction with extracts containing pronuclei or other interphase nuclei induces NE breakdown, chromosome condensation, and spindle assembly, even though the recipient extracts initially are able to support only pronuclear formation. Par-

OH –   H +

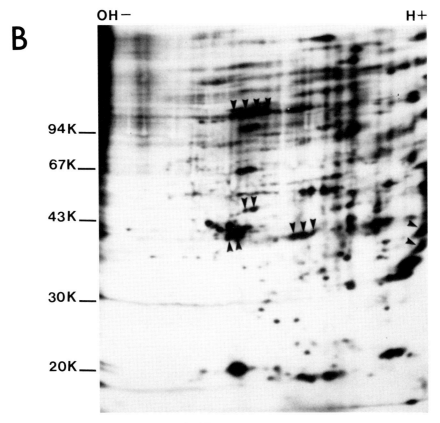

Fig. 16. (*continued*)

tially purified preparations of MPF induce similar changes. Interphase nuclei re-form when Ca^{2+} ions are added to extracts containing condensed chromosomes and spindles. These results suggest that Ca^{2+} ions and MPF activity are involved in the control of the ability of the cell-free extracts to support NE assembly and chromosome decondensation on the one hand, or NE breakdown, chromosome condensation, and spindle formation, on the other.

Protein phosphorylation may play an important role in this control mechanism. The addition of App(NH)p, a nonhydrolzable analog of ATP, to the cytoplasmic extracts prevents NE breakdown, chromosome condensation, and spindle formation, but not NE assembly. In contrast, γS-ATP, a hydrolyzable analog, does not affect NE breakdown, chromosome condensation, or spindle formation, but prevents NE formation. These results suggest that protein phosphorylation is required for NE breakdown, chromosome condensation, and spindle formation. However, since thiophosphorylated proteins are poor substrates for protein phosphatases, dephosphorylation may be necessary for NE assembly and chromatin decondensation. In addition, two-dimensional polyacrylamide gel electrophoresis of proteins phosphorylated in the cell-free extracts has shown that at least six phosphoproteins are found only in extracts that can induce NE breakdown and the other associated changes, but are absent in extracts that induce pronuclear formation. The phosphorylation of these proteins is greatly reduced or absent when the unactivated egg extracts are converted, by the addition of Ca^{2+} ions, to ones that induce pronuclear formation. Thus, the cell-free system described here may provide the basis for the analysis of the molecular mechanisms responsible for the control of NE assembly–breakdown and chromosome condensation–decondensation.

ACKNOWLEDGMENTS

We thank R. Villadiego of the Electron Microscope Laboratory, Department of Zoology, University of Toronto for sectioning the material for ultrastructural examination and Jo Erikson for comments on the manuscript. Portions of the work described here were supported by grants from NIH (GM 26743), the American Cancer Society (CD-279), and the March of Dimes Birth Defects Foundation (1-922). M. J. L. is a Research Fellow of the National Cancer Institute of Canada, and J. L. M. is an Established Investigator of the American Heart Association.

REFERENCES

Arms, K. (1968). Cytonucleoproteins in cleaving eggs of *Xenopus laevis*. *J. Embryol. Exp. Morphol.* **20**, 367–374.

Bajer, A., and Molè-Bajer, J. (1969). Formation of spindle fibers, kinetochore orientation and behavior of the nuclear envelope during mitosis in endosperm. Fine structural and *in vitro* studies. *Chromosoma* **27**, 448–484.

Barry, J. M., and Merriam, R. W. (1972). Swelling of hen erythrocyte nuclei in cytoplasm from *Xenopus* eggs. *Exp. Cell Res.* **71**, 90–96.

Bataillon, E. (1928). Etudes analytiques sur la maturation des oeufs de batraciens. *C. R. Hebd. Seances Acad. Sci.* **187**, 520–523.

Bataillon, E., and Tchou-Su. (1934). L'analyse expérimentale de la fécondation et sa définition par les processus cinétiques. *Ann. Sci. Nat., Zool. Biol. Anim.* [10] **17**, 19–36.

Belanger, A. M., and Schuetz, A. W. (1975). Precocious activation of amphibian oocytes by divalent cation ionophore A23187. *Dev. Biol.* **45**, 378–381.

Benavente, R., Krohne, G., and Franke, W. W. (1985). Cell type-specific expression of nuclear lamina proteins during development of *Xenopus laevis*. *Cell (Cambridge, Mass.)* **41**, 177–190.

Benbow, R. M., and Ford, C. C. (1975). Cytoplasmic control of nuclear DNA synthesis during early development of *Xenopus laevis:* A cell-free assay. *Proc. Natl. Acad. Sci. U.S.A.* **72**, 2437–2441.

Benbow, R. M., Pestell, R. Q. W., and Ford, C. C. (1975). Appearance of DNA polymerase activities during early development of *Xenopus laevis*. *Dev. Biol.* **43**, 159–174.

Bloch, D. P. (1969). A catalog of sperm histones. *Genetics* **61**, Suppl. 93–111.

Bonner, W. M. (1975a). Protein migration into nuclei. I. Frog oocyte nuclei *in vivo* accumulate microinjected histones, allow entry to small proteins and exclude large proteins. *J. Cell Biol.* **64**, 421–430.

Bonner, W. M. (1975b). Protein migration into nuclei. II. Frog oocyte nuclei accumulate a class of microinjected oocyte nuclear proteins and exclude a class of microinjected oocyte cytoplasmic proteins. *J. Cell Biol.* **64**, 431–437.

Bonner, W. M. (1978). Protein migration and accumulation in nuclei. *In* "The Cell Nucleus" (H. Busch, ed.), Vol. 6, Part C, pp. 97–148. Academic Press, New York.

Borle, A. B. (1981). Control, modulation, and regulation of cell calcium. *Rev. Physiol. Biochem. Pharmacol.* **90**, 13–153.

Burke, B. and Gerace, L. (1986). A cell free system to study assembly of the nuclear envelope at the end of mitosis. *Cell (Cambridge, Mass)* **44**, 639–652.

Busa, W. B., and Nuccitelli, R. (1985). An elevated free cytosolic Ca^{2+} wave follows fertilization in eggs of the frog, *Xenopus laevis*. *J. Cell Biol.* **100**, 1325–1329.

Calarco, P., Donahue, R. P., and Szollosi, D. (1972). Germinal vesicle breakdown in the mouse oocyte. *J. Cell Sci.* **10**, 369–385.

Calarco-Gilliam, P. D., Siebert, M. C., Hubble, R., Mitchison, T., and Kirschner, M. (1983). Centrosome development in early mouse embryos as defined by an autoantibody against pericentriolar material. *Cell (Cambridge, Mass.)* **35**, 621–629.

Chai, L. S., Weinfeld, H., and Sandberg, A. A. (1974). Ultrastructural changes in the nuclear envelope during mitosis of Chinese hamster cells: A proposed mechanism of nuclear envelope reformation. *J. Natl. Cancer Inst. (U.S.)* **53**, 1033–1048.

Chang, J. P., and Gibley, C. W., Jr. (1968). Ultrastructure of tumor cells during mitosis. *Cancer Res.* **28**, 521–534.

Cox, J. V., Schenck, E. A., and Olmsted, J. B. (1983). Human anticentromere antibodies. Distribution, characterization of antigens, and effect on microtubule organization. *Cell (Cambridge, Mass.)* **35**, 331–339.

Cuthbertson, K. S. R., Whittingham, D. G., and Cobbold, P. H. (1981). Free Ca^{2+} increases in exponential phases during mouse oocyte activation. *Nature (London)* **294**, 754–757.

DeRobertis, E. M., Longthorne, R., and Gurdon, J. B. (1978). Intracellular migration of nuclear proteins in *Xenopus* oocytes. *Nature (London)* **272**, 254–256.

deRoeper, A., Smith, J. A., Watt, R. A., and Barry, J. M. (1977). Chromatin dispersion and DNA synthesis in G_1 and G_2 HeLa cell nuclei injected into *Xenopus* eggs. *Nature (London)* **265**, 469–470.

Dingwall, C., and Allan, J. (1984). Accumulation of the isolated carboxy-terminal domain of histone H1 in the *Xenopus* oocyte nucleus. *EMBO J.* **3**, 1933–1937.

Dingwall, C., Sharnick, S. V., and Laskey, R. A. (1982). A polypeptide domain that specifies migration of nucleoplasmin into the nucleus. *Cell (Cambridge, Mass.)* **30**, 449–458.

Dorée, M., Peaucellier, G., and Picard, A. (1983). Activity of the maturation-promoting factor and the extent of protein phosphorylation oscillate simultaneously during meiotic maturation of starfish oocytes. *Dev. Biol.* **99**, 489–501.

Earnshaw, W. C., and Heck, M. S. (1985). Localization of topoisomerase II in mitotic chromosomes. *J. Cell Biol.* **100**, 1716–1725.

Earnshaw, W. C., Halligan, N., Cooke, C., and Rothfield, N. (1984). The kinetochore is part of the metaphase chromosome scaffold. *J. Cell Biol.* **98**, 352–357.

Earnshaw, W. C., Halligan, B., Cooke, C. A., Heck, M. M. S., and Liu, L. F. (1985). Topoisomerase II is a structural component of mitotic chromosome scaffolds. *J. Cell Biol.* **100**, 1706–1716.

Elinson, R. P. (1977). Fertilization of immature frog eggs: Cleavage and development following subsequent activation. *J. Embryol. Exp. Morphol.* **37**, 187–201.

Erikson, E., and Maller, J. L. (1985). A protein kinase from *Xenopus* eggs specific for ribosomal protein S6. *Proc. Natl. Acad. Sci. U.S.A.* **82**, 742–746.

Fankhauser, G., and Moore, C. (1941). Cytological and experimental studies of polyspermy in the newt, *Triturus vividescens* I. Normal fertilization. *J. Morphol.* **68**, 347–386.

Feldherr, C. M. (1975). The uptake of endogenous proteins by oocyte nuclei. *Exp. Cell Res.* **93**, 411–419.

Feldherr, C. M., Cohen, R. J., and Ogburn, J, A. (1983). Evidence for mediated protein uptake by amphibian oocyte nuclei. *J. Cell Biol.* **96**, 1486–1490.

Feldherr, C. M., Kallenbach, E., and Schultz, N. (1984). Movement of a karyophilic protein through the nuclear pores of oocytes. *J. Cell Biol.* **99**, 2216–2222.

Forbes, D. J., Kirschner, M. W., and Newport, J. W. (1983). Spontaneous formation of nucleus-like structures around bacteriophage DNA microinjected into *Xenopus* eggs. *Cell (Cambridge, Mass.)* **34**, 13–23.

Franke, W. W. (1974). Structure, biochemistry, and functions of the nuclear envelope. *Int. Rev. Cytol., Suppl.* **4**, 71–236.

Franke, W. W., and Scheer, U. (1970). The ultrastructure of the nuclear envelope of amphibian oocytes: a reinvestigation. The mature oocyte. *J. Ultrastruct. Res.* **30**, 289–316.

Gerace, L., and Biobel, G. (1980). The nuclear envelope lamina is reversibly depolymerized during mitosis. *Cell (Cambridge, Mass.)* **19**, 277–287.

Gerhart, J., Wu, M., and Kirschner, M. (1984). Cell cycle dynamics of an M-phase-specific cytoplasmic factor in *Xenopus laevis* oocytes and eggs. *J. Cell Biol.* **98**, 1247–1255.

Gilkey, J. C., Jaffe, L. F., Ridgway, E. B., and Reynolds, G. T. (1978). A free calcium wave transverses the activating egg of the medaka, *Oryzias latipes*. *J. Cell Biol.* **76**, 448–466.

Gordon, K., Brown, D. B., and Ruddle, F. H. (1985). *In vitro* activation of human sperm induced by amphibian egg extract. *Exp. Cell Res.* **157**, 409–418.

Graham, C. F. (1966). The regulation of DNA synthesis and mitosis in multinucleate frog eggs. *J. Cell Sci.* **1**, 363–374.

Graham, C. F., Arms, K., and Gurdon, J. B. (1966). The induction of DNA synthesis by frog egg cytoplasm. *Dev. Biol.* **14**, 349–381.

Guerrier, P., Moreau, M., and Dorée, M. (1977). Hormonal control of meiosis in starfish: Stimulation of protein phosphorylation induced by 1-methyladenine. *Mol. Cell. Endocrinol.* **7**, 137–150.

Gulyas, B. J. (1972). The rabbit zygote. III. Formation of the blastomere nucleus. *J. Cell Biol.* **55**, 533–541.

Gurdon, J. B. (1967). On the origin and persistence of a cytoplasmic state inducing nuclear DNA synthesis in frogs' eggs. *Proc. Natl. Acad. Sci. U.S.A.* **58**, 545–552.

Gurdon, J. B. (1968). Changes in somatic cell nuclei inserted into growing and maturing amphibian oocytes. *J. Embryol. Exp. Morphol.* **20**, 401–414.

Gurdon, J. B. (1970). Nuclear transplantation and the control of gene activity in animal development. *Proc. R. Soc. London, Ser. B* **176**, 303–314.

Gurley, L. R., Tobey, R. A., Walters, R. A., Hildebrand, C. E. Hohmann, P. G., D'Anna, J. A., Barham, S. S., and Deaven, L. L. (1978). Histone phosphorylation and chromatin structure in synchronized mammalian cells. *In* "Cell Cycle Regulation" (J. R. Jeter, Jr., I. L. Cameron, G. M. Padilla, and A. M. Zimmerman, eds.), pp. 37–60. Academic Press, New York.

Halleck, M. S., Reed, J. A., Lumley-Spanski, K., and Schlegel, R. A. (1984). Injected mitotic extracts induce chromosome condensation of interphase chromatin. *Exp. Cell Res.* **153**, 561–569.

Harland, R. M., and Laskey, R. A. (1980). Regulated replication of DNA microinjected into eggs of *X. laevis. Cell (Cambridge, Mass.)* **21**, 761–771.

Harris, P. (1961). Electron microscope study of mitosis in sea urchin blastomeres. *J. Biophys. Biochem. Cytol.* **11**, 413–419.

Harris, P. (1962). Some structural and functional aspects of the mitotic apparatus in sea urchin embryos. *J. Cell Biol.* **14**, 475–489.

Harris, P. (1975). The role of membranes in the organization of the mitotic apparatus. *Exp. Cell Res.* **94**, 409–425.

Harris, P. (1978). Triggers, trigger waves and mitosis. A new model. *In* "Cell Cycle Regulation," (J. R. Jeter, Jr., I. L. Cameron, G. M. Padilla, and A. M. Zimmerman, eds.), pp. 75–104. Academic Press, New York.

Hepler, P. K., and Wolniak, S. M. (1984). Membranes in the mitotic apparatus: Their structure and function. *Int. Rev. Cytol.* **90**, 169–238.

Hinegardner, R. T., Rao, B., and Feldman, D. E. (1964). The DNA synthetic period during early development of the sea urchin egg. *Exp. Cell Res.* **36**, 53–61.

Hoffner, N. J., and DiBerardino, M. A. (1977). The acquisition of egg cytoplasmic non-histone proteins by nuclei during nuclear reprogramming. *Exp. Cell Res.* **108**, 421–427.

Huchon, D., Crozet, N., Cantenot, N., and Ozon, R. (1981). Germinal vesicle breakdown in the *Xenopus laevis* oocyte: Description of a transient microtubular structure. *Reprod. Nutr. Dev.* **21**, 135–148.

Ikegami, S., Tagushi, T., Ohaski, M., Oguro M., Nagano, H., and Mano, Y. (1978). Aphidi-colin prevents mitotic cell division by interfering with the activity of DNA polymerase-α. *Nature (London)* **275**, 458–460.

Itoh, S., Dan, K., and Goodenough, D. (1981). Ultrastructure and ^{3}H-thymidine incorpora-tion in chromosome vesicles in sea urchin embryos. *Chromosoma* **83**, 441–453.

Iwao, Y., and Katagiri, C. (1984). *In vitro* induction of sperm nucleus decondensation by cytosol from mature toad eggs. *J. Exp. Zool.* **230**, 115–124.

Izant, J. G. (1983). The role of calcium ions during mitosis. Calcium participates in the anaphase trigger. *Chromosoma* **88**, 1–10.

Karsenti, E., Newport, J., Hubble, R., and Kirschner, M. (1984). Interconversion of meta-phase and interphase microtubule arrays, as studied by the injection of centrosomes and nuclei into *Xenopus* eggs. *J. Cell Biol.* **98**, 1730–1745.

Katagiri, C., and Moriya, M. (1976). Spermatozoan response to the toad egg matured after removal of germinal vesicle. *Dev. Biol.* **50**, 235–241.

Kiehart, D. P. (1981). Studies on the *in vivo* sensitivity of spindle microtubules to calcium ions and evidence for a vesicular calcium-sequestering system. *J. Cell Biol.* **88**, 604–617.

Kishimoto, T., and Kanatani, H. (1976). Cytoplasmic factor responsible for germinal vesicle breakdown and meiotic maturation in starfish oocyte. *Nature (London)* **260**, 321–322.

Kishimoto, T., Kuriyama, R., Kondo, H., and Kanatani, H. (1982). Generality of the action of various maturation-promoting factors. *Exp. Cell Res.* **137**, 121–126.

Kishimoto, T., Yamazaki, K., Kato, Y., Koide, S. S., and Kanatani, H. (1984). Induction of starfish oocyte maturation by maturation-promoting factor of mouse and surf clam oocytes. *J. Exp. Zool.* **231**, 293–295.

Kleinschmidt, J. A., and Franke, W. W. (1982). Soluble acidic complexes containing histones H3 and H4 in nuclei of *Xenopus laevis* oocytes. *Cell (Cambridge, Mass.)* **29**, 799–809.

Kleinschmidt, J. A., Fortkamp, E., Krohne, G., Zentgraf, H., and Franke, W. W. (1985). Co-existence of two different types of soluble histone complexes in nuclei of *Xenopus laevis* oocytes. *J. Biol. Chem.* **260**, 1166–1176.

Kunkle, M., Magun, B. E., and Longo, F. J. (1978). Analysis of isolated sea urchin nuclei incubated in egg cytosol. *J. Exp. Zool.* **203**, 381–390.

Lewis, C. D., and Laemmli, U. K. (1982). Higher order metaphase chromosome structure: Evidence for metalloprotein interactions. *Cell (Cambridge, Mass.)* **17**, 849–858.

Lewis, C. D., Lebkowski, J. S., Daly, A. K., and Laemmli, U. K. (1984). Interphase nuclear matrix and metaphase scaffolding structures. *J. Cell Sci., Suppl.* **1**, 103–122

Lohka, M. J., and Maller, J. L. (1985). Induction of nuclear envelope breakdown, chromosome condensation and spindle formation in cell-free extracts. *J. Cell Biol.* **101**, 518–523.

Lohka, M. J., and Maller, J. L. (1987). M-phase protein phosphorylation in *Xenopus laevis* eggs. *Mol. Cell Biol.* (In press.)

Lohka, M. J., and Masui, Y. (1983a). Formation *in vitro* of sperm pronuclei and mitotic chromosomes induced by amphibian ooplasmic components. *Science* **220**, 719–721.

Lohka, M. J., and Masui, Y. (1983b). The germinal vesicle material required for sperm pronuclear formation is located in the soluble fraction of egg cytoplasm. *Exp. Cell Res.* **148**, 481–491.

Lohka, M. J., and Masui, Y. (1984a). Roles of cytosol and cytoplasmic particles in nuclear envelope assembly and sperm pronuclear formation in cell-free preparations from amphibian eggs. *J. Cell Biol.* **98**, 1222–1230.

Lohka, M. J., and Masui, Y. (1984b). Effects of Ca^{2+} ions on the formation of metaphase chromosomes and sperm pronuclei in cell-free preparations from unactivated *Rana pipiens* eggs. *Dev. Biol.* **103**, 434–442.

Longo, F. J. (1972). An ultrastructural analysis of mitosis and cytokinesis in the zygote of the sea urchin *Arbacia punctulata*. *J. Morphol.* **138**, 207–238.

Longo, F. J. (1973). Fertilization: A comparative ultrastructural review. *Biol. Reprod.* **9**, 149–215.

Longo, F. J. (1976). Derivation of the membrane comprising the male pronuclear envelope in inseminated sea urchin eggs. *Dev. Biol.* **49**, 347–368.

Longo, F. J. (1981). Regulation of pronuclear development. In "Bioregulators of Reproduction" (G. Jagiello and H. J. Vogel, eds.), pp. 529–557. Academic Press, New York.

Longo, F. J., and Kunkle, M. (1978). Transformations of sperm nuclei upon insemination. *Curr. Top. Dev. Biol.* **12**, 149–184.

Maller, J. L. (1985). Regulation of amphibian oocyte maturation. *Cell Differ.* **16**, 211–221.

Maller, J. L., and Smith, D. S. (1985). Two-dimensional polyacrylamide gel analysis of changes in protein phosphorylation during maturation of *Xenopus* oocytes. *Dev. Biol.* **109**, 150–156.

Maller, J. L., Wu, M., and Gerhart, J. C. (1977). Changes in protein phosphorylation accompanying maturation of *Xenopus laevis* oocytes. *Dev. Biol.* **58**, 295–312.

Masui, Y. (1974). A cytostatic factor in amphibian oocytes: Its extraction and partial characterization. *J. Exp. Zool.* **187**, 141–147.

Masui, Y. (1982). Oscillating activity of maturation promoting factor (MPF) in extracts of *Rana pipiens* eggs. *J. Exp. Zool.* **224**, 389–399.

Masui, Y., and Clarke, H. J. (1979). Oocyte maturation. *Int. Rev. Cytol.* **57**, 185–282.

Masui, Y., and Markert, C. L. (1971). Cytoplasmic control of nuclear behavior during meiotic maturation of frog oocytes. *J. Exp. Zool.* **177**, 129–146.

Masui, Y., Meyerhof, P. G., Miller, M. A., and Wasserman, W. J. (1977). Roles of divalent cations in maturation and activation of vertebrate oocytes. *Differentiation* **9**, 49–57.

Masui, Y., Meyerhof, P. G., and Ziegler, D. H. (1979). Control of chromosome behavior during progesterone induced maturation of amphibian oocytes. *J. Steroid Biochem.* **11**, 715–722.

Masui, Y., Lohka, M. J., and Shibuya, E. (1984). Roles of Ca ions and ooplasmic factors in the resumption of metaphase-arrested meiosis in *Rana pipiens* oocytes. *Symp. Soc. Exp. Biol.* **38**, 45–66.

Meijer, J., Paul, M., and Epel, D. (1982). Stimulation of protein phosphorylation during fertilization-induced maturation of *Urechis caupo* oocytes. *Dev. Biol.* **94**, 62–70.

Merriam, R. (1969). Movement of cytoplasmic proteins into nuclei induced to enlarge and initiate DNA or RNA synthesis. *J. Cell. Sci.* **5**, 333–349.

Meyerhof, P. G., and Masui, Y. (1977). Ca and Mg control of cytostatic factors from *Rana pipiens* oocytes which cause metaphase and cleavage arrest. *Dev. Biol.* **61**, 214–229.

Meyerhof, P. G., and Masui, Y. (1979a). Properties of a cytostatic factor from *Xenopus* eggs. *Dev. Biol.* **72**, 182–187.

Meyerhof, P. G., and Masui, Y. (1979b). Chromosome condensation activity in *Rana pipiens* eggs matured *in vitro* and in blastomeres arrested by cytostatic factor (CSF). *Exp. Cell Res.* **123**, 345–353.

Miake-Lye, R., and Kirschner, M. W. (1985). Induction of early mitotic events in a cell-free system. *Cell (Cambridge, Mass.)* **41**, 165–175.

Miake-Lye, R., Newport, J., and Kirschner, M. (1983). Maturation-promoting factor induces nuclear envelope breakdown in cycloheximide-arrested embryos of *Xenopus laevis*. *J. Cell Bio.* **97**, 81–91.

Mills, A. D., Laskey, R. A., Black, P., and DeRobertis, E. M. (1980). An acidic protein which assembles nucleosomes *in vitro* is the most abundant protein in *Xenopus* oocyte nuclei. *J. Mol. Biol.* **139**, 561–568.

Moriya, M., and Katagiri, C. (1976). Microinjection of toad sperm into oocytes undergoing maturation division. *Dev., Growth Differ.* **18**, 349–356.

Nelkin, B., Nichols, C., and Vogelstein, B. (1980). Protein factor(s) from mitotic CHO cells induce meiotic maturation in *Xenopus laevis* oocytes. *FEBS Lett.* **109**, 233–238.

Newmeyer, D. D., Lucocq, J. M., Bürglin, J. R., and DeRoberts, E. M. (1986). Assembly *in vitro* of nuclei active in nuclear transport: ATP is required for nucleoplasmin accumulation. *EMBO J.* **5**, 501–510.

Newport, J. W., and Kirschner, M. (1982). A major developmental transition in early *Xenopus* embryos. 1. Characterization and timing of cellular changes at the midblastula stage. *Cell (Cambridge, Mass.)* **30**, 675–686.

Newport, J. W., and Kirschner, M. (1984). Regulation of the cell cycle during early *Xenopus* development. *Cell (Cambridge, Mass.)* **37**, 731–742.

Ottaviano, Y., and Gerace, L. (1985). Phosphorylation of the nuclear lamins during interphase and mitosis. *J. Biol. Chem.* **260**, 624–632.

Peaucellier, G., Dorée, M., and Damaille, J. G. (1982). Stimulation of endogenous protein phosphorylation in oocytes of *Sabellaria alveolata* (polychaete annelid) at meiosis initiation induced by protease, fertilization or ionophore A23187. *Gamete Res.* **5,** 115–123.

Peaucellier, G., Dorée, M., and Picard, A. (1984). Rise and fall of protein phosphorylation during meiotic maturation in oocytes of *Sabellaria alveolata* (polychaete annelid). *Dev. Biol.* **106,** 267–274.

Petzelt, C., and Auel, D. (1978). Purification and some properties of the mitotic Ca^{2+}-ATPase. *In* "Cell Reproduction: In Honor of Daniel Mazia" (E. R. Dirksen, D. M. Prescott, and C. F. Fox, eds.), pp. 487–494. Academic Press, New York.

Poenie, M., Alderton, J., Tsien, R. Y., and Steinhardt, R. A. (1985). Changes of free calcium levels with stages of the cell division cycle. *Nature (London)* **315,** 147–149.

Porter, K. R., and Machado, R. D. (1960). Studies of the endoplasmic reticulum. IV. Its form and distribution during mitosis in cells of onion root tip. *J. Biophys. Biochem. Cytol.* **7,** 167–180.

Rao, P. N., and Johnson, R. T. (1974). Induction of chromosome condensation in interphase cells. *Adv. Cell Mol. Biol.* **3,** 135–189.

Ridgway, F. B., Gilkey, J. C., and Jaffe, L. F. (1977). Free calcium increases explosively in activated medaka eggs. *Proc. Natl. Acad. Sci. U.S.A.* **74,** 623–627.

Roos, U. P. (1973). Light and electron microscopy of rat kangaroo cells in mitosis. I. Formation and breakdown of the mitotic apparatus. *Chromosoma* **40,** 43–82.

Salmon, E. D., and Segall, R. R. (1980). Calcium-labile mitotic spindles isolated from sea urchin eggs (*Lytechinus variegatus*). *J. Cell Bio.* **86,** 355–365.

Scheer, U. (1973). Nuclear pore flow rate of ribosomal RNA and chain growth rate of its precursor during oogenesis of *Xenopus laevis*. *Dev. Biol.* **30,** 13–28.

Shibuya, E. K., and Masui, Y. (1982). Sperm-induced cell cycle activities in blastomeres arrested by the cytostatic factor of unfertilized eggs in *Rana pipiens*. *J. Exp. Zool.* **220,** 381–385.

Silver, R. B., Cole, R. D., and Cande, W. Z. (1980). Isolation of mitotic apparatus containing vesicles with calcium sequestration activity. *Cell (Cambridge, Mass.)* **19,** 505–516.

Simmel, E. B., and Karnofsky, D. A. (1961). Observations on the uptake of tritiated thymidine in the pronuclei of fertilized sand dollar embryos. *J. Biophys. Biochem. Cytol.* **10,** 59–65.

Skoblina, M. N. (1976). Role of karyoplasm in the emergence of capacity of egg cytoplasm to induce DNA synthesis in transplanted sperm nuclei. *J. Embryol. Exp. Morphol.* **36,** 67–72.

Smith, L. D., and Ecker, R. E. (1971). The interactions of steroids with *Rana pipiens* oocytes in the induction of maturation. *Dev. Biol.* **25,** 233–247.

Sorensen, R. A., Cyert, M. S., and Pedersen, R. A. (1985). Active maturation-promoting factor is present in mature mouse oocytes. *J. Cell Biol.* **100,** 1637–1640.

Steinhardt, R. A., Epel, D., Carrol, E. J., and Yanigamachi, R. (1974). Is calcium ionophore a universal activator of eggs? *Nature (London)* **252,** 41–43.

Stick, R., and Hausen, P. (1985). Changes in the nuclear lamina composition during early development in *Xenopus laevis*. *Cell (Cambridge, Mass.)* **41,** 191–200.

Stick, R., and Krohne, G. (1982). Immunological localization of the major architectural protein associated with the nuclear envelope of the *Xenopus laevis* oocyte. *Exp. Cell Res.* **138,** 319–330.

Subtelny, S., and Bradt, C. (1963). Cytological observations on the early developmental stages of activated *Rana pipiens* eggs receiving a transplanted blastula nucleus. *J. Morphol.* **112,** 45–59.

Sunkara, P. S., Wright, D. A., and Rao, P. N. (1979). Mitotic factors from mammalian cells induce germinal vesicle breakdown and chromosome condensation in amphibian oocytes. *Proc. Natl. Acad. Sci. U.S.A.* **76**, 2799–2802.

Sunkara, P. S., Wright, D. A., Adlakha, R. C., Sahasrabuddhe, C. G., and Rao, P. N. (1982). Characterization of chromosome condensation factors of mammalian cells. *In* "Premature Chromosome Condensation: Application in Basic, Clinical and Mutation Research" (P. N. Rao, R. T. Johnson, and K. Sperling, eds.), pp. 233–251. Academic Press, New York.

Szollosi, D. (1966). Time and duration of DNA synthesis in rabbit eggs after sperm penetration. *Anat. Rec.* **154**, 209–212.

Szollosi, D., Calarco, P. G., and Donahue, R. P. (1972a). The nuclear envelope: its breakdown and fate in mammalian oogonia and oocytes. *Anat. Rec.* **174**, 325–340.

Szollosi, D., Calarco, P., and Donahue, R. P. (1972b). Absence of centrioles in the first and second meiotic spindles of mouse oocytes. *J. Cell Sci.* **11**, 521–541.

Tanphaichitr, N., Moore, K. C., Granner, D. K., and Chalkley, R. (1976). Relationship between chromosome condensation and metaphase lysine-rich histone phosphorylation. *J. Cell Biol.* **69**, 43–50.

Wakimoto, B. T. (1979). DNA synthesis after polyspermic fertilization in the axolotl. *J. Embryol. Exp. Morphol.* **52**, 39–48.

Wasserman, W. J., and Masui, Y. (1976). A cytoplasmic factor promoting oocyte maturation: Its extraction and preliminary characterization. *Science* **191**, 1266–1268.

Wasserman, W. J., and Smith, L. D. (1978). The cyclic behavior of a cytoplasmic factor controlling nuclear membrane breakdown. *J. Cell Biol.* **78**, R15–R22.

Weintraub, H., Buscaglia, M., Ferrez, M., Weiller, S., Boulet, A., Fabré, R., and Baulieu, E. E. (1982). Mise en évidence d'une activité "MPF" chez *Saccharomyces cerevisiae*. *C. R. Seances Acad. Sci. Paris Ser. III* **295**, 787–790.

Weisenberg, R. (1972). Microtubule formation *in vitro* in solutions containing low calcium ions. *Science* **177**, 1196–1197.

Weissbach, A. (1979). The functional roles of mammalian DNA polymerases. *Arch. Biochem. Biophys.* **198**, 386–396.

Wilson, E. B. (1896). "The Cell In Development and Heredity," pp. 132–135. Macmillan, New York (reprinted: Johnson Reprint Corp., New York, 1966).

Woodland, H. R., and Adamson, E. D. (1977). The synthesis and storage of histones during oogenesis of *Xenopus* oocytes. *Dev. Biol.* **57**, 118–135.

Wu, M., and Gerhart, J. C. (1980). Partial purification and characterization of the maturation-promoting factor from eggs of *Xenopus laevis*. *Dev. Biol.* **79**, 465–477.

Younglai, E. V., Godeau, F., Mulvihill, B., and Baulieu, E. E. (1982). Effects of cholera toxin and actinomycin on synthesis of ^{35}S methionine labeled proteins during progesterone-induced maturation of *Xenopus laevis* oocytes. *Dev. Biol.* **91**, 36–42.

Ziegler, D., and Masui, Y. (1973). Control of chromosome behavior in amphibian oocytes. I. The activity of maturing oocytes inducing chromosome condensation in transplanted brain nuclei. *Dev. Biol.* **35**, 283–292.

Ziegler, D., and Masui, Y. (1976). Control of chromosome behavior in amphibian oocytes. II. The effect of inhibitors of RNA and protein synthesis on the induction of chromosome condensation. *J. Cell Biol.* **68**, 620–628.

Zierler, M. K., Marini, N. J., Stowers, D. J., and Benbow, R. M. (1985). Stockpiling of DNA polymerases during oogenesis and embryogenesis in the frog, *Xenopus laevis*. *J. Biol. Chem.* **260**, 974–981.

Zimmerman, A. M., and Zimmerman, S. (1967). Action of colcemid in sea urchin eggs. *J. Cell Biol.* **34**, 483–488.

4

Role of Protein Phosphorylation in *Xenopus* Oocyte Meiotic Maturation

RENÉ OZON, ODILE MULNER, JEANNE BOYER, AND ROBERT BELLE

Laboratoire de Physiologie de la Reproduction
Université Pierre et Marie Curie
75230 Paris Cedex 05, France

I. INTRODUCTION

The phosphorylation–dephosphorylation of proteins is a reversible covalent modification that regulates a large number of cellular processes. The relative concentration of the phosphorylated and the nonphosphorylated forms of protein substrates is controlled by two enzymes: a protein kinase and a protein phosphatase. Protein kinase transfers a phosphoryl group from ATP (or GTP) to an amino acid residue, generally a serine or a threonine, and, less frequently, to a tyrosine residue (Krebs and Beavo, 1979).

Protein phosphorylation is now recognized to play a major role in the regulation of cell growth and division. The amphibian oocyte offers a unique system that can be used to study how phosphorylated proteins are implicated in the regulation of the cell cycle. In the ovary, the oocyte is arrested at the end of the first meiotic prophase, the diplotene stage, during the entire growing period. A steroid hormone, such as progesterone, releases the prophase block and synchronously triggers the first meiotic cell division of full grown oocytes.

111

MOLECULAR REGULATION OF NUCLEAR EVENTS
IN MITOSIS AND MEIOSIS

Maturation *in vitro* requires 4–8 hr to complete at 20°C, following stimulation of defolliculated *Xenopus* oocytes by progesterone. During progression from the diplotene stage to metaphase II, a critical biochemical step takes place concomitant with the morphological event of germinal vesicle breakdown (GVBD). During this period, maturation-promoting factor (MPF) appears in the cytoplasm; it can directly induce maturation in about 2 hr when injected into untreated oocytes (see Chapter 1, this volume). The formation of MPF does not appear to require genomic function, since it can be induced in enucleated oocytes. MPF activity remains at a high level until the oocyte is fertilized or activated. When MPF activity can first be detected, an increase in protein phosphorylation occurs (Maller *et al.*, 1977; Bellé *et al.*, 1978, 1979; Boyer *et al.*, 1980, 1983). This discovery that the biological activity of MPF is invariably associated *in ovo* with a burst of protein phosphorylation provides the first biochemical indication that a regulatory cytoplasmic factor may control, via phosphorylation–dephosphorylation reactions, the transition between prophase and metaphase.

These fundamental observations raise numerous fascinating questions: (1) What are the mechanisms that maintain the prophase block of the diplotene oocyte? Why, in such oocytes, is MPF activation prevented until steroid hormone stimulation? (2) What are the mechanisms governing the genesis of MPF activity? (3) Is MPF itself a phosphorylating–dephosphorylating enzyme? (4) What are the biochemical links between MPF, phosphorylation of proteins, GVBD, chromosome condensation, and metaphase spindle organization?

The first part of this chapter summarizes current knowledge on the role of protein phosphorylation in the prophase arrest of *Xenopus* oocytes and the mechanism for the release of oocytes from this block by progesterone. The second part deals with the accumulating body of evidence that MPF is regulated by a phosphorylation–dephosphorylation cascade.

II. MATURATION-INHIBITING PHOSPHOPROTEINS (Mp-P) AND cAMP

A. *In Vivo* Evidence for Mp-P

In an elegant series of microinjection experiments, Maller and Krebs (1977) provided the first clear and convincing demonstration that cAMP-dependent protein kinase controls prophase arrest in the *Xenopus* oocyte. They showed that microinjection of the regulatory (R) subunit of the enzyme into oocytes provokes maturation in the absence of hormonal stimulation and that microinjection of the catalytic (C) subunit inhibits steroid-induced maturation. These results were confirmed by microinjection of the pure heat-stable protein kinase inhibitor (PKI) of the C subunit, which also releases the prophase block (Maller and Krebs, 1977; Ozon *et al.*, 1978; Huchon *et al.*, 1981b).

Microinjection of pure PKI (1.5 μM) into the water-soluble compartment of the *Xenopus* oocyte triggers 100% maturation faster than progesterone but slower than MPF (Fig. 1); even at low concentration, PKI (0.15 μM) may induce a biological response but with much slower kinetics (Huchon *et al.*, 1981b). If microinjected PKI is quantitatively bound to endogenous C subunit, it may be calculated that a decrease in the concentration of the free C subunit to 1 μM or less is necessary to release the prophase block. Because cholera toxin pretreatment, which increases oocyte cAMP levels (see below), does not modify the effectiveness of PKI, cAMP-dependent protein kinase can be assumed to be fully dissociated in quiescent oocytes with the free C subunit at a maximum level of 1 μM.

The most simple and logical interpretation of these experiments is that a phosphorylated protein(s), Mp-P, which participates in the maintenance of the prophase arrest at the diplotene stage is the substrate of cAMP-dependent protein kinase. These results also suggest that a drop in the

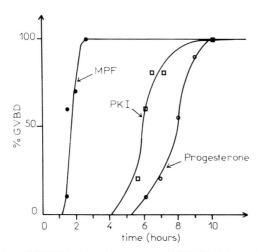

Fig. 1. Kinetics of GVBD induced by transfer of MPF (50 nl from donor progesterone-matured oocyte), microinjection of PKI (1.5 μM into oocyte), and by progesterone incubation (1 μM).

phosphorylation level of Mp-P is sufficient for the resumption of maturation. The dephosphorylation of a phosphoprotein may be regulated either
by a kinase or by a phosphatase. It was therefore of interest to know if
protein phosphatase(s) may be involved in the regulation of Mp-P and
therefore in oocyte maturation. To test this possibility the heat-stable
inhibitor-1 (Huchon *et al.*, 1981a) and inhibitor-2 (Foulkes and Maller,
1982) of protein phosphatase-1 were microinjected into *Xenopus* oocytes.
Both were found to delay progesterone-induced maturation. Inhibitor-1
(15 μM in the oocyte) in its phosphorylated form is more efficient than
inhibitor-2 since in some oocytes it totally prevented maturation induced
by progesterone and also inhibited PKI-induced maturation.

Taken together, these data clearly show that the prophase arrest of the
Xenopus oocyte is maintained by a hypothetical maturation phosphoprotein, Mp-P, whose phosphorylated state is controlled *in ovo* by cAMP-
dependent protein kinase and protein phosphatase-1. In agreement with
these conclusions is the recent finding of Bellé *et al.* (1984) that γ-S-ATP
is a potent inhibitor of steroid-induced maturation when it is microinjected into *Xenopus* oocytes (0.1 mM in the cell). γ-S-ATP is a substrate
in vitro (Gratecos and Fischer, 1974) and *in ovo* (see below) for protein
kinases but a poor substrate for protein phosphatases. A probable interpretation for the *in ovo* effect of γ-S-ATP is that it maintains the maturation protein in a thiophosphorylated form that inhibits maturation induced
by progesterone.

B. Search for Mp-P: *In Vivo* Phosphorylation of Proteins

In order to identify the putative phosphoprotein, Mp-P, the incorporation of ³²P into *Xenopus* oocyte proteins was analyzed by electrophoresis
and autoradiography. Numerous oocyte proteins were phosphorylated
following incubation with ortho[³²P]phosphate or microinjection of
[³²P]ATP (Boyer *et al.*, 1983). However, the problem is how to identify
among the numerous phosphoproteins those that satisfy the two physiological criteria of Mp-P, i.e., they must serve as a substrate *in ovo* for both
cAMP-dependent protein kinase and protein phosphatase-1, and they
must inhibit the prophase–metaphase transition of the oocyte.

A characterization of the phosphoproteins that are substrates for the C
subunit was undertaken first. Since the C subunit appears to be fully
dissociated in the resting diplotene oocyte (see above and Huchon *et al.*,
1981b; Bellé *et al.*, 1979), the only way to modulate its intracellular concentration is to microinject PKI or the C subunit. As shown in Fig. 2, two
major proteins, 20K and 32K, are more phosphorylated in oocytes mi-

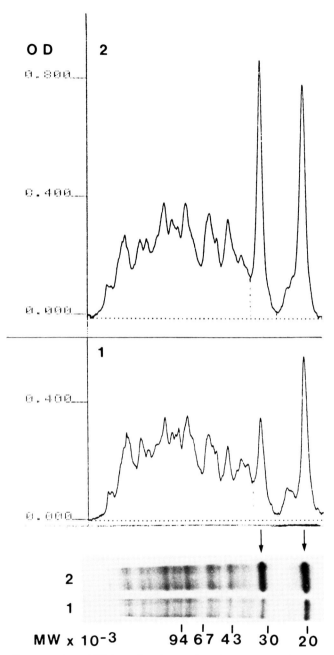

Fig. 2. Incorporation of ^{32}P into phosphoproteins after microinjection of [^{32}P]ATP (1 μCi) (1) or [^{32}P] ATP (1 μCi) and C subunit (0.6 pmoles) (2) into oocytes as shown by autoradiography and densitometric scanning. (From Boyer *et al.*, 1986.)

croinjected with the C subunit than in control oocytes. Furthermore, the level of phosphorylation of the 32K phosphoprotein is increased after microinjection of inhibitor-1 of protein phosphatase-1 (Boyer *et al.*, 1986).

Since γ-S-ATP may exert its inhibitory effect on maturation through the thiophosphorylation of Mp-P, [γ-^{35}S]ATP was microinjected into diplotene oocytes. The results of this experiment (Fig. 3) show first that [γ-^{35}S]ATP is a poor substrate *in ovo* for protein kinases as compared to [^{32}P]ATP, and second that two major proteins, one of which is 20K, are preferentially thiophosphorylated.

When the partially purified 20K phosphoprotein (after heat and 1% trichloroacetic acid treatment) was microinjected into oocytes, it was found to delay progesterone-induced maturation (unpublished results).

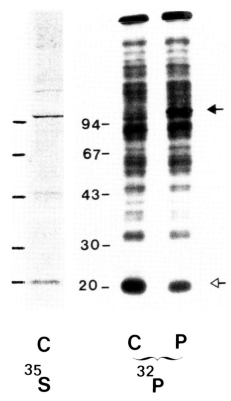

Fig. 3. Incorporation of [^{35}S]thiophosphate and [^{32}P]phosphate into phosphoproteins. Autoradiography after microinjection of [γ-^{35}S]ATP (1 μCi) (left) and after microinjection of [^{32}P]ATP (1 μCi) into control (C) and maturing (P) oocytes 1 hr after progesterone (1 μM) stimulation (right).

Together these results provide the first direct evidence that two phospho-proteins, designated Mp20 and Mp32, may correspond to the hypothetical maturation-inhibiting proteins. One may therefore expect that the levels of phosphorylation of these proteins would decrease when oocytes were induced to mature. As shown in Fig. 3, progesterone did induce a significant drop in the phosphorylation level of Mp20. The effect of progesterone was amplified when nonradioactive γ-S-ATP was microinjected before the addition of progesterone. Under these conditions, therefore, progesterone decreases the levels of phosphorylation of both Mp20 and Mp32 (Boyer *et al.*, 1986).

C. Regulation of Mp-P

How are the levels of phosphorylation of Mp-P regulated in order to maintain the prophase block? A priori two mechanisms may be operative, one involving the C subunit of cAMP-dependent protein kinase and the other involving protein phosphatase-1.

1. Regulation of cAMP-Dependent Protein Kinase Activities

a. cAMP Levels. The first suggestion, that cAMP is implicated in the control of the prophase arrest of amphibian oocytes, was reported by O'Connor and Smith (1976) who showed that theophylline, a phospho-diesterase inhibitor, inhibits progesterone-induced maturation. This finding was subsequently confirmed by a number of investigators. All treatments known to increase intracellular cAMP consistently inhibit the prophase–metaphase transition induced by steroids. Incubation of oocytes in the presence of phosphodiesterase inhibitors, IBMX or papaverine (Ozon *et al.*, 1979; Bravo *et al.*, 1978), or in the presence of cyclase activators such as cholera toxin, *E. coli* enterotoxin, or forskolin (Godeau *et al.*, 1978; Mulner *et al.*, 1979; Schorderet-Slatkine and Baulieu, 1982) abolishes subsequent progesterone-induced maturation. It was therefore proposed that an early event during the course of progesterone-induced maturation is a decrease in the cAMP content of the oocyte.

A number of experiments were then conducted in order to study the cAMP fluctuations in the oocyte after progesterone stimulation. Results were largely contradictory in terms of the occurrence of the drop in the level of cAMP, its time course, or its amplitude (Drury and Schorderet-Slatkine, 1975; Pays de Schutter *et al.*, 1975; O'Connor and Smith, 1976; Schorderet-Slatkine *et al.*, 1978; Bravo *et al.*, 1978; Maller *et al.*, 1979). These discrepancies probably result from the elevated endogenous ex-

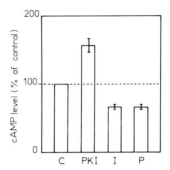

Fig. 4. cAMP levels (percentage of control [C]) in oocytes microinjected with PKI (1.5 μM), or with inhibitor-1 of protein phosphatase-1 (I) (15 μM), or incubted in the presence of progesterone (P) (1 μM). (From Mulner *et al.*, 1983a.)

tractable pool of cAMP in *Xenopus* oocytes, which can mask subtle changes in the "free" pool, and also from the variability in the basal levels of nucleotides in oocytes from different females.

Under experimental conditions that amplify cAMP fluctuation (cholera toxin and IBMX pretreatment), progesterone reproducibly induces a 30% decrease in the cAMP level of oocytes in 1 hr (Thibier *et al.*, 1982; Schorderet-Slatkine *et al.*, 1982; Fig. 4). Recent studies by Cicirelli and Smith (1985) conclusively show that under normal conditions a 20% decrease in the cAMP content of *Xenopus* oocytes occurs during the first 2–50 min following progesterone stimulation.

b. Phosphodiesterase Activity. Phosphodiesterase activities are present in oocytes in both soluble and membrane-bound forms (O'Connor and Smith, 1976; Allende *et al.*, 1977; Mulner *et al.*, 1980). A calmodulin-dependent phosphodiesterase activity has been identified in the soluble fraction *in vitro* (Miot and Erneux, 1982; Echeverria *et al.*, 1981; Orellana *et al.*, 1981, 1984). This activity, however, seems to play a minor role in the degradation of cAMP *in ovo* (Allende and Allende, 1982). Furthermore, it was shown that progesterone does not modify, either *in ovo* or *in vitro*, the phosphodiesterase activities of the oocyte (Mulner *et al.*, 1980; Baltus *et al.*, 1981). Therefore, an activation of phosphodiesterase, either directly or via calmodulin, does not seem to be a major mechanism for the decrease in cAMP content induced by progesterone.

c. Adenylate Cyclase Activity. Progesterone inhibits oocyte cyclase activity *in ovo* (Mulner *et al.*, 1979) as well as *in vitro* in rough membrane pellets (Finidori-Lepicard *et al.*, 1981; Jordana *et al.*, 1981; Sadler and

Maller, 1981). The mechanism of inhibition of adenylate cyclase by progesterone in the oocyte does not involve the classical guanyl nucleotide inhibitory subunit Ni (Goodhardt *et al.*, 1984; Olate *et al.*, 1984; Sadler *et al.*, 1984; Mulner *et al.*, 1985).

Although the Ca–calmodulin complex was reported to inhibit cyclase activity, progesterone action was shown to be independent of the level of endogenous calmodulin (Mulner *et al.*, 1983b). Therefore, progesterone acts on oocyte adenylate cyclase by a not-yet elucidated mechanism, either directly on the catalytic subunit or indirectly via a putative inhibitory subunit distinct from Ni. Blondeau and Baulieu (1985) have recently identified a 48-kDa phosphoprotein *in vitro* whose phosphorylation is rapidly decreased by progesterone and which is present in the same membrane fraction of the oocyte as the cyclase. It would be of interest to know if this protein is involved in the transduction mechanism of progesterone action.

d. Evidence for a Feedback Regulation of the cAMP Level. We have recently established that the cAMP content of the oocyte may be regulated by protein phosphorylation (Mulner *et al.*, 1983a). Microinjection of PKI into oocytes induced a 58% increase in the cAMP content (Fig. 4). Alternatively, microinjection of inhibitor-1 of protein phosphatase-1 induced a 30% decrease in the cAMP level (Fig. 4). A similar inhibition was obtained after C subunit microinjection. These results indicate that oocyte cAMP concentration is regulated by an intracellular feedback mechanism involving a regulatory phosphoprotein substrate of cAMP-dependent protein kinase and of protein phosphatase-1. In its dephosphorylated form, this regulatory protein is capable of stimulating adenylate cyclase activity; in its phosphorylated form it is capable of inhibiting adenylate cyclase activity. This feedback mechanism is already present in the immature growing oocyte. The role of such a mechanism would be to maintain a constant cAMP level and, consequently, to participate in the maintenance of the prophase block. The possibility that proteins Mp20 and Mp32 (see Fig. 2) are the proteins implicated in the regulation of cAMP levels is currently under investigation in our laboratory.

2. Protein Phosphatase-1

Protein phosphatase-1 is strongly inhibited by two thermostable proteins, inhibitor-1 and inhibitor-2 (Cohen, 1982). It is of special interest that inhibitor-1 can inhibit protein phosphatase-1 only after phosphorylation on a threonine residue by cAMP-dependent protein kinase. Since microinjection of inhibitor-1 prevents maturation, it was suggested that it corresponds to a maturation-inhibiting protein (Huchon *et al.*, 1981a). The finding that Mp20 is a heat-stable protein whose level of phosphorylation is decreased after induction of maturation favors the view that it may be an endogenous inhibitor of protein phosphatase-1. We are now undertaking a more extensive purification of Mp20 in order to test this hypothesis directly.

It is of interest to note that calcineurin is a phosphatase that acts *in vitro* on inhibitor-1 (Stewart *et al.*, 1983; Nimmo and Cohen, 1984). Calcineurin is a Ca-dependent enzyme that is stimulated by Ca^{2+} ions and calmodulin in the presence of Mg^{2+} (Tallant and Cheung, 1984). However, it is known that different divalent cations, such as Ca^{2+}, Mg^{2+}, and Mn^{2+} (microinjected or introduced by incubation in the presence of the ionophore A23187) are capable of releasing the prophase block in *Xenopus* oocytes (Masui *et al.*, 1977; Kofoid *et al.*, 1979; Bellé *et al.*, 1986). An attractive hypothesis is that a possible target for these cations is a phosphoprotein phosphatase, probably a calcineurin-like enzyme.

3. Mechanism(s) of Progesterone Action on Mp-P

Progesterone triggers maturation by a mechanism that does not involve gene transcription but probably does involve membrane phenomena. The cellular localization of the initial site of action of the steroid hormone (plasma membrane or not) and the transduction mechanism remain the objects of numerous studies (Maller, 1983). The effect of the steroid at the level of adenylate cyclase has been examined the most extensively (see above). Both the discovery of an intracellular feedback mechanism regulating cAMP levels and the probable existence of an endogenous inhibitor-

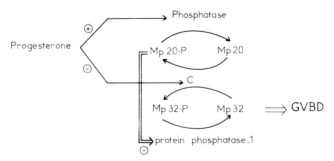

Fig. 5.   Hypothetical scheme for the regulation of maturation-inhibiting phosphoproteins (Mp20 and Mp32) by the C subunit of cAMP-dependent protein kinase and two protein phosphatases (phosphatase-1 and a putative calcineurin-like phosphatase). See text for sites of progesterone effect.

1 of protein phosphatase-1 (Mp20) may suggest another possibility for the effect of the steroid. As shown in Fig. 5, progesterone may also directly regulate the activity of an inhibitor-1 phosphatase (calcineurin-like enzyme). However, until now, there has been no definitive experimental evidence to support the choice between the two possible mechanisms of steroid action; a reasonable conclusion would be that the hormone operates at both levels.

III. PROTEIN PHOSPHORYLATION AND MATURATION-PROMOTING FACTOR (MPF) ACTIVITY

A. Increase in *in Ovo* Protein Phosphorylation When MPF Is Present

1. Characteristics of the Enzymatic Activities

A burst in protein phosphorylation invariably occurs when MPF appears in maturing oocytes (Maller *et al.*, 1977; Bellé *et al.*, 1978, 1979; Boyer *et al.*, 1980, 1983). This burst is not due to changes in the specific activity of the phosphate or ATP pools (Maller *et al.*, 1977), and it occurs in enucleated oocytes as well as in oocytes treated with cycloheximide (Maller *et al.*, 1977), indicating that it results mainly from an increase in protein kinase and/or a decrease in protein phosphatase activities located in the cytoplasmic compartment of the oocyte. The enzymes involved here have not yet been identified, although it has been shown that cAMP-dependent protein kinase is not implicated since microinjection of the thermostable PKI of cAMP-dependent protein kinase into oocytes does not affect the phosphorylation burst (Boyer *et al.*, 1980). Using an *in situ* phosphorylation assay following nondenaturing gel electrophoresis, Halleck *et al.* (1984) showed new protein kinase activity in mature oocytes; it is not yet known, however, if this activity can account for the phosphorylation burst.

To characterize further the properties of the enzymes involved in the increase in protein phosphorylation associated with MPF activity, the amino acids that become phosphorylated have been analyzed after acid hydrolysis of the phosphoproteins. Boyer *et al.* (1983) find a relative increase in phosphothreonine as compared to phosphoserine. The level of phosphotyrosine is at the limit of detection, i.e., less than 0.01% of the phosphoamino acids. It should be noted, however, that more recently, Spivack *et al.* (1984) reported a higher level of phosphotyrosine in control oocytes that does not increase in matured oocytes (between 0.2 and 2% of the total phosphoamino acids). If this result is confirmed, the presence of phosphotyrosine indicates that a tyrosine kinase and a tyrosine phosphatase are active in *Xenopus* oocytes, as is the case in sea urchin eggs (Ribot *et al.*, 1984), but that their activities are not modified during the phosphorylation burst. It can therefore be suggested that a threonine-specific phosphorylating enzyme or a serine-specific dephosphorylating activity is mainly responsible for the phosphorylation burst.

2. Analysis of the Protein Substrates

The proteins that are phosphorylated during maturation are nonyolk proteins (Maller *et al.*, 1977; Bellé *et al.*, 1978) and are found both in the cytosol and particulate fraction (Bellé *et al.*, 1978). Autoradiograms of the proteins separated by gel electrophoresis show that the level of phosphorylation of many proteins is increased (Boyer *et al.*, 1983). Interestingly, the same radioactive banding patterns are found in both progesterone- and PKI-matured oocytes (Boyer *et al.*, 1983). Among the numerous phosphoproteins of matured oocytes, three were analyzed with particular attention. First, the phosphorylation level of a protein of 105 kDa was shown to increase 1 hr after the induction of maturation and several hours before the overall phosphorylation burst (Boyer *et al.*, 1983; Fig. 3). That the phosphorylation of the 105-kDa protein occurs after PKI microinjection indicates that this result is a consequence of the dephosphorylation of Mp-P. Also, it takes place in the cytoplasmic compartment since it can be detected in enucleated oocytes (Boyer *et al.*, 1983). The 105-kDa protein represents an early molecular link between the dephosphorylation of Mp-P and MPF appearance. The second substrate is the ribosomal protein S6 (Hanocq-Quertier and Baltus, 1981; Nielsen *et al.*, 1982). Its phosphorylation state increases before GVBD. The role of S6 phosphorylation in the regulation of oocyte protein synthesis is now being actively studied (Erikson and Maller, 1985). The third substrate is a protein of 47 kDa which is invariably detected in mature oocytes when MPF is present. The phosphorylation of the 47-kDa protein occurs on threonine residues and is alkali resistant (Asselin *et al.*, 1984). The phosphorylation of the 47-kDa protein takes place in enucleated progesterone-treated oocytes and is detectable 15 min after MPF transfer into recipient oocytes (Fig. 6).

B. *In Vitro* MPF Activity

Although MPF activity was first discovered in 1971 by Masui and Markert (see Chapter 1, this volume), all attempts to purify MPF bio-

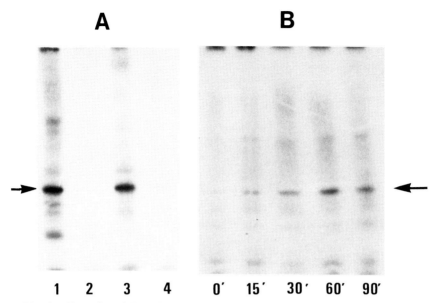

Fig. 6.   Detection of the 47-kDa alkali-resistant phosphoprotein (arrows) in ^{32}P-labeled oocytes. (A) (1, 2), progesterone-treated and control oocytes, respectively; (3, 4), enucleated progesterone-treated and enucleated control oocytes, respectively. Progesterone-treated oocytes were homogenized at the time of GVBD. (B) Time course of the phosphorylation of the 47-kDa protein after MPF transfer into ^{32}P-labeled oocytes.

chemically have so far been unsuccessful. MPF extraction from mature oocytes or mitotic cells (Wasserman and Masui, 1976; Drury, 1978; Sunkara *et al.*, 1979; Wu and Gerhart, 1980; Nelkin *et al.*, 1980; Weintraub *et al.*, 1982; Kishimoto *et al.*, 1982; Hermann *et al.*, 1983) must be performed in the presence of Mg^{2+} (above 5 mM), the Ca^{2+} chelator ethylene glycol bis (β-aminoethyl ether)N,N'-tetraacetic acid) (EGTA) (10 mM), and phosphatase inhibitors such as 2-glycerophosphate (generally 50 mM). The necessity of the phosphatase inhibitors and Mg^{2+} ions for MPF activity as well as for the stabilization of MPF extracts by ATP (Wu and Gerhart, 1980) or its analog γ-S-ATP (Hermann *et al.*, 1984) are compatible with the hypothesis that MPF could be a phosphoprotein.

One major problem with the purification of MPF resides in the biological test necessary to measure its activity, i.e., induction of maturation after microinjection of fractionated extracts. Hermann *et al.* (1983) reported that MPF activity is protected against Ca^{2+} inactivation following precipitation by 5% polyethylene glycol. This procedure allowed the omission of EGTA from the redissolution buffers and led to the observation that the presence of 2-glycerophosphate not only is absolutely required in the buffer (Hermann *et al.*, 1983) but also interferes *in ovo* with MPF activity (Hermann *et al.*, 1984, see below). Furthermore, microinjection of Mg^{2+} ions alone is capable of inducing a low percentage of maturation; the efficiency of Mg^{2+} microinjection is increased by the presence of 2-glycerophosphate (Bellé *et al.*, 1986). The maturation induced by Mg^{2+} and 2-glycerophosphate occurs with a time course close to induction by transfer of MPF (1 hr and 20 min to 2 hr). Therefore, great caution must be taken in the analysis of MPF activity in cellular extracts since most of the extracts contain both Mg^{2+} ions and 2-glycerophosphate.

C. *In Ovo* Regulation of MPF Activity

When 2-glycerophosphate is microinjected into oocytes, a considerable reduction in the time course of progesterone-induced maturation is observed (Hermann. *et al.*, 1984; Fig. 7). Based on substrate specificity, the *in ovo* effect of 2-glycerophosphate can be interpretated as an inhibition of protein phosphatases. On the other hand, microinjection of alkaline phosphatase at low concentration (200 nM) inhibits progesterone-induced maturation (Fig. 7) as well as MPF-induced maturation. Since alkaline phosphatase was shown to be a phosphotyrosyl–protein phosphatase at neutral pH and at low concentration (Swarup *et al.*, 1981), a possible

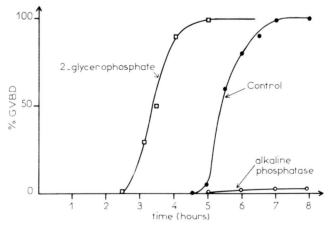

Fig. 7.   Facilitation of GVBD by 2-glycerophosphate and inhibition by alkaline phosphatase. Control oocytes were induced with progesterone. 2-Glycerophosphate was injected at 50 mM into oocytes 1 hr before hormonal stimulation. Alkaline phosphatase was injected at 200 nM 3 hr before hormonal stimulation.

explanation of these results is that protein phosphorylation on tyrosine residues could be implicated in MPF activity (Hermann *et al.*, 1984).

Neither 2-glycerophosphate nor alkaline phosphatase modify quantitatively the phosphorylation level of proteins in the prophase-arrested *Xenopus* oocyte; the pattern of protein phosphorylation remains identical in control-matured and 2-glycerophosphate-matured oocytes. Furthermore, the burst of phosphorylation induced by MPF does not represent a burst in tyrosine phosphorylation (Boyer *et al.*, 1983; Spivack *et al.*, 1984). Taken together these results indicate that tyrosine phosphorylation, if involved in MPF genesis, corresponds to a discrete change in the phosphorylation level of only a few tyrosine residues.

Interestingly, in a recent investigation Spivack *et al.* (1984) showed that the product of the *src* gene of *Rous sarcoma* virus, the protein pp60^v-src^ , induces a biological response similar to that of 2-glycerophosphate. This protein is a phosphotyrosine kinase (Erikson *et al.*, 1980) which provokes a twofold increase in the phosphotyrosine content of the *Xenopus* oocyte (Spivack *et al.*, 1984).

The biological action of alkaline phosphatase (Hermann *et al.*, 1984) as well as the effect of the tyrosine kinase activity of pp60^v-src^ (Spivack *et al.*, 1984) strongly suggest that the phosphorylation of a hypothetical protein, Rp, at the level of tyrosine residues plays a regulatory role in MPF formation. The identification of this hypothetical phosphotyrosine protein has yet to be done.

IV. CONCLUSIONS

The *Xenopus* oocyte has provided fundamental insights into the role of phosphorylation–dephosphorylation of proteins during MPF formation at the transition between prophase and metaphase of the cell cycle. In fact, experimental results indicate unequivocally that three distinct phosphorylation–dephosphorylation mechanisms are involved in MPF regulation.

1. A protein, Mp-P, substrate of cAMP-dependent protein kinase and of protein phosphatase-1, must be dephosphorylated (at the level of serine/threonine residues) before active MPF becomes detectable.

2. A second regulatory protein, Rp, substrate of an alkaline phosphatase-like enzyme and of a tryosine kinase, must be maintained in a phosphorylated state for MPF activity.

3. A consequence of MPF appearance at the time of GVBD is a burst of protein phosphorylation; it is cAMP independent and also tyrosine kinase

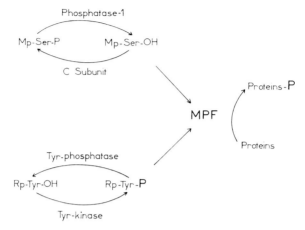

Fig. 8. Model proposed for the concerted control of MPF activity. The model postulates that three different phosphorylation–dephosphorylation enzymatic systems are involved in MPF formation (left) and activity (right) (see text)

independent. It is not yet clear whether MPF itself is the kinase responsible for this burst.

Figure 8 represents a speculative scheme that shows the possible links between the three phosphorylating systems. It indicates that MPF is a consequence of two necessary events: the dephosphorylation of Mp-P and the phosphorylation of Rp. An intriguing possibility is that the direct transfer of active MPF results in the transfer of both phosphorylating systems. It would explain why all attempts to purify MPF have not been successful. Molecular identification of the regulatory phosphoproteins Mp and Rp, whose existences are based only on indirect evidence, would indicate if this scheme is valid.

ACKNOWLEDGMENTS

This work was supported by grants from the Centre National de la Recherche Scientifique and from the Institut National de la Sante et de la Recherche Medicale.

REFERENCES

Allende, C. C., and Allende, J. E. (1982). Calmodulin-dependent cyclic nucleotide phosphodiesterase of amphibian oocytes is inhibited in vivo. *Biochem. Int.* **5,** 91–96.

Allende, C. C., Bravo, R., and Allende, J. E. (1977). Comparison of in vivo and in vitro properties of cyclic adenosine 3′, 5′-monophosphate phosphodiesterase of amphibian oocytes. *J. Biol. Chem.* **252,** 4662–4666.

Asselin, J., Bellé, R., Boyer, J., Mulner, O., and Ozon, R. (1984). Une protéine phosphorylée de 45 kDa, resistante à un traitement alcalin, apparait à l'époque de la rupture de l'enveloppe nucléaire au cours de la première division méiotique de l'ovocyte de Xenope. *C. R. Seances Acad. Sci., Ser. III* **299,** 127–129.

Baltus, E., Hanocq-Quertier, J., and Guyaux, M. (1981). Adenylate cyclase and cyclic AMP-phosphodiesterase activities during the early phase of maturation in *Xenopus laevis* oocytes. *FEBS Lett.* **123,** 37–40.

Bellé, R., Boyer, J., and Ozon, R. (1978). Endogenous protein phosphorylation in *Xenopus laevis* oocytes. Quantitative and qualitative changes during progesterone-induced maturation. *Biol. Cell.* **32,** 97–102.

Bellé, R., Boyer, J., and Ozon, R. (1979). Endogenous phosphorylated proteins during maturation in *Xenopus laevis* oocytes. *Gamete Res.* **2,** 137–145.

Bellé, R., Boyer, J., and Ozon, R. (1984). ATP-gamma-S [Adenosine 5′-0-(3-thiotriphosphate)] blocks progesterone-induced maturation of the *Xenopus* oocyte. *J. Exp. Zool.* **231,** 131–136.

Bellé, R., Mulner-Lorillon, O., Marot, J., and Ozon, R. (1986) A possible role for Mg²⁺ ions in the induction of meiotic maturation of *Xenopus* oocyte. *Cell. Differ.* (in press).

Blondeau, J. P., and Baulieu, E. E. (1985). Progesterone-inhibited phosphorylation of an unique Mr 48,000 protein in the plasma membrane of *Xenopus laevis* oocytes. *J. Biol. Chem.* **260,** 3617–3625.

Boyer, J., Bellé, R., Huchon, D., and Ozon, R. (1980). In ovo protein kinase activity during progesterone-induced maturation of *Xenopus laevis* oocytes. *In* "Steroids and their Mechanism of Action in Nonmammalian Vertebrates" (G. Delrio and J. Brachet, eds.), pp. 85–92. Raven Press, New York.

Boyer, J., Bellé, R., and Ozon, R. (1983). Early increase of a 105,000-dalton phosphoprotein during meiotic maturation of *Xenopus laevis* oocytes. *Biochimie* **65,** 15–23.

Boyer, J., Asselin, J., Bellé, R., and Ozon, R. (1986). The level of phosphorylation of two proteins (Mr 20,000 and 32,000) is regulated by progesterone and cAMP-dependent protein kinase in *Xenopus* oocytes. *Dev. Biol.* **113,** 420–428.

Bravo, R., Otero, C., Allende, C. C., and Allende, J. (1978). Amphibian oocyte maturation and protein synthesis: Related inhibition by cyclic AMP, theophylline and papaverine. *Proc. Natl. Acad. Sci. U.S.A.* **75,** 1242–1246.

Cicirelli, M. F., and Smith, L. D. (1985). Cyclic AMP levels during the maturation of *Xenopus* oocytes. *Dev. Biol.* **57,** 305–316.

Cohen, P. (1982). The role of protein phosphorylation in neural and hormonal control of cellular activity. *Nature (London)* **296,** 613–620.

Drury, K. C. (1978). Method for the preparation of active maturation promoting factor (MPF) from in vitro matured oocytes of *Xenopus laevis*. *Differentiation* **10,** 181–186.

Drury, K. C., and Schorderet-Slatkine, S. (1975). Effect of cycloheximide on the "autocatalytic" nature of the maturation promoting factor (MPF) in oocytes of *Xenopus laevis*. *Cell (Cambridge, Mass.)* **4,** 296–274.

Echeverria, M., Orellana, O., Jedlicki, E., Plaza, M., Allende, C., and Allende, J. (1981). Regulation of a cyclic nucleotide phosphodiesterase from *Xenopus laevis* ovary by calmodulin and calcium. *Biochem. Int.* **2,** 539–546.

Erikson, E., and Maller, J. (1985). A protein kinase from *Xenopus* eggs specific for ribosomal protein S6. *Proc. Natl. Acad. Sci. U.S.A.* **82,** 742–746.

Erikson, R. L., Purchio, A. F., Erikson, E., Collet, M. S., and Brugge, J. S. (1980).

Molecular events in cells transformed by *Rous sarcoma* virus. *J. Cell Biol.* **87**, 319–325.

Finidori-Lepicard, J., Schorderet-Slatkine, S., Hanoune, J., and Baulieu, E. E. (1981). Progesterone inhibits membrane-bound adenylate cyclase. *Nature (London)* **292**, 255–257.

Foulkes, J. C., and Maller, J. L. (1982). In vivo actions of protein phosphatase inhibitor-2 in *Xenopus* oocytes. *FEBS Lett.* **150**, 155–160.

Godeau, F., Boquet, P., Schorderet, M., Schorderet-Slatkine, S., and Baulieu, E. E. (1978). Inhibition par l'enterotoxine de Vibrio cholerae de la reinitiation meiotique de l'ovocyte de *Xenopus* laevis induite in vitro par la progesterone. *C. R. Hebd. Seances Acad. Sci.* **286**, 685–688.

Goodhardt, M., Ferry, N., Buscaglia, M., Baulieu, E. E. and Hanoune, J. (1984). Does the guanine nucleotide regulatory protein Ni mediate progesterone inhibition of *Xenopus* oocyte adenylate cyclase? *EMBO J.* **3**, 2653–2657.

Gratecos, D., and Fischer, E. H. (1974). Adenosine 5'-0-(3-thiophosphate) in the control of phosphorylase phosphatase activity. *Biochem. Biophys. Res. Commun.* **58**, 960–967.

Halleck, M. S., Lumley-Sapanski, K., Reed, J. A., Iyer, A. P., Mastro, A. M., and Schlegel, R. A. (1984). Characterization of protein kinases in mitotic and meiotic cell extracts. *FEBS Lett.* **167**, 193–198.

Hanocq-Quertier, J., and Baltus, E. (1981). Phosphorylation of ribosomal proteins during maturation of *Xenopus laevis* oocytes. *Eur. J. Biochem.* **120**, 351–355.

Hermann, J., Bellé, R., Tso, J., and Ozon, R. (1983). Stabilization of the maturation promoting factor (MPF) from *Xenopus laevis* oocytes. Protection against calcium ions. *Cell Differ.* **13**, 143–148.

Hermann, J., Mulner, O., Bellé, R., Marot, J., Tso, J., and Ozon, R. (1984). In vivo effects of microinjected alkaline phosphatase and its low molecular weight substrates on the first meiotic cell division in *Xenopus laevis* oocytes. *Proc. Natl. Acad. Sci. U.S.A.* **81**, 5130–5154.

Huchon, D., Ozon, R., and Demaille, J. G. (1981a). Protein phosphatase-1 is involved in *Xenopus* oocyte maturation. *Nature (London)* **294**, 358–359.

Huchon, D., Ozon, R., Fischer, E. H., and Demaille, J. G. (1981b). The pure inhibitor of cAMP-dependent protein kinase initiates *Xenopus laevis* meiotic maturation. A four-step scheme for meiotic maturation. *Mol. Cell. Endocrinol.* **22**, 211–222.

Jordana, X., Allende, C. C., and Allende, J. E. (1981). Guanine nucleotides are required for progesterone inhibition of amphibian oocyte adenylate cyclase. *Biochem. Int.* **3**, 527–532.

Kishimoto, T., Kuriyama, R., Kondo, H., and Kanatani, H. (1982). Generality of the action of various maturation-promoting factors. *Exp. Cell Res.* **137**, 121–126.

Kofoid, E. C., Knauber, D. C., and Allende, J. E. (1979). Induction of amphibian oocyte maturation by polyvalent cations and alkaline pH in the absence of potassium ions. *Dev. Biol.* **72**, 374–380.

Krebs, E. G., and Beavo, J. A. (1979). Phosphorylation-dephosphorylation of enzymes. *Annu. Rev. Biochem.* **48**, 923–959.

Maller, J. L. (1983). Interaction of steroids with the cyclic nucleotide system in amphibian oocytes. *Adv. Cyclic Nucleotide Res.* **15**, 295–336.

Maller, J. L., and Krebs, E. G. (1977). Progesterone stimulated meiotic cell division in *Xenopus laevis* oocytes. Induction by regulatory subunit and inhibition by catalytic subunit of adenosine 3', 5' monophosphate dependent protein kinase. *J. Biol. Chem.* **252**, 1712–1718.

Maller, J. L., Wu, M., and Gerhart, J. C. (1977). Changes in protein phosphorylation accompanying maturation of *Xenopus laevis* oocytes. *Dev. Biol.* **58**, 295–312.

Maller, J. L., Butcher, F. R., and Krebs, E. G. (1979). Early effect of progesterone on levels of cyclic adenosine 3', 5' monophosphate in *Xenopus* oocytes. *J. Biol. Chem.* **254,** 579–582.

Masui, Y., and Markert, C. L. (1971). Cytoplasmic control of nuclear behavior during meiotic maturation of frog oocytes. *J. Exp. Zool.* **177,** 129–146.

Masui, Y., Meyerhof, P. G., Miller, M. A., and Wasserman, W. J. (1977). Roles of divalent cations in maturation and activation of verterbrate oocytes. *Differentiation* **9,** 49–57.

Miot, F., and Erneux, C. (1982). Characterization of a cyclic nucleotide phosphodiesterase from *Xenopus laevis* oocytes. Evidence for a calmodulin-dependent enzyme. *Biochim. Biophys. Acta* **701,** 253–259.

Mulner, O., Huchon, D., Thibier, C., and Ozon, R. (1979). Cyclic AMP synthesis in *Xenopus laevis* oocytes: Inhibition by progesterone. *Biochim. Biophys. Acta* **582,** 179–184.

Mulner, O., Cartaud, A., and Ozon, R. (1980). Cyclic AMP phosphodiesterase activities in *Xenopus laevis* oocytes. *Differentiation* **16,** 31–39.

Mulner, O., Bellé, R., and Ozon, R. (1983a). cAMP dependent protein kinase regulates in ovo cAMP level of the *Xenopus* oocyte: Evidence for an intracellular feed-back mechanism. *Mol. Cell. Endocrinol.* **31,** 151–160.

Mulner, O., Tso, J., Huchon, D., and Ozon, R. (1983b). Calmodulin modulates the cyclic AMP level in *Xenopus* oocytes. *Cell Differ.* **12,** 211–218.

Mulner, O., Megret, F., Alouf, J. E., and Ozon, R. (1985). Pertussis toxin facilitates the progesterone-induced maturation of *Xenopus* oocytes: Possible role for protein phosphorylation. *FEBS Lett.* **181,** 397–402.

Nelkin, B., Nichols, C., and Vogelstein, B. (1980). Protein factor(s) from mitotic CHO cells induces meiotic maturation in *Xenopus laevis* oocytes. *FEBS Lett.* **109,** 233–238.

Nielsen, P. J., Thomas, G., and Maller, J. L. (1982). Increased phosphorylation of ribosomal protein S6 during meiotic maturation of *Xenopus* oocytes. *Proc. Natl. Acad. Sci. U.S.A.* **79,** 2937–2941.

Nimmo, G. A., and Cohen, P. (1984). The regulation of glycogen metabolism: phosphorylation of inhibitor-1 from rabbit skeletal muscle and its interaction with protein phosphatase II and III. *Eur. J. Biochem.* **87,** 353–365.

O'Connor, C. M., and Smith, L. D. (1976). Inhibition of oocyte maturation by theophylline: Possible mechanism of action. *Dev. Biol.* **52,** 318–322.

Olate, J., Allende, C. C., Allende, J. E., Sekura, R. D., and Birmbaumer, L. (1984). Oocyte adenylate cyclase contains Ni, yet the guanine nucleotide-dependent inhibition by progesterone is not sensitive to pertussis toxin. *FEBS Lett.* **175,** 25–30

Orellana, O., Allende, C. C., and Allende, J. E. (1981). Trypsin activates the calmodulin sensitive cyclic nucleotide phosphodiesterase of amphibian oocytes. *Biochem. Int.* **3,** 663–668.

Orellana, O., Jedlicki, E., Allende, C. C., and Allende, J. E. (1984). Properties of a cyclic nucleotide phosphodiesterase of amphibian oocytes that is activated by calmodulin and calcium, by tryptic proteolysis, and by phospholipids. *Arch. Biochem. Biophys.* **231,** 345–354.

Ozon, R., Bellé, R., Huchon, D., and Marot, J. (1978). cAMP-dependent protein kinase and the control of progesterone induced maturation in amphibian oocytes. *Ann. Biol. Anim., Biochim., Biophys.* **18,** 91–95.

Ozon, R., Marot, J., and Huchon, D. (1979). Progesterone stimulated meiotic maturation in *Xenopus laevis*: Inhibition by methylxanthines. *Res. Steroids* **8,** 259–263.

Pays de Schutter, A., Kram, R., Hubert, E., and Brachet, J. (1975). Cyclic nucleotides and amphibian development. *Exp. Cell Res.* **96,** 7–14.

Ribot, H. D., Jr., Eisenman, E. A., and Kinsey, W. H. (1984). Fertilization results in increased tyrosine phosphorylation of egg proteins. *J. Biol. Chem.* **259,** 5333–5338.

René Ozon *et al.*

Sadler, S. E., and Maller, J. L. (1981). Progesterone inhibits adenylate cyclase in *Xenopus* oocytes: Action of the guanine nucleotide regulatory protein. *J. Biol. Chem.* **256,** 6368–6373.

Sadler S. E., Maller, J. L., and Cooper, D. M. F. (1984). Progesterone inhibition of *Xenopus* oocyte adenyl cyclase is not mediated via the *Bordetella pertussis* toxin substrate. *Mol. Pharmacol.* **26,** 526–531.

Schorderet-Slatkine, S., and Baulieu, E. E. (1982). Forskolin increases cAMP and inhibits progesterone induced meiosis reinitiation in *Xenopus laevis* oocytes. *Endocrinology (Baltimore)* **111,** 1385–1387.

Schorderet-Slatkine, S., Schorderet, M., Boquet, P., Godeau, F., and Baulieu, E. E. (1978). Progesterone-induced meiosis in *Xenopus laevis* oocytes: A role for cAMP at the "maturation-promoting factor" level. *Cell (Cambridge, Mass.)* **15,** 1269–1275.

Schorderet-Slatkine, S., Schorderet, M., and Baulieu, E. E. (1982). Cyclic AMP control of meiosis: Effect of progesterone, cholera toxin, and membrane-active drugs in *Xenopus laevis* oocytes. *Proc. Natl. Acad. Sci U.S.A.* **79,** 850–854.

Spivak, J. G., Erikson, R. L. and Maller, J. L. (1984). Microinjection of pp 60[v-src] into *Xenopus* oocytes increases phosphorylation of ribosomal protein S6 and accelerates the rate of progesterone-induced meiotic maturation. *Mol. Cell. Biol.* **4,** 1631–1634.

Stewart, A. A., Ingebritsen, T. S., and Cohen, P. (1983). The protein phosphatases involved in cellular regulation. 5. Purification and properties of a Ca^{2+}/calmodulin-dependent protein phosphatase (2B) from rabbit skeletal muscle. *Eur. J. Biochem.* **132,** 289–295.

Sunkara, P. S., Wright, D. A., and Rao, P. N. (1979). Mitotic factors from mammalian cells induce germinal vesicle breakdown and chromosome condensation in amphibian oocytes. *Proc. Natl. Acad. Sci. U.S.A.* **76,** 2799–2802.

Swarup, G., Cohen, S., and Garbers, D. L. (1981). Selective dephosphorylation of proteins containing phosphotyrosine by alkaline phosphatase. *J. Biol. Chem.* **256,** 8197–8201.

Tallant, E. A., and Cheung, W. Y. (1984). Characterization of bovine brain calmodulin-dependent protein phosphatase. *Arch. Biochem. Biophys.* **232,** 269–279.

Thibier, C., Mulner, O., and Ozon, R. (1982). In vitro effects of progesterone and estradiol-17B on choleragen activated *Xenopus* oocyte adenylate cyclase. *J. Steroid Biochem.* **17,** 191–196.

Wasserman, W. J., and Masui, Y. (1976). A cytoplasmic factor promoting oocyte maturation. Its extraction and preliminary characterization. *Science* **191,** 1266–1268.

Weintraub, H., Buscaglia, M., Ferrez, S., Weiler, A., Boulet, A., Fabre, F., and Baulieu, E. E. (1982). Mise en évidence d'une activité MPF chez *Saccharomyces cerevisiae. C. R. Seances Acad. Sci., Ser 2* **295,** 787–790.

Wu, M., and Gerhart, J. C. (1980). Partial purification and characterization of the maturation promoting factor from eggs of *Xenopus laevis. Dev. Biol.* **79,** 465–477.

5

Maintenance of Oocyte Meiotic Arrest by Follicular Fluid Factors

IVAN L. CAMERON

Department of Cellular and Structural Biology
The University of Texas Health Science Center at San Antonio
San Antonio, Texas 78284

I. INTRODUCTION

The oocytes of most animals normally arrest at prophase of the first meiotic or maturation division (Masui and Clarke, 1979). In mammals this event takes place at about the time of birth. As the female reaches reproductive maturity, one (in monotocous species) or a few (in polytocous species) of the primordial oocytes is called on at each reproductive cycle to grow until it is larger than other cells of the adult. Thus, most primordial oocytes are kept in reserve and only a small number of oocytes grow to full size. The reason that only one or a few small oocytes grow and that the rest of the oocytes do not grow is of considerable interest but is not the subject of this chapter.

The fully grown oocyte has a single large nucleus, called a germinal vesicle, (GV), with a double complement of chromosomes and is maintained or arrested at full size until it is signaled to complete its first maturation division. The controlling factor or factors that maintain arrest of the fully grown oocyte at this late stage of meiosis is the subject of this chapter.

The term oocyte maturation is applied to the process by which the fully

MOLECULAR REGULATION OF NUCLEAR EVENTS
IN MITOSIS AND MEIOSIS

grown oocyte is caused to complete its first meiotic or maturation division and to progress to an arrest point at the second metaphase stage. This maturation process produces a small polar body (cell) and a large fertilizable egg. The first step in the oocyte maturation process must logically involve a stimulatory signal, the removal of an inhibitory signal or some combination of these two possibilities. Early observations by Pincus and Enzmann (1935) and Chang (1955) provided evidence that there is an inhibitor present in the ovarian follicle of mammals which either antagonizes oocyte maturation prior to the preovulatory surge of gonadotropin, or else these inhibitors are diminished in the follicle by the preovulatory surge of gonadotropin. For example, Pincus and Enzmann showed that oocytes isolated from rabbit follicles underwent spontaneous maturation when cultured free of follicular fluid. Chang noticed that addition of rabbit follicular fluid to cultured rabbit oocytes partially inhibited the spontaneous maturation.

In 1969, Foote and Thibault showed that cocultivation of pig granulosa cells with immature pig oocytes prevented spontaneous maturation. This observation provided evidence that granulosa cells may be the source of the oocyte maturation inhibitor (OMI) found in normal follicular fluid. Tsafriri and Channing (1975a,b), Hillensjö et al. (1979a), as well as Gwatkin and Anderson (1976), showed clearly that pig and beef follicular fluid was inhibitory to cumulus-enclosed pig and hamster oocytes, respectively. Although there has been a great deal of data to support the findings of oocyte maturation inhibitor in follicular fluid (as discussed throughout this chapter), there have been some reports of failure to find OMI in follicular fluid (Sato and Ishibashi, 1977; Leibfried and First, 1980; Racowsky and McGaughey, 1982). Tsafriri et al. (1982) and Chari et al. (1983) suggest that these failures to find follicular OMI may be due to differences in the condition of the follicles from which fluid or oocytes were obtained. Channing et al. (1982), as discussed in Section II, provide some support for the idea that follicular fluid contains both OMI as well as a stimulator of oocyte maturation which may also help explain the failure to find OMI in follicular fluid.

In summary, the observations that support the existence of OMI in follicular fluid are as follows: (1) Mammalian oocytes removed from follicular fluid undergo spontaneous maturation when cultured in the absence of follicular fluid; (2) coculture of follicular granulosa cells with immature mammalian oocyte–cumulus cell complexes maintains meiotic arrest; (3) follicular fluid or fractions thereof can inhibit oocyte maturation. The conceptional need for follicular OMI has been questioned on a theoretical basis by Biggers and Powers (1979). However, no experimental test of the Biggers and Powers concept has yet appeared in the literature.

II. CHARACTERIZATION OF FOLLICULAR FLUID OOCYTE MATURATION INHIBITOR (OMI) BY USE OF ISOLATED MAMMALIAN OOCYTES

In order to test for OMI, Tsafriri and Channing (1975a,b) and Stone *et al.* (1978) developed an assay using cumulus-enclosed oocytes obtained from freshly killed pigs. Although the oocytes were obtained from follicles ranging in diameter from 1 to 20 mm, those from the smaller and larger size follicles were reported to mature at the same rate (Channing *et al.*, 1978, 1982). The oocytes were cultured for 43–46 hr in groups of 10–15 per well in the presence of control medium (#199 plus 15% pig serum, 2.5 mM lactate, 0.03 mM pyruvate, and 12.5 mU/ml of insulin) plus inhibitor. At the end of incubation the oocytes were fixed, stained with acetoorcein, and scored as immature or mature (including metaphase I and II with a polar body). Results were expressed as percentage of inhibition, calculated as the percentage of mature oocytes in the control solution minus the percentage of mature oocytes in the experimental test solution divided by the percentage of mature oocytes in the control solution multiplied by 100. During purification of OMI, a dose–response curve was performed on each fraction using different dilutions of the test fraction.

The use of this assay procedure has resulted in the following findings: addition of 50% pig follicular fluid exerts 50% inhibition of cumulus-enclosed pig oocyte maturation; leutinizing hormone (LH) overcame this inhibition; an Amicon PM10 membrane filtrate (less than 10,000 MW) of follicular fluid contained the OMI, while an Amicon UM2 filter (with a 2000-MW cutoff) retarded the OMI (Tsafriri, *et al.*, 1976); freezing and thawing of follicular fluid had no effect on OMI activity (Stone *et al.*, 1978); replacement of follicular fluid with culture medium caused the reversal of OMI effects; pig follicular fluid from smaller follicles displayed more OMI activity than follicular fluid from larger follicles (Stone *et al.*, 1978); and pig oocytes from small, medium, and large pig follicles were equally sensitive to a PM10 (<10,000 MW) filtrate of pig follicular fluid. Tsafriri *et al.* (1976) conclude that follicular OMI is protease-sensitive but heat-stable and that cumulus-enclosed oocytes but not pig or rat oocytes denuded of cumulus cells are inhibited by addition of OMI (Hillensjö *et al.*, 1979a,b; Channing *et al.*, 1982; Tsafriri *et al.*, 1982). Human follicular fluid was also found to be inhibitory to spontaneous maturation of pig oocytes (Hillensjö *et al.*, 1979a,b).

The pig oocyte assay system has now been used to identify and quantify OMI during its purification by Channing *et al.* (1982), who report that OMI purified from larger pig follicles by chromatography on Sephadex G-

25 followed by CM-Sephadex chromatography yielded greater than 100% of the starting OMI activity. These results were interpreted to suggest that even the fluid from large follicles contains OMI but that it is "masked" by a stimulator of oocyte maturation. The presence of an oocyte maturation stimulator may therefore help explain the failure of some workers to measure OMI in follicular fluid, as indicated earlier.

The most recent and as yet unpublished purification of OMI from pig follicular fluid (S. H. Pomerantz and C. P. Channing, personal communication) is by QAE-Sephadex and reversed phase high performance liquid chromatography (HPLC) of an Amicon PM10 filtrate. Two main fractions of OMI activity were observed from QAE-Sephadex separation. These two fractions were purified further by use of HPLC with a gradient of trifluoroacetic acid–acetonitrite–water. The specific activity of one fraction was 15×10^3 times greater than that in the PM10 filtrate and was 15×10^4 times greater than that in the pig follicular fluid. Thus, at least two major fractions of OMI activity were found in the PM10 filtrate of follicular fluid.

In summary, the published characterization of pig follicular fluid OMI by use of the pig cumulus-enclosed oocyte assay indicates that OMI is less than 10,000 in molecular weight, is heat-stable, and is a peptide. Partial purification and fractionation of pig follicular fluid produces more than one fraction which contains OMI activity.

III. HYPOXANTHINE IN PIG FOLLICULAR FLUID AS THE PRINCIPAL INHIBITOR OF MOUSE OOCYTE MATURATION AND THE PREVENTION OF FOLLICLE STIMULATING HORMONE (FSH) REVERSAL OF cAMP-MAINTAINED MEIOTIC ARREST

Downs and Eppig have continued to study the putative inhibitor of oocyte maturation in porcine follicular fluid (PFF) as reported in a series of recent publications (Eppig and Downs, 1984; Downs and Eppig, 1984, 1985; Downs et al., 1985; Schultz et al., 1983a). They found that an Amicon PM10 filtrate of PFF produced a transient inhibition of maturation in cumulus-enclosed and in cumulus-denuded mouse oocytes. To denude the mouse oocyte it was passed in and out of a hand-operated Pasteur pipet until free of adherent cumulus cells. Addition of FSH or $N^6,2'$-O-dibutyryl-cAMP (Bt$_2$-cAMP) to the culture medium containing PFF produced a synergistic inhibitory effect on the maturation of cumulus-enclosed oocytes. The putative PFF inhibitor passed through dialysis tubing with a nominal molecular weight cutoff of 1000. Ether extraction, acid hydrolysis, and proteolysis of the PFF-PM10 filtrate did not reduce the inhibitory synergism between Bt_2-cAMP and the PFF filtrate. Treatment of the PFF filtrate with charcoal completely removed its inhibitory properties. Downs and Eppig (1984) concluded from these findings that factor(s) in PFF act synergistically with a cAMP-dependent process to inhibit mouse oocyte maturation in culture. These results also indicate that suppression of maturation by PFF fractions was greater in cumulus-enclosed oocytes than in denuded oocytes and led to the suggestion that the cumulus cells around the oocyte help mediate the inhibitory effects of PFF on oocyte maturation.

In a subsequent report, the low molecular weight fraction (<1000) obtained by dialysis of PFF was subjected to further study (Downs et al., 1985). Four components of the PFF dialysate were separated by ion exchange chromatography, characterized by UV spectroscopy, and tested for OMI activity using mouse oocytes. One of the fractions contained hypoxanthine which was found to be present in PFF at an estimated concentration of 1.4 mM. Both this fraction, as well as commercial preparations of hypoxanthine (a known inhibitor of phosphodiesterase), showed a transient inhibition of oocyte maturation which was enhanced by the presence of FSH and/or Bt_2-cAMP. It was concluded that hypoxanthine is the main low molecular weight (<1000) component of PFF that inhibits mouse oocyte maturation, although in this 1985 publication and in personal communication, Downs does not preclude the possibility that other factors such as peptides play a role in maintenance of meiotic arrest in oocytes, and he was careful to emphasize that other species may exhibit differential sensitivities to various PFF components.

Mechanistic schemes for action of follicular inhibitors of oocyte maturation are discussed in Section VI of this chapter.

IV. ASSAY FOR AND CHARACTERIZATION OF HUMAN FOLLICULAR FLUID OMI BY USE OF OOCYTES FROM THE AMPHIBIAN, *XENOPUS LAEVIS*

That mammalian oocytes, when removed from antrial follicles, undergo spontaneous maturation has provided the basis for bioassay of OMI. Progress in establishing the nature and mechanism of action of OMI has been impeded by the limited number and heterogeneity with respect to stage of the mammalian oocytes used in the bioassay. There was, therefore, the need for a readily available, simple to use, rapid, and unambiguous assay system to help define the chemical properties and action of OMI, as expounded by Schuetz and Rock (1981). My laboratory group, therefore, set out to develop such an assay system. We specifically chose to assay for the presence of OMI in human follicular fluid (hFF) by use of fully grown oocytes from the South African clawed toad, *Xenopus laevis*. A main difference between isolated mammalian oocytes and amphibian oocytes is that the latter do not undergo spontaneous maturation but can be stimulated by progesterone to undergo maturation.

Female adult *X. laevis* with maturable oocytes are commercially available year-round and were obtained from either Nasco (Ft. Atkinson, WI) or from Carolina Biological Supply (Burlington, NC). A booklet describing proper care and breeding of *Xenopus* is provided by the commercial suppliers at the time of purchase. Proven breeder *Xenopus* have been treated with chorionic gonadotropin 2–3 weeks prior to delivery, and oocytes from such toads should be used during the next 2 weeks for best results in the assay described below. If the gravid toads are not used during the first 3 weeks, it becomes important to induce ovulation by two injections of chorionic gonadotropin. The first injection contains 250 IU and a second injection of 500 IU is given 8 hr later. Ovulation usually follows the second injection, and the female is then separated from the eggs and can be used beginning 2 weeks later.

The specific bioassay methods we developed and currently use (Cameron *et al.*, 1983) are as follows: Toads were first anesthesized by being placed in a 0.1% solution of ethyl *m*-aminobenzoate (Tricaine) (Sigma Chemical Co., St. Louis, MO). Both ovaries were removed and treated for 1–1.5 hr in a solution of 0.2% collagenase IA (Sigma). The collagenase-treated oocytes were then placed in amphibian Ringer's (111 mM NaCl, 2 mM KC1, 1.4 mM CaCl$_2$, and 1 mM MgCl$_2$, pH 7.3); fully grown, large oocytes (1.2–1.4 mm) corresponding to Stage VI (Dumont, 1972) were separated using a flamed, wide-mouthed pipette with the aid of a dissecting microscope.

Oocytes were washed and incubated with their respective test solutions in plastic 35-mm tissue culture dishes and stimulated to undergo germinal vesicle breakdown (GVBD) by exposure to progesterone dissolved in ethanol (5 mg/ml) and added to the respective solutions to give a final concentration of 2.5 μg/ml. GVBD was scored by the formation of a distinct white spot on the pigmented animal pole (see Fig. 1) at 12 or 24 hr of incubation and verified by dissection of the formalin-fixed eggs. Data are expressed as the number of oocytes which underwent GVBD divided by the total number scored × 100 within each treatment group. Three dilutions of hFF or other fluids to be tested were used for least squares regression analysis. The dilutions in Ringer's solution were 1 : 1 (50%),

Fig. 1.   Maturation of *Xenopus* oocytes. The photomicrograph on the left shows 1.2 mm diameter oocytes which were shown histologically to possess an intact nucleus (GV). The photograph on the right is of sister oocytes exposed to 2.5 μg/ml of progesterone for 8 hr. Progesterone caused the GV to migrate to the animal pole and disrupt the pigment layer giving a white spot. Histology confirmed GVBD in these oocytes. Thus white spots serve as a marker of the first meiotic maturation.

1:4 (75%), and 1:8 (87.5%). At least 15 oocytes were used for each dilution on each patient's sample.

Follicular fluid was obtained from human patients enrolled in an *in vitro* fertilization protocol by transabdominal aspiration of the dominant follicles via endoscopic visualization under general anesthesia. One hour prior to the operative procedure, the presence of at least one follicle larger than 2.1 cm in its greatest dimension was ascertained by sector scanner ultrasound (Model number MK100, kindly loaned by ATL Ultrasound of Bellevue, WA). Within 1 hr after collection, the follicular fluid was passed through a 0.45-μm filter (Millipore Corp., Bedford, MA), stored at −20°C and analyzed within 24 hr. Blood samples were collected by venipuncture from the same patients at the time of surgery. The blood was allowed to clot, and the serum was separated from the cellular components by centrifugation, stored at −20°C, and analyzed within 24 hr.

Figure 2 shows the results of a typical dose–response curve for hFF. Our bioassay showed that hFF, but not serum from the same patients, contained OMI. The negative intercept value of the linear dose–response slopes of all six patients tested indicates that the undiluted follicular fluid contains more than enough OMI to inhibit the oocytes. The OMI factor of hFF was found to be heat labile and more recently (unpublished) we have shown the OMI factor to be destroyed by protease (trypsin) digestion.

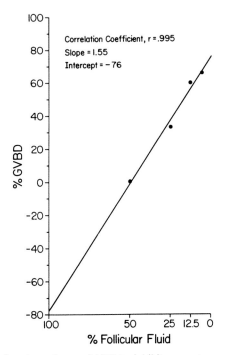

Fig. 2. Ability of various doses of hFF to inhibit progesterone-stimulated *Xenopus* oocytes from undergoing their first meiotic maturation as determined by GVBD. At least 15 oocytes were used to estimate each point on the dose–response curve. The negative intercept (−76% GVBD) at 100% hFF indicates that there is more than enough OMI present in hFF to arrest meiosis.

The inhibitory effects of hFF OMI were reversible once the OMI was washed from the oocytes. The follicular cells that surround the isolated *Xenopus* oocytes were manually removed with watchmaker forceps to show that hFF OMI was indeed inhibitory to oocyte maturation in the absence of follicular cells. Fluid from ovarian cysts of eight of eight patients also showed significant OMI activity in the *Xenopus* oocyte assay, but the concentration of OMI present in the cyst fluid was significantly less than it was in the fluid from the fully developed Graafian follicles. Because such cysts often reach a volume range of from 100 ml to more than 1 liter, cyst fluid could provide a source of large quantities of human OMI for chemical characterization and studies of its mode of action.

Subsequent to our original published characterizations of hFF–OMI, we have performed further experiments to show that an Amicon PM10 filtrate contained OMI activity but that a UMI filtrate (MW <1000) did not contain OMI activity. Thus, OMI in our studies has a molecular weight of between 1,000 and 10,000. We also found that charcoal extraction of hFF fluid did not remove OMI activity.

S. H. Pomerantz at the University of Maryland School of Medicine (personal communication) has used a *Xenopus* assay similar to that described previously in concert with the pig oocyte assay to test Amicon PM10 filtrates of PFF for OMI activity. The *Xenopus* oocyte assay detected two of the same peaks of OMI activity from QAE-Sephadex fractions as did the pig oocyte assay. However, one additional OMI activity peak was found with the *Xenopus* oocyte assay, which was not seen with the pig oocyte assay.

From these results it appears that the low molecular weight OMI–peptide fractions found in human and pig follicular fluid using pig oocytes are also inhibitory to maturation of progesterone-stimulated *Xenopus* oocytes. Thus, the *Xenopus* oocyte bioassay has proved to be a simple, reliable and independent means for detection and characterization of follicular OMI. The *Xenopus* oocyte assay can, we believe, also provide a simple method for testing normal and abnormal follicular development.

V. FAILURE TO INHIBIT *XENOPUS* OOCYTE MATURATION BY USE OF EXOGENOUS HYPOXANTHINE IN COMBINATION WITH $N^6,2'$-O-DIBUTYRYL-cAMP (Bt$_2$-CYCLIC AMP)

Based on the report of Downs *et al.* (1985) that hypoxanthine in PFF is the principle OMI for mouse oocyte maturation, we became curious to test hypoxanthine as a potential inhibitor of *Xenopus* oocyte maturation. We decided to test hypoxanthine even though we knew of several facts that might suggest that hFF–hypoxanthine was not the main OMI factor in the *Xenopus* oocyte system. For example, we were aware that Downs and Eppig had found OMI activity in PFF dialyzed filtrate with a molecular weight cutoff of 1000 and that we had not found OMI activity in an Amicon UMI filtrate with a molecular weight cutoff of 1000. Also, our OMI activity was heat-labile, protease-digestible, and not extracted by charcoal.

The results of the *Xenopus* oocyte assay system for inhibitory action of hypoxanthine and of Bt$_2$-cAMP, when used at the same concentrations that were found to be inhibitory in the mouse oocyte assay of Downs *et al.*

TABLE I

Effect of Hypoxanthine, Bt₂-cAMP and Human Follicular Fluid on Maturation of Progesterone-Stimulated *Xenopus* Oocytes[a]

	Xenopus oocyte assay			
	GVBD/oocytes scored		Average GVBD (%)	Mouse oocyte assay,[b] average GVBD
Compounds added	Animal 1	Animal 2		
Progesterone	14/14	8/14	79	—
+ (2 mM) Hypoxanthine	14/15	8/15	73	92
+Bt₂-cAMP (100 μM)	15/15	10/15	83	68
+ (2 mM) Hypoxanthine + Bt₂-cAMP (100 μM)	13/15	9/14	76	5
+ 50% Human follicular fluid	—	0/15	0	—
No progesterone	0/13	0/14	0	98

[a] Oocytes were cultured for 1 hr in amphibian Ringer's solution in the presence of hypoxanthine and/or Bt₂-cAMP prior to addition of progesterone to a final concentration of 4 μg/ml. The human follicular fluid was added immediately prior to the progesterone. A trypan blue dye exclusion test run at the end of the experiment showed oocyte viability to be 95% or greater for every combination of compounds tested.

[b] Data from Downs *et al.* (1985). Cumulus cell-enclosed mouse oocytes were cultured in minimum essential medium, but as progesterone is not needed for stimulation of mouse oocyte maturation, it was not added in any of these tests.

(1985), are summarized in Table I. Hypoxanthine was not found to be a significant inhibitor of *Xenopus* oocyte maturation when the oocytes were cultured either in the absence or in the presence of Bt₂-cAMP. For comparison purposes data from the mouse oocyte assay system of Downs *et al.* (1985) are also summarized in Table I. That the *Xenopus* oocytes used in this assay were indeed sensitive to OMI factor present in hFF was shown by its inhibition of oocyte maturation (Table I). These data allow us to conclude that the progesterone-stimulated *Xenopus* oocytes were not maintained in meiotic arrest by concentrations of hypoxanthine and by concentrations of Bt₂-cAMP which maintained meiotic arrest in the mouse oocytes. Thus, the OMI activity of hFF which maintains meiotic arrest in *Xenopus* oocytes must consist of a factor or factors other than hypoxanthine and/or cAMP. The role of intraoocyte concentrations of cAMP in oocyte maturation is discussed further in Section VI.

VI. HOW DO THE INHIBITORY FACTORS PRESENT IN FOLLICULAR FLUID MAINTAIN MEIOTIC ARREST IN OOCYTES?

How many follicular factors are involved in meiotic arrest? Taken together the evidence presented above indicates that at least three factors inhibitory to oocyte maturation are present in follicular fluid. The work of Downs and Eppig (1985) and Downs *et al.* (1985) establishes that a fraction of PFF significantly potentiates the inhibitory effect of cAMP on oocyte maturation in the mouse. This factor was found to have a molecular weight of less than 1000, was heat-stable and charcoal-extractable, and was not destroyed by protease digestion. They identify this factor as hypoxanthine. These workers were careful to point out that these findings "cannot preclude the possibility that other types of molecules such as peptides play a role in the control of meiotic arrest, whether present in follicular fluid or restricted to the follicle cell/oocyte syncytium."

Past reports (Channing *et al.*, 1982; Cameron *et al.*, 1983; Lum and Cameron, 1983; S. H. Pomerantz, personal communication, discussed in Section IV) when taken together do indicate that there are at least two and probably more than two peptides of <10,000 MW in follicular fluids that contain OMI activity. Because one of the PM10 filtrate fractions from QAE-Sephadex separations inhibited *Xenopus* oocyte maturation but did not inhibit pig oocyte maturation, one can assume that there are species differences in sensitivities to various follicular fluid components. Clearly the presence in follicular fluid of more than one peptide component with OMI activity now seems established.

How do the follicular OMI factors work? An earlier publication indicated that cumulus-enclosed pig oocytes were necessary to demonstrate the maturation-inhibiting action of pig follicular fluid (Hillensjö *et al.*, 1979a,b). In contrast, Eppig and Downs (1984) report the PFF PM10 filtrate alone produced a transitory inhibition of maturation in cumulus cell-denuded mouse oocytes. Several possible explanations for these differences can be suggested: There are species differences, the act of PM10 filtration removed something that antagonized the inhibitory effect of OMI in the filtered follicular fluid, or else Hillensjö *et al.* (1979a) did not notice the transitory inhibition due to the infrequency of scoring for GVBD. Our observations on manually defolliculated (naked) *Xenopus* oocytes show clearly that an active OMI peptide had a direct inhibitory effect on oocyte maturation (Cameron *et al.*, 1983). Without removal of the cumulus cells from mammalian oocytes, one cannot differentiate a

direct inhibitory effect of an OMI on the oocyte proper from an indirect
effect that acts through the attached cumulus cells. Because denuded
mouse oocytes have proved responsive to a synergistic effect with hypoxanthine and exogenous Bt_2-cAMP, and because denuded *Xenopus* oocytes are responsive to hFF peptide, it seems likely that naked oocytes
are the target for both classes of meiotic inhibitors. This does not, of

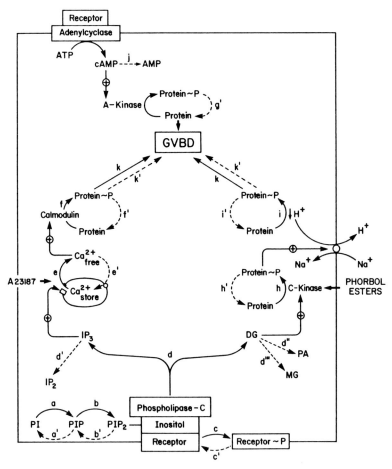

Fig. 3. Proposed pathways to explain meiotic arrest and the release from meiotic arrest
in oocytes. This scheme was adapted and modified from Berridge (1984). The positive
aspects of the pathways (represented by the solid lines) illustrate the way in which inositol
triphosphate (IP_3), diacylglycerol (DG), and cyclic AMP (cAMP) could contribute to GVBD
and oocyte maturation. The dashed lines represent pathways involved in maintenance of
meiotic arrest. Abbreviations: PI, phosphatidylinositol; PIP, phosphatidylinositol 4-phos-

course, rule out a role for cumulus cells in the maintenance of meiotic arrest, but allows one to study and define the action of OMI factors on the oocyte proper. As much as possible, the following discourse concentrates on the effects of OMI on the oocyte proper.

Eppig and Downs (1984) have previously reviewed the role of cAMP in maintenance of meiotic arrest. Briefly, analogs of cAMP such as Bt_2-cAMP which can enter cells, or permeable phosphodiesterase inhibitors, such as hypoxanthine, theophylline, and 3-isobutyl-1-methylxanthine (IBMX), inhibit the spontaneous maturation of denuded mammalian oocytes. Elevated levels of cAMP overcome the inhibitory effects of IBMX. A decrease in oocyte cAMP precedes GVBD in both mammals and amphibians (Maller and Krebs, 1980: Schultz *et al.*, 1983c).

Is a decrease in cAMP essential for GVBD? Recent data (Schultz *et al.*, 1983b,c) link the reduced cAMP level in the oocyte to a protein dephosphorylation step. This dephosphorylation step may involve the cAMP-dependent protein kinase (A-kinase). However, as shown in Fig. 3, at least two other pathways can control the phosphorylation of specific proteins in target tissues: (1) the calcium–calmodulin-dependent protein kinase and (2) the calcium–phospholipid-dependent protein kinase (C-kinase). In fact, Urner and Schorderet-Slatkine (1984) have already shown that the phorbol esters (phorbol 12,13-dibutyrate or phorbol 12-myris-

---

phate; PIP_2, phosphatidylinositol 4,5-bisphosphate; ATP, adenosine triphosphate; IP_2, inositol 1,4-bisphosphate; PA, phosphatidic acid; MG, monoacylglycerol; A23187, calcium ionophore.

Enhance GVBD	Arrest GVBD
a. PI kinase	a'. PIP phosphomonoesterase
b. PIP kinase	b'. PIP_2 phosphomonoesterase
c. Tyrosine kinase	c'. Phosphotyrosyl–protein phosphatase
d. PIP_2 phospholipase C	d'. Inositol trisphosphatase
e. IP_3-sensitive Calcium release	d''. DG kinase
	d'''. DG lipase
f. Ca^{2+}/calmodulin-dependent protein kinase	e'. Calcium pump
h. C-kinase	f'. Protein phosphatase
	g'. Protein phosphatase
i. pH-sensitive protein phosphorylation	h'. Protein phosphatase
j. Nuclear proteins	i'. Protein phosphatase
	j'. Phosphodiesterase

k or k'. Unspecified whether the phosphorylated protein (protein ~P) or the protein enhances or arrests GVBD.

tate,13-acetate) when added to the culture medium of denuded mouse oocytes inhibit their spontaneous GVBD. Because such active phorbol esters can replace phospholipid or Ca^{2+} as activators of C-kinase (Gschwendt *et al.*, 1983), it is suggested that arrest of oocyte maturation may occur via an alternate pathway not involving cAMP. It therefore seems that a decrease in intracellular cAMP can bring about GVBD but that a decrease in cAMP is not of itself essential for GVBD because alternate pathways not dependent on cAMP can be used to control oocyte maturation (see Fig. 3).

The scheme presented in Fig. 3 is intended to illustrate that putative inhibitory factors present in follicular fluid might potentially intervene at multiple points along the pathways controlling oocyte maturation. For example, the data of Downs *et al.* (1985) indicate involvement of cAMP; on the other hand, our failure to inhibit *Xenopus* oocyte maturation with hypoxanthine and Bt_2-cAMP suggest that one or more of the alternate pathways shown in Fig. 3 was used. Likewise, numerous workers have provided evidence for the involvement of a surge of free Ca^{2+} in oocyte maturation (Wasserman *et al.*, 1980; Moreau *et al.*, 1980). Indeed, Schultz *et al.* (1983b), Cartaud *et al.* (1981), and Wasserman and Smith (1981) have provided evidence that the Ca^{2+} effect on oocyte maturation is mediated by calmodulin. Thus a Ca^{2+}–calmodulin-dependent step is also involved in oocyte maturation. Work from our laboratory has shown that drugs which inhibit the Na^+/H^+ antiport at the oocyte surface (benzamil and amiloride) inhibit *Xenopus* oocyte maturation (Cameron *et al.*, 1982). Thus, according to the scheme shown in Fig. 3, such treatment would prevent the intracellular alkaline surge and would cause a change in protein phosphorylation. It is not as yet clear if such protein phosphorylation, as has been reported in fully grown *Xenopus* oocytes, is by itself sufficient to trigger GVBD (Wasserman and Houle, 1983).

Regarding a site and mechanism of action for the peptide factors from follicular fluid it seems logical to think that they exert their inhibitory action at the oocyte surface. The rationale for this statement is that other peptide factors which are growth regulators act on receptors at the cell surface and that the cell membrane is not readily permeable to peptides.

VII. CONCLUSIONS

From the foregoing review it is concluded: (1) that follicular fluid contains more than one factor inhibitory to oocyte maturation; and (2) alternate metabolic pathways appear to be involved in regulation of oocyte maturation. It is proposed that the multiple inhibitory peptide factors and the nonpeptide factors that have been identified in follicular fluid act on one or more of the proposed metabolic pathways illustrated in Fig. 3.

ACKNOWLEDGMENTS

I wish to dedicate this chapter to the memory of Dr. Cornelia P. Channing. Dr. Channing died of metastatic cancer on April 8, 1985. She was a foremost leader in the field of ovarian physiology. Her original contributions on the role on nonsteroidal regulators in control of oocyte and follicular maturation are recognized as elegant landmarks in the field. She will be missed by her many friends and colleagues.

I wish to thank Seymour Pomerantz for providing preprints of his papers and for helpful discussion. I am grateful to Virginia Ord and Keithley Hunter for their skilled and dedicated assistance in performing the experimental work and to Jean B. Lum who contributed greatly to the original research. The work was supported in part by NIH-CA36372.

REFERENCES

Berridge, M. J. (1984). Oncogenes, inositol lipids and cellular proliferation. *Bio/Technology* **2,** 541–546.

Biggers, J. D., and Powers, R. D. (1979). Comments on the control of meiotic maturation in mammals. *In* "Ovarian Follicular Development and Function" (A. R. Midgley and W. W. Sadler, eds.), pp. 365–373. Raven Press, New York.

Cameron, I. L., Hunter, K. E., and Cragoe, E. J., Jr. (1982). A stage-specific inhibitory effect of benzamil on *Xenopus* oocyte maturation located at the cell surface. *Exp. Cell Res.* **139,** 455–457.

Cameron, I. L., Lum, J. B., Nations, C., Asch, R. H., and Silverman, A. Y. (1983). Assay for characterization of human follicular oocyte maturation inhibitor using *Xenopus* oocytes. *Biol. Reprod.* **28,** 817–822.

Cartaud, A., Huchon, D., Marot, J., Ozon, R., and Demaille, J. G. (1981). Calmodulin is involved in the first step of oocyte maturation: Effects of the antipsychohotic drug fluphenazine and of anticalmodulin antibodies on the progesterone-induced maturation of *Xenopus laevis* oocyte. *Cell Differ.* **10,** 357–363.

Chang, M. C. (1955). The maturation of rabbit oocytes in culture and their maturation, activation, fertilization and subsequent development in the fallopian tubes. *J. Exp. Zool.* **28,** 379–405.

Channing, C. P., Stone, S. L., Kripner, S. C., and Pomerantz, S. H. (1978). Studies on an oocyte maturation inhibitor present in porcine follicular fluid. *In* "Novel Aspects of Reproductive Physiology" (C. H. Spilman and J. W. Wilks, eds.), pp. 37–59. S. P. Med. Sci. Books, New York.

Channing, C. P., Anderson, L. D., Hoover, D. J., Kolena, J., Osteen, K. G., Pomerantz, H., and Tanabe, K. (1982). The role of nonsteroidal regulators in control of oocyte and follicular maturation. *Recent Prog. Hor. Res.* **38,** 331–408.

Chari, S., Hillensjö, T., Magnusson, C., Sturm, G., and Daume, E. (1983). In vitro inhibition of rat oocyte meiosis by human follicular fluid fractions. *Arch. Gynecol.* **233**, 155–164.

Downs, S. M., and Eppig, J. J. (1984). Cyclic adenosine monophosphate and ovarian follicular fluid act synergistically to inhibit mouse oocyte maturation. *Endocrinology (Baltimore)* **114**, 418–427.

Downs, S. M., and Eppig, J. J. (1985). A follicular fluid component prevents gonadotropin reversal of cyclic adenosine monophosphate-dependent meiotic arrest in murine oocytes. *Gamete Res.* **11**, 83–97.

Downs, S. M., Coleman, D. L., Ward-Bailey, P. F., and Eppig, J. J. (1985). Hypoxanthine is the principal inhibitor of murine oocyte maturation in a low molecular weight fraction of porcine follicular fluid. *Proc. Natl. Acad. Sci. U.S.A.* **82**, 454–458.

Dumont, J. N. (1972). Oogenesis in *Xenopus laevis*. 1. Stages of oocyte development in laboratory maintained animals. *J. Morphol.* **136**, 153–180.

Eppig, J. J., and Downs, S. M. (1984). Chemical signals that regulate mammalian oocyte maturation. *Biol. Reprod.* **30**, 1–11.

Foote, W. D., and Thibault, C. (1969). Recherches expérimentales sur la maturation in vitro des oocytes de truie et de veau. *Ann. Biol. Anim., Biochim., Biophys.* **3**, 329–349.

Gschwendt, M., Horn, F., Kittstein, W., Fürstenberger, G., and Marks, F. (1983). Soluble phorbol ester binding sites and phospholipid- and calcium-dependent protein kinase activity in cytosol of chick oviduct. *FEBS Lett.* **162**, 147–150.

Gwatkin, R. B. L., and Anderson, O. F. (1976). Hamster oocyte maturation in vitro: inhibition by follicular components. *Life Sci.* **19**, 527–536.

Hillensjö, T., Botta, S. K., Schwartz-Kripner, A., Wentz, A. C., Sulewski, V., and Channing, C. P. (1978). Inhibitory effect of human follicular fluid upon the maturation of porcine oocytes in culture. *J. Clin. Endocrinol. Metab.* **47**, 1332–1335.

Hillensjö, T., Channing, C. P., Pomerantz, S. H., and Schwartz-Kripmer, A. (1979a). Intrafollicular control of oocyte maturation in the pig. *In Vitro* **15**, 32–39.

Hillensjö, T., Schwartz-Kripner, A., Pomerantz, S. H., and Channing, C. P. (1979b). Action of Porcine follicular fluid oocyte maturation inhibitor *in vitro:* Possible role of the cumulus cells. *In* "Ovarian Follicular and Corpus Luteum Function" (C. P. Channing, J. M. March, and W. Z. Sadler, eds.), pp. 283–291. Plenum, New York.

Leibfried, L., and First, N. L. (1980). Effect of bovine and porcine follicular fluid and granulosa cells on maturation of oocytes in vitro. *Biol. Reprod.* **23**, 699–704.

Lum, J. B., and Cameron, K. L. (1983). Characterization of human oocyte maturation inhibitor (OMI) using the *Xenopus* oocyte assay. *J. Cell Biol.* **97**, 22a.

Maller, J. L., and Krebs, E. G. (1980). Regulation of oocyte maturation. *Curr. Top. Cell Regul.* **16**, 272–311.

Masui, Y., and Clarke, H. J. (1979). Oocyte maturation. *Int. Rev. Cytol.* **57**, 185–223.

Moreau, M., Vilain, J. P., and Guerrier, P. (1980). Free calcium changes associated with hormone action in amphibian oocytes. *Dev. Biol.* **78**, 201–214.

Pincus, G., and Enzmann, E. V. (1935). The comparative behavior of mammalian eggs in vivo and in vitro. I. The activation of ovarian eggs. *J. Exp. Med.* **62**, 665–675.

Racowsky, C., and McGaughey, R. W. (1982). Further studies of the effects of follicular fluid and membrana granulosa cells on the spontaneous maturation of pig oocytes. *J. Reprod. Fertil.* **66**, 505–512.

Sato, E., and Ishibashi, T. (1977). Meiotic arresting action of the substance obtained from cell surface of porcine granulosa cells. *Jpn. J. Zootech. Sci.* **48**, 22–26.

Schuetz, A. W., and Rock, J. (1981). Stimulatory and inhibitory effects of human follicular fluid on amphibian oocyte maturation and ovulation *in vitro*. *Differentiation* **21**, 41–44.

Schultz, R. M., Montgomery, R. R., Ward-Bailey, P. F., and Eppig, J. J. (1983a). Regulation of oocyte maturation in the mouse: Possible roles of intercellular communication, cAMP, and testosterone. *Dev. Biol.* **95,** 294–304.

Schultz, R. M., Heller, D. T., and Baklad, N. (1983b). Implication of a calmodulin-dependent step in mouse oocyte maturaton. *Biol. Reprod.* **28,** Suppl., 138.

Schultz, R. M., Montgomery, R. R., and Belanoff, J. R. (1983c). Regulation of mouse meiotic maturation: Implication of a decrease in oocyte cAMP and protein dephosphorylation in commitment to resume meiosis. *Dev. Biol.* **97,** 264–273.

Stone, S. L., Pomerantz, S. H., Schwartz-Kripner, A., and Channing, C. P. (1978). Inhibitor of oocyte maturation from porcine follicular fluid: Further purification and evidence for reversible action. *Biol. Reprod.* **19,** 585–592.

Tsafriri, A., and Channing, C. P. (1975a). An inhibitory influence of granulosa cells and follicular fluid upon porcine oocyte meiosis *in vitro Endocrinology (Baltimore)* **96,** 922–927.

Tsafriri, A., and Channing, C. P. (1975b). Influence of follicular maturation and culture conditions upon porcine oocyte meiosis *in vitro. J. Reprod. Fertil.* **43,** 149–152.

Tsafriri, A., Pomerantz, S. H., and Channing, C. P. (1976). Porcine follicular fluid inhibitor of oocyte meiosis: Partial purification and characterization. *Biol. Reprod.* **14,** 511–516.

Tsafriri, A., Dekel, N., and Bar-Ami, S. (1982). The role of oocyte maturation inhibitor in follicular regulation of oocyte maturation. *J. Reprod. Fertil.* **64,** 541–551.

Urner, F., and Schorderet-Slatkine, S. (1984). Inhibition of denuded mouse oocyte meiotic maturation by tumor-promoting phorbol esters and its reversal by retinoids. *Exp. Cell Res.* **154,** 600–605.

Wasserman, W. J., and Houle, J. G. (1983). An increase in intracellular pH, S-6 phosphorylation and protein synthesis are not sufficient to trigger GVBD in Stage IV *Xenopus* oocytes. *J. Cell Biol.* **97,** 22a.

Wasserman, W. J., and Smith, L. D. (1981). Calmodulin triggers the resumption of meiosis in amphibian oocytes. *J. Cell Biol.* **89,** 389–394.

Wasserman, W. J., Pinto, L. H., O'Connor, C. M., and Smith, L. D. (1980). Progesterone induces a rapid increase in Ca of *Xenopus laevis* oocytes. *Proc. Natl. Acad. Sci. U.S.A.* **77,** 1534–1536.

6

Regulation of Chromatin Condensation and Decondensation in Sea Urchin Pronuclei

DOMINIC POCCIA

Department of Biology
Amherst College
Amherst, Massachusetts 01002

I. INTRODUCTION

It is a long-established principle that the state of chromosome compaction is cytoplasmically controlled. For example, in 1922, Brachet established that sperm nuclei entering sea urchin oocytes undergoing meiotic maturation condensed into chromosomes, whereas those entering mature eggs decondensed into pronuclei. More recent work, based on nuclear transplantation and cell fusion techniques, has not only provided much confirmation of that principle, but also introduced several experimental systems for studying such phenomena (reviewed by Gurdon and Woodland, 1968; Johnson and Rao, 1971).

A deep understanding of the control of chromatin condensation will require knowledge of the nature and timing of the cytoplasmic conditions, as well as a description of the biochemical and physical alterations in the responding chromatin. In particular, it will be necessary to distinguish causative events from permissive or merely coincidental ones.

Several generalizations have emerged from the study of chromosome condensation–decondensation. Meiotic and mitotic condensation factors are fundamentally similar (Sunkara *et al.*, 1982; Halleck *et al.*, 1984).

149

MOLECULAR REGULATION OF NUCLEAR EVENTS
IN MITOSIS AND MEIOSIS

Condensation factors are not species-specific (Johnson *et al.*, 1970; von der Haar *et al.*, 1981). The morphology of the chromosomal response to condensation conditions varies throughout interphase (Johnson and Rao, 1971), supporting the concept of a continuous condensation–decondensation cycle (Mazia, 1963).

We have explored in my laboratory essentially the same system studied by Brachet (1922), attempting to relate biochemical and morphological transitions associated with chromosome condensation–decondensation phenomena. Using polyspermically fertilized sea urchin eggs, it is possible to isolate sufficient quantities of chromatin for biochemical analyses of the male chromosomal proteins. In virtually all respects, male chromatin from polyspermic and monospermic eggs behaves identically (Poccia *et al.*, 1978, 1981). This chapter presents experiments on three morphological transitions: mitotic chromosome condensation; premature chromosome condensation; and male chromatin decondensation following fertilization. In each case, the timing of the cytoplasmic conditions promoting the transitions are described, the requirements for protein or RNA synthesis, if any, are discussed, and analyses of changes in histone phosphorylation states of the responding chromatins are presented.

II. MITOTIC AND PREMATURE CHROMOSOME CONDENSATION

A. Chromosome Condensing Conditions

The first cell cycle in monospermic or moderately polyspermic *Stronglyocentrotus purpuratus* eggs lasts 90 min at 15°C from fertilization to metaphase (Hinegardner, 1967; Poccia *et al.*, 1978). Supernumerary male pronuclear chromosomes condense in synchrony with the female pronuclear chromosomes (Mazia, 1974; Krystal and Poccia, 1979).

Conditions promoting chromosome condensation in the fertilized sea urchin egg have been investigated by initiating the first cell cycle with ammonium chloride, and fertilizing at successively later times (Krystal and Poccia, 1979). Condensation of the male chromosomes reveals the development of these conditions, even when the female chromosomes are lacking, as in enucleated egg halves (merogons). The condensation of male chromosomes which are introduced as interphase chromatin into the egg during late stages of the maternal cell cycle is termed premature chromosome condensation (PCC), by analogy to PCC in heterophasic homokaryons produced by cell fusion (Johnson and Rao, 1971; Rao *et al.*, 1982).

Using the PCC assay, conditions promoting male chromosome condensation were shown to develop and decay in parallel with the mitotic period of the female pronucleus (Krystal and Poccia, 1979). The male nuclei undergo a transformation from a state with a thin layer of decondensed chromatin around the periphery, to an intermediate state in which only a small central region is still not decondensed while recondensation into chromosomes is occurring at the periphery, to a state where a full haploid complement of thin (unreplicated) chromosomes can be seen (Fig. 1) (Mazia, 1974; Krystal and Poccia, 1979; Longo, 1983).

Induction of PCC is an all-or-none phenomenon within a given egg. As many as 50 pronuclei per egg may undergo PCC, resulting in the condensation of approximately 1100 chromosomes per cell, suggesting that whatever factors are responsible for condensation, they either are present in

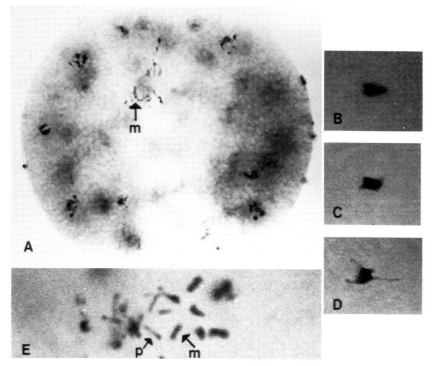

Fig. 1.   Premature chromosome condensation. (A) Premature chromosome condensation 30 min following polyspermic fertilization of a prometaphase egg; (B–D) Male pronucleus at (B) 7, (C) 14, and (D) 20 min following fertilization of a prometaphase egg; (E) prematurely condensed chromosomes intermingled with maternal chromosomes at 33 min following fertilization of a prometaphase egg; p = prematurely condensed paternal, m = maternal chromosomes. (From Krystal and Poccia, 1979.)

excess or act catalytically. Since pronuclear fusion is not necessary for PCC induction, condensation must be cytoplasmically mediated.

To assess the role of the maternal pronucleus, enucleated egg halves can be prepared either by ultracentrifugation or manual bisection (Krystal and Poccia, 1979). Enucleated merogons are capable of developing chromosome condensing conditions at the correct time after ammonium chloride activation, demonstrating that the female pronucleus or chromosomes are not required for development of condensing conditions. Nonetheless, the enucleates never show the full response seen in nucleated halves or whole eggs. The lifetime of condensing conditions is shorter, and fewer eggs at any given time show PCC (Fig. 2). These observations were interpreted as indicating an absence of stabilizing effects of either

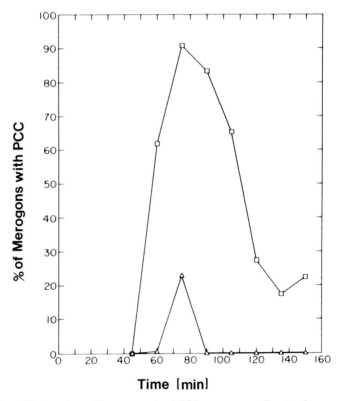

Fig. 2.   Lifetime of conditions promoting PCC in merogons. Aliquots of merogons from ammonia-activated cultures were fertilized at successive intervals and cultured for an additional 30 min at which time they were fixed and scored for PCC (200 merogons scored for each time point). □, Nucleated; △, enucleated merogons. (From Krystal and Poccia, 1979.)

TABLE I

Stability of the Conditions Inducing PCC in Merogons Produced from Mid-prophase Eggs[a]

Time of fertilization after removal from gradient (min)	% PCC[b]	
	Nucleated halves	Enucleated halves
Exp. 1		
0	79	25
6	88	16
12	85.5	6.5
Exp. 2		
0	78	25
7	87	6
14	76	0.5
Exp. 3		
0	99	10.5
10	97	0
20	94	—
30	80	—
40	64	—
50	35	—

[a] Eggs from cultures showing 80–90% maternal pronuclear envelope breakdown were layered on density gradients and merogons produced by centrifugation. After the merogons were washed free of sucrose, aliquots were fertilized at the specified intervals, cultured for an additional 30 min, fixed, and scored for PCC. (From Krystal and Poccia, 1979.)

[b] Two hundred eggs counted for each point.

the maternal pronucleus or chromosomes on the chromosome condensing conditions. The instability of chromosome condensing conditions was directly assayed in merogons produced by density ultracentrifugation from midprophase cells. Enucleated merogons began to lose the ability to condense chromosomes while the nucleated merogons were still developing their maximal response (Table I).

Development of condensing conditions is independent of transcription as shown by adding actinomycin D or ethidium bromide from 90 min before fertilization until mitosis at concentrations sufficient to inhibit 95% of RNA synthesis (Krystal and Poccia, 1979). Protein synthesis may be required for condensation, but after late S or early G_2 (50 min postactivation), exposure to emetine, which inhibits protein synthesis by $> 98\%$,

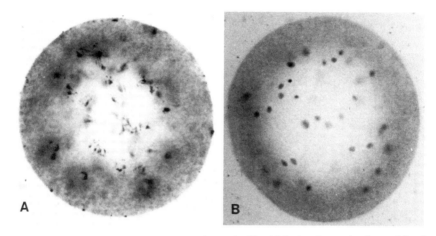

Fig. 3. Appearance of emetine-treated eggs fertilized 85 min after activation and fixed 30 min later. Emetine (10^{-4} M) added (A) 60 min or (B) 20 min after activation. (From Krystal and Poccia, 1979.)

has no effect (Figs. 3 and 4). The results suggest that any proteins needed for chromosome condensation are synthesized prior to G_2 and rule out a significant dependence of condensation on transcription of maternal cytoplasmic or pronuclear genomes. The results also eliminate any contribution of mRNA carried by the sperm, since translation can be blocked prior to fertilization without effect on PCC.

B. Cleavage-Stage (CS) H1 Phosphorylation

The phosphorylation state of histone H1 was investigated in mitotic and prematurely condensed chromosomes in the first cell cycle (Krystal and Poccia, 1981). The only H1 species in the chromatin during condensation is cleavage-stage H1 (CS H1). CS H1 is an unusually large H1, with a molecular weight of approximately 34,000 (G. R. Green and D. Poccia, unpublished). It exists as a phosphoprotein in unfertilized eggs and enters the chromatin in a phosphorylated state within minutes after fertilization (Green and Poccia, 1985). It becomes progressively more phosphorylated through the cell cycle, reaching a maximum at metaphase (Krystal and Poccia, 1981). The degree of phosphorylation as judged by mobility on long acid–urea polyacrylamide gels is not a function of the amount of time that the male chromatin is exposed to cytoplasm, but of the stage of the cytoplasm to which it is exposed. Male pronuclei introduced into the activated egg cytoplasm for a total of 37–38 min during G_1–S, G_2–early

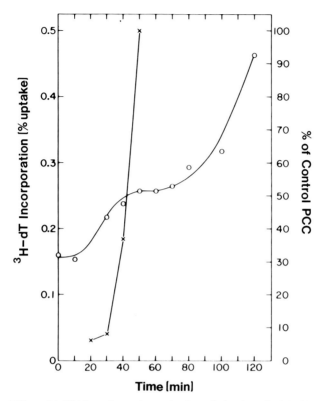

Fig. 4. Effect of inhibition of protein synthesis on induction of PCC. Six parallel cultures of *L. pictus* eggs (from a single female, activated simultaneously) were established. One served as a control. One was labeled with [³H]thymidine and sampled at various times to measure DNA synthesis. Emetine was added to the other four, one at a time, at the times after activation indicated on the abscissa. The emetine-treated cultures and the control were fertilized at 85 min, cultured for an additional 30 min, fixed and scored for PCC (200 eggs for each time point). ○, [³H]thymidine incorporation; X, % PCC. (From Krystal and Poccia, 1979.)

prophase, or prometaphase–metaphase showed increasing amounts of CS H1 phosphorylation (Fig. 5; Krystal and Poccia, 1981).

As noted, chromosome condensation can be blocked by the protein synthesis inhibitor emetine. A comparison of CS H1 in condensed (metaphase) and uncondensed (emetine-inhibited) male chromatin reveals little if any difference in its degree of phosphorylation (Krystal and Poccia, 1981). Assuming functionally equivalent phosphorylation sites, it would appear that the high degree of phosphorylation of CS H1 is not a sufficient condition for induction of chromosome condensation. The experiment

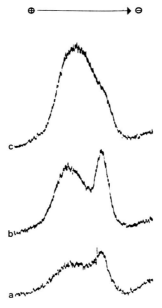

Fig. 5. Cell cycle specificity of CS H1 phosphorylation. CS H1 was extracted from male pronuclei residing in egg cytoplasm for approximately 35 min at the specified times after activation and electrophoresed on long acid–urea polyacrylamide gels, pH 2.8. Densitometer tracings were made of the CS H1 region. Phosphorylated proteins are retarded on this gel system. (a) 7–45 min (maternal G_1-S period); (b) 41–78 min (G_2-early prophase); (c) 86–124 min (maternal prometaphase; prematurely condensed male chromosomes). (From Krystal and Poccia, 1981.)

also shows that all of the protein factors needed for the phosphorylation of CS H1 and its normal timing are present in the unfertilized egg. It also suggests that the high degree of phosphorylation of CS H1 is not simply a result of increased accessibility of phosphorylation sites made available during chromosome condensation.

C. Comparison to Other Systems

Premature chromosome condensation of sea urchin male pronuclear chromatin is analogous to PCC in, for example, heterophasic homokaryons (Johnson and Rao, 1971). However, the interphase chromosomes in the hybrid cells rarely become completely condensed. Sperm nuclei, like erythrocyte nuclei, may be considered heterochromatic, arrested in G_1 or G_0 phase. In this sense, male pronuclear PCC is more formally analogous to fusion of bovine sperm or mature avian erythrocytes with mitotic cells (Johnson *et al.*, 1970), or to transplantation of sperm nuclei into meiotically arrested frog eggs (Gurdon and Woodland, 1968). Sperm nuclei undergo decondensation prior to recondensing into chromosomes. Male pronuclear recondensation (Fig. 1; Longo, 1983) is similar morphologically to events that occur during "prophasing" as defined by Matsui *et al.* (1982).

The apparent requirement for protein synthesis but not RNA synthesis in G_2 for development of chromosome condensing conditions in the sea urchin egg was also observed for mammalian PCC (Matsui *et al.*, 1971). Although originally a distinction was made between the behavior of mitotic and prematurely condensed chromosomes, it was subsequently suggested that both condensations result from an accumulation of common factors synthesized during G_2 (Matsui *et al.*, 1982). At least some proteins synthesized during G_2 become associated with mitotic and prematurely condensed chromosomes (Rao and Johnson, 1974).

Factors from mammalian cells which promote chromosome condensation can be assayed indirectly by microinjection into frog oocytes or embryos (Sunkara *et al.*, 1979; Miake-Lye *et al.*, 1983; Halleck *et al.*, 1984). Activity begins to increase during G_2 and is maximal during mitosis (Sunkara *et al.*, 1979), when it is found both in the cytoplasm and associated with the chromosomes (Adlakha *et al.*, 1982). It was suggested that as the factors are synthesized in G_2 they associate with the nuclei, and as they continue to be synthesized during M, the excess resides in the cytoplasm (Sunkara *et al.*, 1982). If the rate of degradation of these factors in the cytoplasm is greater than in the chromatin, the observation that chromosome condensing conditions developing in enucleated egg halves are less stable than in nucleated halves might be explained (Poccia *et al.*, 1978; Krystal and Poccia, 1979). Differential stability would also explain the difficulty of inducing PCC upon fusion of enucleated mitotic cytoplasts with interphase cells (Sunkara *et al.*, 1977, 1980).

Phosphorylation of histone H1 has been suggested as having a role in chromosome condensation (Lake and Salzman, 1972; Balhorn *et al.*, 1975; Bradbury *et al.*, 1973, 1974; Gurley *et al.*, 1974; Inglis *et al.*, 1976). Such a mechanism is made attractive because H1 is believed to have a role in higher order chromatin structure (McGhee and Felsenfeld, 1980; McGhee *et al.*, 1980), its phosphorylation is reversible, and it as well as histone H3 become maximally phosphorylated at mitosis (Gurley *et al.*, 1978; Matsumoto *et al.*, 1980). These observations were extended to PCC for H1 in sea urchins (Krystal and Poccia, 1981) and for H1 and H3 in mammalian systems (Ajiro *et al.*, 1983; Hanks *et al.*, 1983).

The demonstration that H1 phosphorylation can proceed when chromosome condensation is inhibited in the urchin egg, together with the observation of Tanphaichitr *et al.* (1976) that inhibition of dephosphorylation does not prevent decondensation strongly suggest that H1 phosphorylation–dephosphorylation is not sufficient to drive condensation–decondensation. It seems more likely that phosphorylation of H1 is a permissive event that allows other factors, perhaps newly synthesized proteins (Matsui *et al.*, 1971), Mg^{2+} (Finch and Klug, 1976; Jerzmanowski and Staron, 1980), nonhistone chromosomal proteins (Sahasrabuddhe *et al.*, 1984), or other histone modifications, such as H3 phosphorylation (Gurley *et al.*, 1978) or H2A ubiquitination (Matsui *et al.*, 1979; this volume Chapter 11), to drive condensation. Dephosphorylation may be necessary to reset the chromatin for the next cell cycle.

Since CS H1 is phosphorylated in eggs blocked in protein synthesis, all proteins necessary for H1 phosphorylation and its timing must be stored in the egg. These may be activated simultaneously with the chromosome condensing conditions described previously.

III. MALE PRONUCLEAR DECONDENSATION

A. Chromatin Decondensing Conditions

The chromatin of the sea urchin sperm nucleus is highly and uniformly condensed (Longo and Anderson, 1969; Zentgraf *et al.*, 1980). Its DNA concentration is approximately the same as in mitotic chromosomes (Green and Poccia, 1985). Immediately after fertilization the sperm nucleus loses its nuclear envelope and its chromatin begins to decondense, in a process that proceeds from the periphery toward the center, resulting after 10–12 min in a rather uniformly decondensed, euchromatic nucleus with a new nuclear envelope (Longo and Anderson, 1968; Poccia *et al.*, 1978; Kunkle, 1982; Longo, 1983). The conditions for decondensation persist for a time into embryogenesis (Sugiyama, 1951; Longo, 1980)

The sea urchin egg is fertilized in the ootid stage, i.e., after completion of meiosis. The conditions that promote decondensation appear to be absent from previtellogenic or vitellogenic oocytes and only begin to appear during meiotic maturation divisions (Brachet, 1922; Longo, 1978). In the mature egg, maintenance of these conditions is not dependent on the presence of the female pronucleus, nuclear or mitochondrial RNA synthesis, or protein synthesis (Krystal and Poccia, 1979, 1981).

B. Sperm (Sp) H1 and Sp H2B Phosphorylation

During the first 10 min following fertilization, the only changes in *S. purpuratus* sperm nuclear proteins detected by gel electrophoresis occur in the H1 and H2B classes (Poccia *et al.*, 1981; Green and Poccia, 1985). These include the conversion of the sperm histone H1 variant (Sp H1) to a phosphorylated form called N; the replacement of N by CS H1; and the phosphorylation of the two Sp H2B variants to proteins called O and P. The CS H1 variant (Newrock *et al.*, 1978) is synthesized during oogenesis and the first few cell cycles after fertilization (Newrock *et al.*, 1978; Herlands *et al.*, 1982), is stored in the egg as a phosphoprotein (Green and Poccia, 1985), and exists in the chromatin in a phosphorylated form. The phosphorylation of Sp H1 and Sp H2B occurs within 1–2 min after fertilization, before detectable decondensation of the male pronuclear chromatin has taken place (Green and Poccia, 1985; Kunkle, 1982). At 5 min postfertilization, Sp H1, Sp H2B, and CS H1 are the only major phosphohistones in the male chromatin (Fig. 6). As decondensation proceeds, phosphorylated Sp H1 is lost and CS H1 accumulates. In contrast, phosphorylated Sp H2B remains in the chromatin (Poccia *et al.*, 1984; Green and Poccia, 1985). The data suggest that phosphorylation of Sp H1 and Sp H2B might be a prerequisite for decondensation. Phosphorylation events following fertilization might be considered in some sense a reversal of

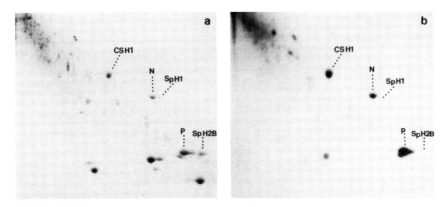

Fig. 6. Pronuclear phosphoproteins 5 min postfertilization. Eggs were prelabelled with ortho [^{32}P] phosphate for 4 hr and polyspermically fertilized. Pronuclear chromatin was isolated and subjected to two-dimensional gel electrophoresis (acid–urea–Triton X-100 vs. SDS). The gel was stained (a), dried, and exposed to X-ray film (b). (From Green and Poccia, 1985.)

dephosphorylation events known to occur during late stages of sper-
miogenesis (Louie *et al.*, 1974; Kennedy and Davies, 1981; see Poccia,
1986).

C. An Unusual Tetrapeptide

The restriction of the early phosphorylation events to sperm histones
H1 and H2B implies a possible underlying structural or functional similarity in these molecules. Both molecules are longer than their embryonic
counterparts (Fig. 7), mostly because of extensions at their N-terminal
ends (von Holt *et al.*, 1984). These extensions in *Parechinus angulosus*,
Echinolampus crassa, *Spherechinus granulosus*, and *Psammechinus mi-*

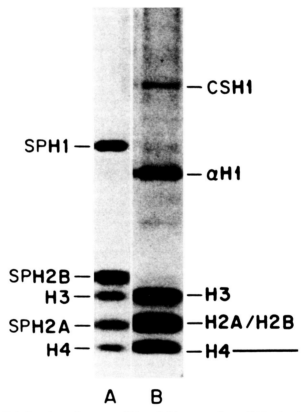

Fig. 7.   Relative sizes of sperm and blastula histone variants. SDS gels of (A) sperm and
(B) actinomycin D-treated blastula histones. (From Salik *et al.*, 1981.)

liaris are composed predominantly of unusual amino acid sequences forming repeating short peptides (M. S. Strickland *et al.*, 1977, 1978a,b; W. N. Strickland *et al.*, 1977, 1980, 1982a,b; von Holt *et al.*, 1979; Busslinger and Barberis, 1985). The repeat is a tetrapeptide for Sp H1 and either of two pentapeptides for Sp H2B (Table II).

The relationship of the Sp H2B pentapeptide sequences to each other as well as to other pentapeptides delimited by Pro and Ser (or Thr) and including three intervening amino acids, one or more of which are basic, has been stressed by von Holt and colleagues as a general feature of a variety of DNA binding proteins. Potential evolutionary relationships of these proteins have been discussed (von Holt *et al.*, 1984). In particular, it was suggested that Sp H2B histones were precursors to the protamines, and attention was called to the potential protamine-like properties of these histones.

Based on the common phosphorylation behavior of Sp H1 and Sp H2B following fertilization, we have emphasized the common aspects of their sequences. If the sequences are grouped slightly differently, all known H1 and H2B repeat sequences from sea urchin sperm can be simply related to one another by considering them as variants of the tetrapeptide sequences Ser-Pro-basic-basic or basic-basic-Ser-Pro, where basic is either Lys or Arg (Table II). Of particular importance is that viewed in this way, they match sequences constituting known phosphorylation sites in trout testis H1 at Ser-145,-161, and -182 (Ser-Pro-Lys-Lys or Lys-Lys-Ser-Pro-Lys-Lys) (Macleod *et al.*, 1977) and in chicken erythrocyte H5 at Ser-104 (Lys-Arg-Ser-Pro) and Ser-148 (Ser-Pro-Lys-Lys) (Sung and Freedlender, 1978; Briand *et al.*, 1980). Analogous mammalian phosphorylation sites have also been mapped (see Hohmann, 1983).

As seen in Table III, such tetrapeptides are found in all complete H1 and H5 sequences thus far determined, with the single exception of the sperm H1's of a polychaete *Platynereis dumerlii*. Other than the sea urchin sperm H2B, and wheat germ H2A, they are absent from all other known histone sequences, from a wide range of animals and plants, as well as fungi and protists. Even though present in H1 or H5 molecules, the sequences are never seen in the amino terminal regions or the relatively conserved central globular domains of these molecules (Allan *et al.*, 1980). They are, however, common in the C-terminal regions of H1 or H5, usually present three times with Thr sometimes substituting for Ser (Macleod *et al.*, 1977; Yaguchi *et al.*, 1979; Cole, 1975; Briand *et al.*, 1980; Stephenson *et al.*, 1981; Levy *et al.*, 1982; Turner *et al.*, 1983; Strickland *et al.*, 1982a,b; Cole *et al.*, 1984; Sugarman *et al.*, 1983).

The asymmetric distribution of these sequences in *P. angulosus* sperm histones is shown in Fig. 8a. The sequence occurs three to four times in

TABLE II
Repeating Peptide Units in SP H1 and SP H2B

Histone (Organism)	Sequence
Sp H1 (*P. angulosus*)	(X7)AA SPRK SPRK SPKK SPRK ASA SPRR *KA*
Sp H1 (*E. crassa*)	(X7)AA SPRK SPKK SPRK SPKK *K*?? SPRK *R*?
Sp H1 (*S. granulosus*)	(X7)AA SPRK SPRK *G* SPKK SP ??
Sp H2B(1) (*P. miliaris*)	PSQKS PTKRS PTKRS *PQ = PSQKSPT KRSP T KRSP Q*
Sp H2B(1) (*P. angulosus*)	PSQKS PTKRS PTKRS PTKRS *PQ = PSQKSPT KRSP T KRSP T KRSP Q*
Sp H2B(2) (*P. angulosus*)	PRSPAKTS PRKGS PRKGS *PS = PRSPAKT SPRK G SPRK GPS*
Sp H2B(3) (*P. angulosus*)	PRSPAKTS PRKGS PRKGS PRKGS *PS = PRSPAKT SPRK G SPRK G SPRK GSPS*

Phosphorylation sites:

In trout testis H1	SPKK	KKSPKK
In chicken rbc H5	SPKK	KRSP

[a] Only the N-terminal portions of the histones are shown. X7 represents the first seven amino acids of the Sp H1's. Sequences for Sp H2B's are grouped in two alternate ways. Data are taken from references cited in text (Section III, C). Amino acids: S = serine, P = proline, K = lysine, R = arginine, T = threonine, A = alanine, G = glycine, Q = glutamine.

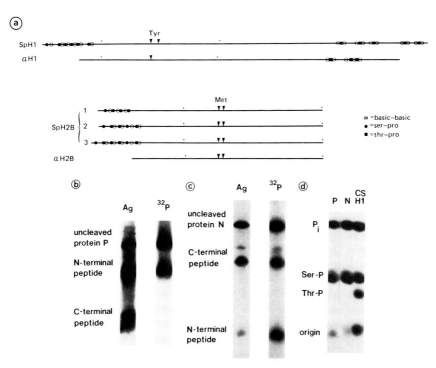

Fig. 8. Localization of the phosphorylation sites in proteins N (Sp H1) and P (Sp H2B). The distribution of selected amino acids in sperm and early embryo H1 and H2B histones is shown. (a) Published amino acid sequences for H1 and the three H2B variants of *P. angulosus* sperm and for the early embryo α H1 and α H2B histones of *S. purpuratus* were used to construct linear representation of the molecules, in which each amino acid is a unit length. Ser, Thr, and Pro residues are indicated only when they occur in the sequence Ser-pro or Thr-pro, and basic amino acids (Lys and Arg) when they occur in pairs adjacent to Ser-Pro or Thr-Pro. The relatively conserved, hydrophobic regions of the molecules are between dots above the lines. Tyr, site of cleavage by *N*-bromosuccinimide, and Met, site of cleavage by cyanogen bromide, are indicated. (b and c) Peptide maps and their autoradiograms of proteins P and N, respectively. P was cleaved with cyanogen bromide and N with *N*-bromosuccinimide, the peptides resolved by SDS gel electrophoresis, silver stained, and autoradiographed. (d) Phosphoamino acid analysis of P, N, and CS H1. *In vivo* labeled proteins were hydrolyzed after elution from gels, and analyzed by thin-layer electrophoresis and autoradiography. (From Green and Poccia, 1985.)

the first 21–32 N-terminal residues of the 143–148 amino acid long Sp H2B variants (M. S. Strickland *et al.*, 1977, 1978b; W. N. Strickland *et al.*, 1977). It appears five times in the first 32 residues and five times in the last 57 amino acids of the 248 amino acid long Sp H1 molecule (Strickland *et al.*, 1980, 1982a).

Analysis of proteins N and P from polyspermic pronuclei shows that

TABLE III

Occurrence of Unusual Tetrapeptide in Known Histone Sequences[a]

Histone	Organism (tissue)	P/N	C/F	Present	Pro position	Sequences
H1	Sea urchin *P. angulosus* (sperm)	P	C	+	11,15,19,23,30,193, 204,213,234,243	SPKK,SPRR,SPRK,KKSP,RRSP,KRSP,RKSP
	Sea urchin *E. crassa* (sperm)	P	F	+	11,15,19,23,28	SPKK,SPRK,KKSP,RKSP
	Sea urchin *S. granulosus* (sperm)	P	F	+	11,15,20	SPKK,SPRK,KKSP,RKSP
	Sea urchin *S. purpuratus* (embryo)	N	C	+	161,175,179	TPKK,KKTP
	Sea urchin *P. miliaris* (embryo)	N	F	−		
	Rabbit (thymus)	P	C	+	137,154,181	TPKK,KKTP
	Toad	N	C	+	145,159,172	SPKK
	Toad	N	F	+	Three times	SPKK
	Chicken	N	C	+	152,170	SPKK,KKSP
	Trout (testis)	P	C	+	142,155,162,183	SPKK,KKSP
	Boar (testis)	P	C	+	182	RKSP
	Newt	N	F	+	At least once	SPKK,KKSP
	Calf (thymus)	P	F	−		
	Fruit fly	N	F	−		
	Polychaete	P	C	−		
H5	Goose (erythrocyte)	P	C	+	106,122,153,190	SPKK,KRSP,RKSP
	Chicken (erythrocyte)	P/N	C	+	105,118,149,186	SPKK,KRSP,RKSP
	Pigeon (erythrocyte)	P	F	−		
H2B	Sea urchin 1 *P. angulosus* (sperm)	P	C	+	11,16,21	KRSP

Sea urchin 2 *P. angulosus* (sperm)	P	C	+	9,14,25	SPRK,SPKR
Sea urchin 3 *P. angulosus* (sperm)	P	C	+	9,14,19,30	SPRK,SPKR
Sea urchin 1 *P. miliaris* (sperm)	P	F	+	11,16	KRSP
Sea urchin 2 *P. miliaris* (sperm)	P	F	+	9,14,19,24	SPRK,SPKR
Sea urchin *E. crassa* (sperm)	P	F	+	9,14	SPRK
Calf, human, chicken, crocodile, toad, brown trout, rainbow trout, limpet, sea urchin embryo, yeast, starfish	P/N	C/F	−	(17 different histones)	
H2A	Wheat (germ)	P	C	+	118,128,137	KKSP,SPKK
	Calf, human, rainbow trout, chicken, rat, sipunculid, sea urchin gonad, wheat germ, sea urchin embryo, *Tetrahymena*, chick embryo, starfish, cuttlefish, fruit fly, yeast	P/N	C/F	−	(19 different histones)	
H3	Calf, mouse, toad, yeast, sea urchin embryo, fruit fly, pea	P/N	C/F	−	(8 different histones)	
H4	Calf, human, pig, rat, toad, rainbow trout, fruit fly, sea urchin embryo, mouse, wheat, pea, yeast, *Tetrahymena*	P/N	C/F	−	(13 different histones)	

[a] P, protein sequences; N, derived from nucleic acid sequence; C, complete sequence; F, fragment. Amino acids: S, serine; P, proline; K, lysine; R, arginine; T, threonine.

the only amino acid phosphorylated in these molecules is Ser (Fig 8d; Green and Poccia, 1985). Chemical cleavage of N with *N*-bromosuccinimide (at Tyr) or of P with cyanogen bromide (at Met) (Figs. 8b and 8c) shows that the labeled phosphate incorporated *in vivo* is approximately equally distributed between the N- and C-terminal fragments for N (H1), but entirely localized in the N-terminal fragment for P (H2B), mirroring the distribution of the putative tetrapeptide phosphorylation sites. (In contrast, phosphorylation of CS H1, whose sequence is not known, was associated with all cyanogen and *N*-bromosuccinimide peptides, and occurred on both Ser and Thr.)

Such a short sequence might be expected to be fairly common. However, it is not. In a computer search of the National Biomedical Research Foundation protein sequence database (2898 protein sequences) done in January 1985, only 1.3% of the sequences contained Ser-Pro flanked by two basic amino acids. Slightly over half of these were found in eukaryotes. Histones, accounting for 2.7% of the total eukaryotic sequences, represented 43% of the eukaryotic sequences containing such tetrapeptides. The only other category of proteins showing a higher than average incidence of the tetrapeptide was viruses. The only nonhistone proteins containing the sequences more than once were galline (rooster sperm nuclear protein) and the nucleocapsid protein of murine caronavirus, each twice, and core antigens from hepatitis B virus, woodchuck hepatitis virus, and ground squirrel hepatitis virus, each four times.

In summary, sea urchin Sp H1 and Sp H2B histones are the only known histones to contain the unusual tetrapeptide in their N-terminal regions. The tetrapeptide is otherwise restricted to the C-terminal regions of H1 or H5 histones, where it can serve as a phosphorylation site, and the C-terminus of wheat germ H2A, whose phosphorylation state is unknown. Sp H2B could be considered an H1-H2B chimeric molecule. Such relatively rare sites might be phosphorylated by a common kinase with simple sequence specificity. A nuclear cyclic AMP-independent kinase capable of phosphorylating such sequences is known and has been partially purified (see Matthews and Huebner, 1984).

D. Chromatin Packing in Sperm

In interpreting the possible consequences of phosphorylation of the terminal domains of Sp H1 and Sp H2B, several other features of sperm histone and chromatin structure are relevant. H1 histones are believed to be bound to linker DNA (Noll and Kornberg, 1977), whereas H2B histones are associated with nucleosomal cores. Sea urchin sperm chromatin has the longest known average nucleosomal repeat length, in *S. purpuratus* about 250 bp as compared to 222 bp in the blastula (Keichline and Wasserman, 1979). Since the DNA associated with the core is the same for sperm and blastula (Keichline and Wasserman, 1979), the additional approximately 30 bp must reside in linker DNA and represent a difference of approximately 60 negative DNA–phosphate charges. Complete DNA charge neutralization is probably of extreme importance in the dense chromatin packing characteristic of the sperm head (Pogany *et al.*, 1981; Balhorn, 1982).

Sp H1, which has 34 more basic amino acids than the major variant at blastula stage (αH1), is therefore probably insufficient to neutralize the additional linker DNA. Each Sp H2B, however, possesses 11–15 more basic residues than an α H2B (Sures *et al.*, 1978). The other histone classes do not differ significantly in basicity between sperm and embryo. Therefore, one Sp H1 and two Sp H2B histones per nucleosomal repeat are sufficient for charge neutralization of the long linker DNA [34 + (2 × 13) = 60], whereas Sp H1 alone can only provide about half this charge. This implies that Sp H2B may bind ionically to linker DNA. Such a role has been suggested previously for Sp H2B (Zalenskaya *et al.*, 1981) and for the short N-terminal arms of other H2B molecules (McGhee *et al.*, 1980; Allan *et al.*, 1983).

A comparison of the chromatin of sea urchin sperm with that of other sperm supports the notion that arms of both H1 and H2B bind to linker DNA. Sea urchin sperm has extended N-termini in both classes of histone and the longest known repeat length (about 250 bp). Starfish sperm chromatin, with an extended H1 but normal H2B's, has a repeat length of 224 bp (Zalenskaya *et al.*, 1981). Goldfish sperm, with apparently typical somatic-type histones, has a repeat length of 205 bp (Muñoz-Guerra *et al.*, 1982). Sperm of the polychaete *P. dumerlii* has normal H2B's. It contains truncated H1's which lack both N- and C-terminal arms, therefore consisting of essentially only the central conserved globular domain. It has a repeat length of only 165 bp (Kmiecik *et al.*, 1985).

The chromatin of the sea urchin sperm nucleus is densely packed. The DNA concentration can be estimated from the nuclear DNA content of *S. purpuratus*, about 0.77 pg (Pikó *et al.*, 1967), and from its volume, about 4.1 μm^3 (see Green and Poccia, 1985; Fig. 33 in Longo and Anderson, 1969). The concentration of DNA, 0.19 pg/μm^3, is similar to that measured for mitotic chromosomes (Bennett *et al.*, 1983). It is about one-quarter that achieved in mouse sperm, but as shown by Pogany *et al.* (1981), such high DNA concentrations are incompatible with nucleosomal

(a)

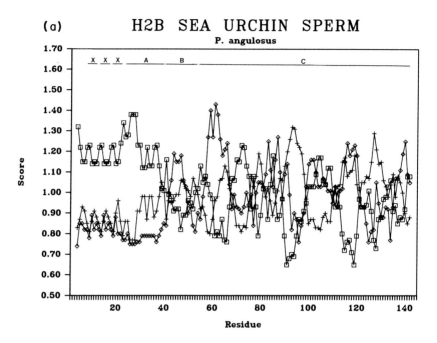

(b)

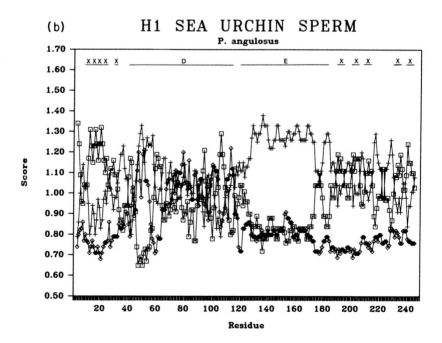

organization, which sets an upper limit of about 0.33 pg/μm^3. Mouse sperm is capable of very high degrees of packing because it has abandoned histones and nucleosomal structure.

It is not known how the nucleosomes of sea urchin sperm chromatin are packed in the sperm head. Physical measurements are consistent with solenoidal packing to achieve a 30-nm fiber (McGhee *et al.*, 1983). A 30-nm solenoid with six nucleosomes per turn and a repeat length of 250 bp has a DNA concentration of about 0.20 pg/μm^3, in close agreement with the concentration of DNA calculated for the entire sperm head. Other models suggesting 50-nm beads as a higher order structure have been proposed (Zentgraf and Franke, 1984). Whatever the details of higher order structure, it is clear that nucleosomes in the sperm head must lie close to one another and to linker DNA of other nucleosomes. This close association must be disrupted for the most part during decondensation when the volume of the male pronucleus increases at least 16 times (assuming a sphere of 5 μm diameter).

E. Secondary Structure of Sp H1 and Sp H2B Arms

Given the short distances involved, it seems likely that the Sp H1 and Sp H2B terminal arms would be able to contact linkers that are many nucleosomes away, but lie nearby. Such contact would allow for crosslinking of the chromatin gel, which could result in stabilization of the structure. Extension of the arms is implied by their unusual sequences, rich in Ser and Pro.

Using the PRPLOT program of the Protein Identification Resource at Georgetown, we have plotted the scores for α-helix, β-sheet, and β-turn formation based on the rules of Chou and Fasman (1978). High scores represent favorable probabilities for formation of secondary structure. As seen in Fig. 9a, Sp H2B is dominated in the first 40 amino acids by predicted β-turn, the entire region preceding the conserved portion of the molecule which begins with a long basic (Arg-rich) sequence (40–53) typi-

---

Fig. 9. Predicted secondary structural propensities and domains of Sp H2B and Sp H1. (a) Sp H2B; (b) Sp H1. Scores are based on rules developed by Chou and Fasman (1978) and are derived from the PRPLOT computer program of the National Biomedical Research Foundation, Georgetown University Medical Center. The window used for averaging was six amino acids. $\square$ = β-turn; + = α-helix; $\diamond$ = β-sheet. Regions indicated above the plot: X = Ser-Pro adjacent to two basic amino acids; A = Gly-rich segment of H2B; B = Arg-rich segment of H2B; C = relatively conserved region of H2B; D = relatively conserved trypsin resistant core of H1; E = Ala-rich basic region of H1.

cal of all H2B's (Isenberg, 1979). The conserved region interacts normally with H2A molecules (Giancotti *et al.*, 1981a), further emphasizing the chimeric nature of these molecules. Note that the region unfavorable for α-helix or β-sheet formation extends well beyond the region containing the tetrapeptides into a glycine-rich segment (from amino acids 22–40).

Sp H1, by such an analysis, appears to have four domains (Fig. 9b). The central conserved globular domain extends approximately from residues 40–114 (Allan *et al.*, 1980) and lies adjacent to a very long Ala/Lys-rich stretch from about 120–185, which is favorable for α-helix formation. The two ends, which contain the tetrapeptides, are unfavorable for α-sheet, but have strong scores for β-turn. The C-terminal end alternates β-turn with α-helix. Both the central and C-terminal domains are known to contain extensive α-helical content (Giancotti *et al.*, 1981b).

W. N. Strickland *et al.* (1977) suggested that the region of Sp H2B containing the repeated structures exists as an extended, kinked helix (see von Holt *et al.*, 1984). Whether the N termini exist in this form or as β-turn structures, the lack of ordered α-helix or β-sheet may allow the arms to follow the path of linker DNA. It has been suggested that the arms of Sp H2B's may crosslink up to three DNA columns in the intact sperm head (von Holt *et al.*, 1984). Upon phosphorylation, the ionic affinity of the arms for DNA should be drastically reduced, since the two positive charges contributed by the basic amino acids will be balanced by the two negative charges of the phosphoserine.

F. Model for Decondensation

Based on the foregoing arguments, we have proposed a model for the decondensation of sea urchin sperm chromatin following fertilization (Green and Poccia, 1985). Sperm chromatin is probably organized as 30-nm fibers and has the longest known length of linker DNA. This extra linker is effectively neutralized by positive charges in the N-terminal extensions of Sp H1 and Sp H2B and the C-terminal extension of Sp H1. These regions allow crosslinking within and between fibers which stabilize the densely packed chromatin. Immediately following fertilization, the extensions are essentially neutralized by phosphorylation of their serine residues, thus lowering their affinities for linker DNA. The chromatin fibers can then move apart, resulting in decondensation. As decondensation proceeds, the phosphorylation of both ends of Sp H1 aids in the removal of the now weakly interacting histone from the chromatin. At the same time, phosphorylated CS H1 from the maternal storage pool becomes associated with the chromatin which is directly exposed to cytoplasm. The Sp H2B molecules remain in the chromatin anchored in nucleosomes by their conserved C-terminal regions. A new nuclear envelope forms and the sperm nucleus has been converted to a male pronucleus.

IV. CONCLUSIONS

Phosphorylation of histones is correlated with two different chromatin structural transitions in fertilized sea urchin eggs, the decondensation of chromatin following fertilization, and the condensation of chromatin at mitosis and in PCC. The work discussed in Section II indicates that H1 histone phosphorylation is not sufficient for chromosome condensation, and the work discussed in Section III indicates that phosphorylation of histones precedes decondensation. Rather than viewing phosphorylation as a cause of these structural changes, an alternative is to consider it as a requisite of such changes. The direction of the condensation–decondensation transition may be controlled by other factors that must act on a chromatin in which the potentially strong interaction of an unmodified highly basic histone is mitigated by the addition of negatively charged phosphates. Such effects may be important not only in gross structural changes such as discussed here, but in more subtle conformational transitions involved in replication, transcription, or recombination. The relatively rare occurrence of the simple tetrapeptide phosphorylation site would allow for control of a variety of unrelated processes through phosphorylation by a common kinase.

ADDENDUM

The deduced amino acid sequence for a SpH2B molecule of one more sea urchin species has recently become available. SpH2B-2 of *L.pictus* contains four N-terminal tetrapeptides of sequence SPKR and SPRK (Lai and Childs (1986).

ACKNOWLEDGMENTS

I am gratefully indebted to two students, Geoff Krystal and Ray Green, for the technical virtuosity of most of the experimental work presented here, as well as for stimulating discussions. Databases were kindly made available by the National Biomedical Research

Foundation. This work was supported by grant No. HD 12982 from the National Institutes of Health.

REFERENCES

Adlakha, R. C., Sahasrabuddhe, C. G., Wright, D. A., Lindsey, W. F., and Rao, P. N. (1982). Localization of mitotic factors on metaphase chromosomes. *J. Cell Sci.* **54**, 193–206.

Ajiro, K., Nishimoto, T., and Takahashi, T. (1983). Histone H1 and H3 phosphorylation during premature chromosome condensation in a temperature-sensitive mutant (tsBN2) of baby hamster kidney cells. *J. Biol. Chem.* **258**, 4534–4538.

Allan, J., Hartman, P. G., Crane-Robinson, C., and Aviles, X. (1980). The structure of H1 and its location in chromatin. *Nature (London)* **288**, 675–679.

Allan, J., Harborne, N., Rau, D. C., and Gould, H. (1983). Participation of core histones "tails" in the stabilization of the chromatin solenoid. *J. Cell Biol.* **93**, 285–297.

Balhorn, R. (1982). A model for the structure of chromatin in mammalian sperm. *J. Cell Biol.* **93**, 295–305.

Balhorn, R., Jackson, V., Granner, D., and Chalkley, R. (1975). Phosphorylation of the lysine-rich histones throughout the cell cycle. *Biochemistry* **14**, 2504–2511.

Bennett, M. D., Heslop-Harrison, J. S., Smith, J. B., and Ward, J. P. (1983). DNA density in mitotic and meiotic metaphase chromosomes of plants and animals. *J. Cell Sci.* **63**, 173–179.

Brachet, A. (1922). Recherches sur la fécondation prematurée de l'oef d'oursin. *Arch. Biol.* **32**, 205–244.

Bradbury, E. M., Inglis, R. J., Matthews, H. R., and Sarner, N. (1973). Phosphorylation of very lysine-rich histone in *Physarum polycephalum*. *Eur. J. Biochem.* **73**, 131–139.

Bradbury, E. M., Inglis, R. J., Matthews, H. R., and Langan, T. A. (1974). Molecular basis of control of mitotic cell division in eukaryotes. *Nature (London)* **249**, 552–556.

Briand, G., Kmiecik, D., Sautiere, P., Wouters, D., Borie-Loy, O., Biserte, G., Mazen, A., and Champagne, M. (1980). Chicken erythrocyte histone H5. IV. Sequence of the carboxy-terminal half of the molecule (96 residues) and complete sequence. *FEBS Lett.* **112**, 147–151.

Busslinger, M., and Barberis, A. (1985). Synthesis of sperm and late histone cDNAs of the sea urchin with a primer complementary to the conserved 3′ terminal palindrome: Evidence for tissue specific and more general histone gene variants. *Proc. Natl. Acad. Sci. U.S.A.* **82**, 5676–5680.

Chou, P. Y., and Fasman, G. D. (1978). Prediction of the secondary structure of proteins from their amino acid sequence. *Adv. Enzymol.* **47**, 45–148.

Cole, K. D., York, R. G., and Kistler, W. S. (1984). The amino acid sequence of boar H1t, a testis-specific H1 histone variant. *J. Biol. Chem.* **259**, 13695–13702.

Cole, R. D. (1975). Special features of the structure of H1 histones. *In* "The Molecular Biology of the Mammalian Genetic Apparatus" (P. T'so, ed.), Vol. 1, pp. 93–104. Elsevier North-Holland Biomedical Press, Amsterdam.

Finch, J. T., and Klug, A. (1976). Solenoidal model for superstructure in chromatin. *Proc. Natl. Acad. Sci. U.S.A.* **73**, 1897–1901.

Giancotti, V., Russo, E., Cosimi, S., Cary, P. D., and Crane-Robinson, C. (1981a). The sea urchin sperm histone H2B readily forms a complex with heterologous H2A despite having an elongated N-terminal domain. *Eur. J. Biochem.* **114**, 629–634.

Giancotti, V., Russo, E., Cosimi, S., Cary, P. D., and Crane-Robinson, C. (1981b). Secondary and tertiary structural differences between histone H1 molecules from calf thymus and sea urchin *(Sphaerechinus granularis)* sperm. *Biochem. J.* **197**, 655–660.

Green, G. R., and Poccia, D. (1985). Phosphorylation of sea urchin sperm H1 and H2B histones precedes chromatin decondensation and H1 exchange during pronuclear formation. *Dev. Biol.* **108**, 235–245.

Gurdon, J. B., and Woodland, H. R. (1968). The cytoplasmic control of nuclear activity in animal development. *Biol. Rev. Cambridge Philos. Soc.* **43**, 233–267.

Gurley, L. R., Walters, R. A., and Tobey, R. A. (1974). Cell cycle specific changes in histone phosphorylation associated with cell proliferation and chromosome condensation. *J. Cell Biol.* **60**, 356–364.

Gurley, L. R., Tobey, R. A., Walters, R. A., Hildebrand, C. E., Hohmann, P. G., D'Anna, J. A., Barham, S. S., and Deaven, L. L. (1978). Histone phosphorylation and chromatin structure in synchronized mammalian cells. *In* "Cell Cycle Regulation" (J. R. Jeter, Jr., I. L. Cameron, G. M. Padilla, and A. M. Zimmerman, eds.), pp. 37–60. Academic Press, New York.

Halleck, M. S., Reed, J. A., Lumley-Sapanski, K., and Schlegel, R. A. (1984). Injected mitotic extracts induce condensation of interphase chromatin. *Exp. Cell Res.* **153**, 561–569.

Hanks, S. K., Rodriquez, L. V., and Rao, P. N. (1983). Relationship between histone phosphorylation and premature chromosome condensation. *Exp. Cell Res.* **148**, 293–302.

Herlands, L., Allfrey, V. G., and Poccia, D. (1982). Translational regulation of histone synthesis in the sea urchin *Strongylocentrotus purpuratus*. *J. Cell Biol.* **94**, 219–223.

Hinegardner, R. T. (1967). Echinoderms. *In* "Methods in Developmental Biology" (F. H. Wilt and N. K. Wessells, eds.), pp. 139–155. Crowell-Collier, New York.

Hohmann, P. (1983). Phosphorylation of H1 histones. *Mol. Cell. Biochem.* **57**, 81–92.

Inglis, R. J., Langan, T. A., Matthews, H. R., Harder, D. G., and Bradbury, E. M. (1976). Advance of mitosis by histone phosphokinase. *Exp. Cell Res.* **97**, 418–425.

Isenberg, I. (1979). Histones. *Annu. Rev. Biochem.* **48**, 159–191.

Jerzmanowski, A., and Staron, K. (1980). Mg2+ as a trigger of condensation-decondensation transition of chromatin during mitosis. *J. Theor. Biol.* **82**, 41–46.

Johnson, R. T., and Rao, P. N. (1971). Nucleocytoplasmic interactions in the achievement of nuclear synchrony in DNA synthesis and mitosis in multinucleate cells. *Biol. Rev. Cambridge Philos. Soc.* **46**, 97–155.

Johnson, R. T., Rao, P. N., and Hughes, S. D. (1970). Mammalian cell fusion: A HeLa cell inducer of premature chromosome condensation active in cells from a variety of animal species. *J. Cell Physiol.* **76**, 151–158.

Keichline, L. D., and Wasserman, P. M. (1979). Structure of chromatin in sea urchin embryos, sperm, and adult somatic cells. *Biochemistry* **18**, 214–219.

Kennedy, B. P., and Davies, P. L. (1981). Phosphorylation of a group of high molecular weight basic nuclear proteins during spermatogenesis in the winter flounder. *J. Biol. Chem.* **256**, 9254–9259.

Kmiecik, D., Sellos, D., Belaiche, D., and Sautiere, P. (1985). Primary structure of the two variants of a sperm-specific H1 from the annelid *Platynereis dumerlii*. *Eur. J. Biochem.* **150**, 359–370.

Krystal, G. W., and Poccia, D. L. (1979). Control of chromosome condensation in the sea urchin egg. *Exp. Cell Res.* **123**, 207–219.

Krystal, G. W., and Poccia, D. L. (1981). Phosphorylation of cleavage stage histone H1 in mitotic and prematurely condensed chromosomes. *Exp. Cell Res.* **134**, 41–48.

Kunkle, M. (1982). Chromatin transitions in the fertilizing sperm nucleus of the sea urchin, *Strongylocentrotus purpuratus. Gamete Res.* **5**, 181–190.

Lai, Z-C., and Childs, G. (1986). *Nucl. Acids Res.* (in press).

Lake, R. S., and Salzman, N. P. (1972). Occurrence and properties of chromatin associated F1 histone phosphokinase in mitotic Chinese hamster cells. *Biochemistry* **11**, 4817–4825.

Levy, S., Sures, I., and Kedes, L. (1982). The nucleotide and amino acid sequence of a gene for H1 histone that interacts with euchromatin. *J. Biol. Chem.* **257**, 9438–9443.

Longo, F. J. (1978). Insemination of immature sea urchin *(Arbacia punctulata)* eggs. *Dev. Biol.* **62**, 271–291.

Longo, F. J. (1980). Reinsemination of fertilized sea urchin *(Arbacia punctulata)* eggs. *Dev. Growth Differ.* **22**, 219–227.

Longo, F. J. (1983). Cytoplasmic and sperm nuclear transformations in fertilized ammonia-activated sea urchin *(Arbacia punctulata)* eggs. *Gamete Res.* **8**, 65–78.

Longo, F. J., and Anderson, E. (1968). The fine structure of pronuclear development and fusion in the sea urchin, *Arbacia punctulata. J. Cell Biol.* **39**, 335–368.

Longo, F. J., and Anderson, E. (1969). Sperm differentiation in the sea urchins *Arbacia punctulata* and *Strongylocentrotus purpuratus. J. Ultrastruct. Res.* **27**, 486–509.

Louie, A. J., Candido, E. M. P., and Dixon, G. H. (1974). Enzymatic modifications and their possible roles in regulating the binding of basic proteins to DNA and in controlling chromosomal structure. *Cold Spring Harbor Symp. Quant. Biol.* **38**, 803–819.

McGhee, J. D., and Felsenfeld, G. (1980). Nucleosome structure. *Annu. Rev. Biochem.* **49**, 1115–1156.

McGhee, J. D., Rau, D., C., Charney, E., and Felsenfeld, G. (1980). Orientation of the nucleosome within the higher order structure of chromatin. *Cell (Cambridge, Mass.)* **22**, 87–96.

McGhee, J. D., Nickol, J. M., Felsenfeld, G., and Rau, D. C. (1983). Higher order structure of chromatin: Orientation of nucleosomes within the 30nm chromatin solenoid is independent of species and spacer length. *Cell (Cambridge, Mass.)* **33**, 831–841.

Macleod, A. R., Wong, N. C. W., and Dixon, G. H. (1977). The amino-acid sequence of trout-testis histone H1. *Eur. J. Biochem.* **78**, 281–291.

Matsui, S., Weinfeld, H., and Sandberg, A. A. (1971). Dependence of chromosome pulverization in virus-fused cells on events in the G_2 period. *J. Natl. Cancer Inst. (U.S.)* **47**, 401–411.

Matsui, S., Seon, B. K., and Sandberg, A. A. (1979). Disappearance of a structural chromatin protein A24 in mitosis: Implications for molecular basis of chromatin condensation. *Proc. Natl. Acad. Sci. U.S.A.* **76**, 6386–6390.

Matsui, S., Weinfeld, H., and Sandberg, A. A. (1982). Factors involved in prophasing and telophasing. *In* "Premature Chromosome Condensation" (P. N. Rao, R. T. Johnson, and K. Sperling, eds.), pp. 207–232. Academic Press, New York.

Matsumoto, Y., Yasuda, H., Mita, S., Marunouchi, T., and Yamada, M. (1980). Evidence for the involvement of H1 histone phosphorylation in chromosome condensation. *Nature (London)* **284**, 181–183.

Matthews, H. R., and Huebner, V. D. (1984). Nuclear protein kinases. *Mol. Cell. Biochem.* **59**, 81–99.

Mazia, D. (1963). Synthetic activities leading to mitosis. *J. Cell. Comp. Physiol.* **62**, *Suppl.* 1, 123–140.

Mazia, D. (1974). Chromosome cycles turned on in unfertilized sea urchin eggs after exposure to NH₄OH. *Proc. Natl. Acad. Sci. U.S.A.* **71**, 690–693.

Miake-Lye, R., Newport, J., and Kirshner, M. (1983). Maturation-promoting factor induces nuclear envelope breakdown in cycloheximide-arrested embyros of *Xenopus laevis*. *J. Cell Biol.* **97**, 81–91.

Muñoz-Guerra, S., Azorin, F., Casas, M. T., Marcet, X., Maristany, M. A., Roca, J., and Subirana, J. A. (1982). Structural organization of sperm chromatin from the fish *Carassius auratus*. *Exp. Cell Res.* **137**, 47–53.

Newrock, K. M., Alfageme, C. R., Nardi, R. V., and Cohen, L. H. (1978). Histone changes during chromatin remodelling in embryogenesis. *Cold Spring Harbor Symp. Quant. Biol.* **42**, 421–431.

Noll, M., and Kornberg, R. D. (1977). Action of micrococcal nuclease on chromatin and the location of histone H1. *J. Mol. Biol.* **109**, 393–404.

Pikó, L., Tyler, A., and Vinograd, J. (1967). Amount, location, priming capacity, and other properties of cytoplasmic DNA in sea urchin eggs. *Biol. Bull. (Woods Hole, Mass.)* **132**, 68–90.

Poccia, D. (1986). Remodeling of nucleoproteins during gametogenesis, fertilization and early development. *Int.Rev.Cytol.* **105**, 1–65.

Poccia, D., Krystal, G., Nishioka, D., and Salik, J. (1978). Control of sperm chromatin structure by egg cytoplasm in the sea urchin. *In* "Cell Reproduction: In Honor of Daniel Mazia" (E. R. Dirksen, D. M. Prescott, and C. F. Fox, eds.), pp. 197–206. Academic Press, New York.

Poccia, D., Salik, J., and Krystal, G. (1981). Transitions in histone variants of the male pronucleus following fertilization and evidence for a maternal store of cleavage-stage histones in the sea urchin egg. *Dev. Biol.* **82**, 287–296.

Poccia, D., Greenough, T., Green, G. R., Nash, E., Erickson, J., and Gibbs, M. (1984). Remodelling of sperm chromatin following fertilization: Nucleosome repeat length and histone variant transitions in the absence of DNA synthesis. *Dev. Biol.* **104**, 274–286.

Pogany, G., C., Corzett, M., Weston, S., and Balhorn, R. (1981). DNA and protein content of mouse sperm. *Exp. Cell Res.* **136**, 127–136.

Rao, P. N., and Johnson, R. T. (1974). Control of proliferation in animal cells. *Cold Spring Harbor Conf. Cell Proliferation* **1**, 785–800.

Rao, P. N., Johnson, R. T., and Sperling, K., eds. (1982). "Premature Chromosome Condensation: Application in Basic, Clinical and Mutation Research." Academic Press, New York.

Sahasrabuddhe, C. G., Adlakha, R. C., and Rao, P. N. (1984). Phosphorylation of non-histone proteins associated with mitosis. *Exp. Cell Res.* **153**, 439–450.

Salik, J., Herlands, L., Hoffmann, H. P., and Poccia, D. (1981). Electrophoretic analysis of the stored histone pool in unfertilized sea urchin eggs: quantification and identification by antibody binding. *J. Cell Biol.* **90**, 385–395.

Stephenson, E. C., Erba, H. P., and Gall, J. G. (1981). Characterization of a cloned histone gene cluster of the newt *Notophthalmus viridescens*. *Nucleic Acids Res.* **9**, 2281–2295.

Strickland, M., Strickland, W. N., Brandt, W., and von Holt, C. (1977). The complete amino-acid sequence of histone H2B(1) from sperm of the sea urchin *Parechinus angulosus*. *Eur. J. Biochem.* **77**, 263–275.

Strickland, M., Strickland, W. N., Brandt, W. F., and von Holt, C. (1978a). The partial amino acid sequences of the two H2B histones from sperm of the sea urchin *Psammechinus miliaris*. *Biochim. Biophys. Acta* **536**, 289–297.

Strickland, M., Strickland, W., Brandt, W., von Holt, C., Wittman-Liebold, B., and

Lehmann, A. (1978b). The complete amino-acid sequence of histone H2B(3) from sperm of the sea urchin *Parechinus angulosus*. *Eur. J. Biochem.* **89**, 443–452.

Strickland, W. N., Strickland, M., Brandt, W., and von Holt, C. (1977). The complete amino-acid sequence of histone H2B(2) from sperm of the sea urchin *Parechinus angulosus*. *Eur. J. Biochem.* **77**, 277–286.

Strickland, W. N., Strickland, M., Brandt, W. F., von Holt, C., Lehmann, A., and Wittmann-Liebold, B. (1980). The primary structure of histone H1 from the sea urchin *Parechinus angulosus*. Sequence of the C-terminal CNBr peptide and the entire primary structure. *Eur. J. Biochem.* **104**, 567–578.

Strickland, W. N., Strickland, M., and von Holt, C. (1982a). A comparison of the amino acid sequences of histone H1 from the sperm of *Echinolampus crassa* and *Parechinus angulosus*. *Biochim. Biophys. Acta* **700**, 127–129.

Strickland, W. N., Strickland, M., von Holt, C., and Giancotti, V. (1982b). A partial structure of histone H1 from sperm of the sea urchin *Sphaerechinus granulosus*. *Biochim. Biophys. Acta* **703**, 95–100.

Sugarman, B. J., Dodgson, J. B., and Engel, J. D. (1983). Genomic organization, DNA sequence, and expression of chicken embryonic histone genes. *J. Biol. Chem.* **258**, 9005–9016.

Sugiyama, M. (1951). Refertilization of the fertilized egg of the sea urchin. *Biol. Bull. (Woods Hole, Mass.)* **101**, 335–338.

Sung, M. T., and Freedlender, E. F. (1978). Sites of *in vivo* phosphorylation of histone H5. *Biochemistry* **17**, 1884–1890.

Sunkara, P. S., Al-Bader, A. A., and Rao, P. N. (1977). Mitoplasts, mitotic cells minus chromosomes. *Exp. Cell Res.* **107**, 444–448.

Sunkara, P. S., Wright, D. A., and Rao, P. N. (1979). Mitotic factors from mammalian cells induce germinal vesicle breakdown and chromosome condensation in amphibian oocytes. *Proc. Natl. Acad. Sci. U.S.A.* **76**, 2799–2802.

Sunkara, P. S., Al-Bader, A. A., Riker, M. A., and Rao, P. N. (1980). Induction of prematurely condensed chromosomes by mitoplasts. *Cell Biol. Int. Rep.* **4**, 1025–1029.

Sunkara, P. S., Wright, D. A., Adlakha, R. C., Sahasrabuddhe, C. G., and Rao, P. N. (1982). Characterization of chromosome condensation factors of mammalian cells. *In* "Premature Chromosome Condensation" (P. N. Rao, R. T. Johnson, and K. Sperling, eds.), pp. 233–251. Academic Press, New York.

Sures, I., Lowry, J., and Kedes, L. (1978). The DNA sequence of sea urchin *(S. purpuratus)* H2A, H2B, and H3 histone coding and spacer. *Cell (Cambridge, Mass.)* **15**, 1033–1044.

Tanphaichitr, N., Moore, K., C., Granner, D., K., and Chalkley, R. (1976). Relationship between chromosome condensation and metaphase lysine-rich histone phosphorylation. *J. Cell Biol.* **69**, 43–50.

Turner, P. C., Aldridge, T. C., Woodland, H. R., and Olds R. W. (1983). Nucleotide sequences of H1 histone genes from *Xenopus laevis*. A recently diverged pair of H1 genes and an unusual H1 pseudogene. *Nucleic Acids Res.* **11**, 4093–4105.

von der Haar, B., Sperling, K., and Gregor, D. V. (1981). Maturing *Xenopus* oocytes induce chromosome condensation in somatic plant nuclei. *Exp. Cell Res.* **134**, 477–481.

von Holt, C., Strickland, W. N., Brandt, W. F., and Strickland, M. S. (1979). More histone structures. *FEBS Lett.* **100**, 201–218.

von Holt, C., deGroot, P., Schwager, S., and Brandt, W. F. (1984). The structure of sea urchin histones and considerations on their function. *In* "Histone Genes" (G. S. Stein, J. L. Stein, and W. F. Marzluff, eds.), pp. 65–105. Wiley, New York.

Yaguchi, M., Roy, C., and Seligy, V. L. (1979). Complete amino acid sequence of goose erythrocyte H5 histone and the homology between H1 and H5 histones. *Biochem. Biophys. Res. Commun.* **90,** 1400–1406.

Zalenskaya, I. A., Pospelov, V. A., Zalensky, A. O., and Vorob'ev, V. I. (1981). Nucleosome structure of sea urchin and starfish chromatin: Histone H2B is possibly involved in determining the length of linker DNA. *Nucleic Acids Res.* **9,** 473–486.

Zentgraf, H., and Franke, W. W. (1984). Differences of supranucleosomal organization in different kinds of chromatin: Cell type-specific globular subunits containing different numbers of nucleosomes. *J. Cell Biol.* **99,** 272–286.

Zentgraf, H., Muller, U., and Franke, W. W. (1980). Supranucleosomal organization of sea urchin sperm chromatin in regularly arranged 40 to 50 nm large granular subunits. *Eur. J. Biochem.* **20,** 254–264.

7

Regulation of Mitosis by Nonhistone Protein Factors in Mammalian Cells

RAMESH C. ADLAKHA AND POTU N. RAO

Department of Medical Oncology
The University of Texas System Cancer Center
M. D. Anderson Hospital and Tumor Institute
Houston, Texas 77030

I. INTRODUCTION

Mitosis is a unique phase in the life cycle of a eukaryotic cell when the cell is involved in a major macromolecular reorganization. Chiefly, the dispersed interphase chromatin is condensed into highly compact metaphase chromosomes, the nuclear membrane breaks down, and the cytoskeletal structure is disassembled and reorganized into the mitotic apparatus with the centrosomes serving as the organizing centers. Furthermore, gene transcription is known to be shut off, and synthesis of proteins and RNA has been shown to be important for the G_2 to mitosis transition (for reviews, see Mazia, 1974; Baserga, 1976, 1981; Prescott, 1976; Pardee et al., 1978; Rao, 1982; Rao and Adlakha, 1985; Adlakha et al., 1985a). What are the factors responsible for this structural reorganization and what are the molecular mechanisms that drive these events in one direction during prophase and metaphase and in the opposite direction during telophase and cytokinesis? This chapter is primarily a review of our studies aimed at answering these questions.

The cell fusion studies performed in the early 1970s (Rao and Johnson, 1970; Johnson and Rao, 1970) clearly showed that within mitotic cells is a

179

mechanism that can induce chromosome condensation and the dissolution of the nuclear membrane in the interphase nuclei following fusion between a mitotic cell and an interphase cell. This phenomenon is called premature chromosome condensation (PCC), and its products are the prematurely condensed chromosomes (PCCs) (for review, see Rao, 1982). In general, the induction of PCC may be regarded as equivalent to the initiation of mitosis. The mitotic factors responsible for the induction of PCC have no species specificity (Rao and Johnson, 1974). Studies from our laboratory have revealed that compounds carrying a net positive charge promote PCC induction, while negatively charged compounds inhibit this process (Rao and Johnson, 1971; Rao et al., 1975). During the induction of PCC, prelabeled proteins from the mitotic cell become associated with the PCCs of the unlabeled interphase cell (Rao and Johnson, 1974), an observation confirmed by using polyclonal antibodies specific to human chromosomes (Davis and Rao, 1982). However, neither the nature of this interaction nor the mechanism by which the chromatin is transformed into chromosomes is understood. In the first part of this chapter, we discuss in detail the progress that has been made in the characterization of the mitotic factors and their role in the induction of mitosis, meiosis, and PCC.

The ability of a mitotic cell to induce PCC in an interphase nucleus has been shown to depend largely on the ratio of mitotic to interphase nuclei in the cell at the time of fusion. Very low frequency or no induction of PCC was observed in tri- or tetranucleates containing one mitotic and two or three G_1 nuclei (Johnson et al., 1971). Furthermore, Obara et al. (1973, 1974a,b) reported that in certain of these multinucleate cells with a higher ratio of interphase cells, a membrane is formed around the metaphase chromosomes. This process, because of its resemblance to the process occurring in normal telophase, has been termed telophasing. Moreover, Rao and Johnson (1970) and Rao et al. (1975) observed that the entry of G_2 cells into mitosis is delayed when G_2 cells are fused with G_1- or S-phase cells until G_1- or S-phase nuclei in the heterophasic binucleate cells have completed DNA synthesis, and subsequently, both nuclei have synchronously entered mitosis. We speculated, therefore, that some factors present in G_1- or S-phase cells were causing decondensation of chromatin in G_2 nuclei, thus blocking them from entering mitosis. In the second part of this chapter, we present experimental evidence to show the presence of factors in G_1 cells that are antoganistic to the action of mitotic factors and are called inhibitors of the mitotic factors (IMF).

In the third part of this chapter, we deal with the studies of molecular mechanisms by which both mitotic factors and the inhibitors of the mitotic factors possibly regulate mitosis. We present experimental evidence that strongly suggests that the phosphorylation and dephosphorylation of non-histone proteins by or of these factors represents a crucial regulatory mechanism for the control of events associated with mitosis in mammalian cells.

II. THE MITOTIC FACTORS

The cell fusion studies referred to earlier clearly demonstrated the existence of certain regulatory factors (proteins) in mitotic cells that are absent from interphase cells. These factors can cause mitosis-like events when introduced into interphase cells by cell fusion. Sunkara *et al.* (1979a,b) demonstrated that extracts from mitotic HeLa cells can induce germinal vesicle breakdown (GVBD) and chromosome condensation when injected into immature oocytes of *Xenopus laevis*. A bioassay was thereby developed for the isolation and characterization of mitotic factors. During the meiotic maturation of *Xenopus* oocytes following stimulation with progesterone, it has been shown that a cytoplasmic factor called maturation-promoting factor (MPF) is synthesized and/or activated. When MPF reaches a critical threshold level, the germinal vesicle breaks down and chromatin condenses into discrete chromosomes (Masui and Markert, 1971; Smith and Ecker, 1971). This process can also be induced by injection of cytoplasmic extracts from matured oocytes (Masui and Markert, 1971; Drury and Schorderet-Slatkine, 1975; Wasserman and Masui, 1976; for reviews, see Masui and Clarke, 1979; Maller and Krebs, 1980). Wasserman and Smith (1978) demonstrated that MPF-like activity was also present in early cleavage-stage amphibian embryos and that the peak of this activity coincided with the time that embryonic nuclei enter mitosis. Because meiotic maturation appears to closely resemble the phenomenon of PCC, it is possible that mitotic factors and MPF may be very similar, if not identical. The activity of mitotic factors from HeLa cells was shown to be dose-dependent and cell cycle-related, initially appearing in G_2 and reaching a peak in mitosis. The maturation-promoting activity (MPA) is not present in extracts from G_1- and S-phase cells. Extracts of cells irreversibly arrested in G_2 by alkylating agents (Al-Bader *et al.*, 1978) did not exhibit MPA (Sunkara *et al.*, 1979a,b, 1982). Similarly, extracts from mitotic Chinese hamster ovary (CHO) cells (Nelkin *et al.*, 1980; Kishimoto *et al.*, 1982), human D98/AH2 cells (Halleck *et al.*, 1984a), and *cdc* mutants of yeast *(Saccharomyces cerevisiae)* arrested in late G_2 or mitosis (Weintraub *et al.*, 1982) have been shown to induce meiotic maturation of *X. laevis* oocytes. Using the oocyte system for bioassay, we have shown that mitotic factors from HeLa cells are heat-labile, Ca^{2+}-sensitive, and Mg^{2+}-dependent nonhistone proteins with a sedimentation value of 4–5 S (Sunkara *et al.*, 1979a,b, 1982).

The bioassay for the mitotic factors in which meiotic cells (oocytes) are used was recently shown to be valid for cells undergoing mitosis (embryonic cells) (Miake-Lye *et al.*, 1983; Newport and Kirschner, 1984). These studies showed that when partially purified MPF from matured *Xenopus* oocytes was injected into cycloheximide-arrested multinucleated *Xenopus* embryos (arrested in a G_2-like state by inhibition of protein synthesis), the nuclear envelope surrounding each nucleus was dispersed within 5 min after injection of MPF. Similar results were obtained when mitotic factors from human D98/AH2 cells were injected into the syncytial frog embryos (Halleck *et al.*, 1984b). We have also been able to successfully adopt and improve this technique in our laboratory by using partially purified mitotic factors from HeLa cells (R. C. Adlakha and P. N. Rao, unpublished data). These results suggest that the factors involved in the initiation of mitosis, meiosis, and in the induction of PCC are very similar, if not identical.

A. Subcellular Localization of Mitotic Factors

The cell fusion experiments of Rao and Johnson (1974) suggested that certain mitotic proteins migrated into interphase nuclei and brought about the condensation of chromatin into PCC. Experiments were designed, therefore, to determine whether the mitotic factors were also localized on metaphase chromosomes. Using a differential extraction technique to obtain cytoplasmic and chromosomal fractions, we observed that both the cytoplasmic extract and the chromosomal fraction exhibited MPA. The specific activity of the chromosomal fraction was at least three to four times greater than that of the cytoplasmic fraction (Table I). Injection of cytoplasmic and nuclear extracts from early- and mid-G_2 phase HeLa cells indicated the presence of MPA in nuclear fractions only. However, in late G_2-phase cells, both the cytoplasmic and nuclear extracts exhibited MPA. These data indicate that a major portion of the mitotic factors is localized on chromosomes and that these factors preferentially bind to chromatin as soon as they are synthesized in G_2. As a cell, in preparation for mitosis, synthesizes more of these factors, increasing amounts of them are retained in the cytoplasm (Adlakha *et al.*, 1982a). It is not clear from these results, however, whether the mitotic factors are synthesized in the cytoplasm and migrate to the nucleus during late-G_2 phase or whether they are synthesized in the nucleus and then are diffused into the cytoplasm as more proteins are synthesized. In either case, these results

TABLE I

Maturation-Promoting Activities of Cytoplasmic and Chromosomal Fraction of Mitotic HeLa Cells

Oocytes injected with:	Extract dilution	Protein injected (ng[a])	No. oocytes injected	No. oocytes showing GVBD	Induction of GVBD (%)
Cytoplasmic	0	585	30	30	100
fraction I	1/2	292	25	25	100
	1/3	195	25	25	100
	1/4	146	25	7	28
	1/8	73	25	0	0
Cytoplasmic	0	69	25	0	0
fraction II					
Chromosomal	0	179	30	30	100
fraction	1/2	90	25	25	100
	1/3	60	25	23	92
	1/4	45	20	5	25
	1/8	22	25	0	0

[a] A total volume of 65 nl of the extracts was injected into each oocyte, and oocytes were scored for GVBD 2–3 hr after injection. A concentration of 40×10^6 cells/ml was used for preparation of these extracts. (From Adlakha et al., 1982a.)

suggest that mitotic factors have a strong affinity to chromatin. It has been suggested that in most cells the synthesis of nuclear proteins occurs primarily, if not exclusively, in the cytoplasm. In certain cases, some special mechanism besides diffusion may be involved in transporting the proteins from the cytoplasmic site of synthesis to their ultimate location in the nucleus (Kuehl, 1974; Kuehl et al., 1980). For mitotic factors, however, no such special mechanism need be invoked. It has already been demonstrated by cell fusion experiments that these factors have free access to the nucleus, presumably through the nuclear pores (Rao and Johnson, 1974).

In a 1982 study by Adlakha et al. (1982b), additional evidence for the association of mitotic factors with chromosomes was provided by partial release of the factors through a mild digestion of the isolated and purified metaphase chromosomes by micrococcal nuclease or DNase II (Table II). Even after extensive digestion of the chromosomes with nucleases, a portion of the mitotic factors remained bound to the chromosomes, indicating that the mitotic factors are probably bound to the DNA in the core particle. However, it is not yet clear whether the binding of these factors to chromatin directly induces the tight packaging of chromatin fibers into

TABLE II

Release of Mitotic Factors by Digestion of Metaphase Chromosomes or Interphase Nuclei with Micrococcal Nuclease and DNase II[a,b]

Oocytes injected with supernatants after:	Micrococcal nuclease digestion[c]		DNase II digestion[c]	
	Protein injected (ng)[d]	GVBD induction (%)[e]	Proteins injected (ng)[d]	GVBD induction (%)[e]
Digestion of metaphase chromosomes	73	80	85	90
Nuclease-digested chromosomes extracted with high salt buffer	111	100	107	100
Digestion of nuclei of G_2 cells	91	50	95	60
Nuclease-digested nuclei from G_2 cells extracted with high salt buffer	97	73	90	80
Digestion of nuclei from S-phase cells	107	0	101	0
Nuclease-digested nuclei from S-phase cells extracted with high salt buffer	123	0	115	0
Chromosomes extracted with digestion buffer without the enzyme	21	0	21	0
Digestion buffer alone	0	0	0	0

[a] Nuclease (200 units/ml); DNase II (100 units/ml). (From Adlakha et al., 1982b.)

[b] Data from several digestion experiments.

[c] The data in this table refer to the maximum release of mitotic factors at the optimum digestion time of 3.5 min in case of micrococcal nuclease and 2 min for the DNase II.

[d] A cell density of 40×10^6 cells/ml was used for preparation of these extracts and 65 nl of the extracts was injected into each oocyte.

[e] An average of 25 oocytes were injected with each extract.

metaphase chromosomes, or whether the mitotic factors become associated with chromatin and then induce a change (e.g., phosphorylation or acetylation) in the structural proteins to cause condensation of chromosomes.

B. Purification of the Mitotic Factors

We have previously shown that mitotic factors extracted from isolated metaphase chromosomes were 3- to 4-fold enriched in specific activity as compared to the cytoplasmic extract (see Table I). Precipitation with $(NH_4)_2SO_4$ (20–40% cut) resulted in another 3- 4-fold enrichment of MPA

(Adlakha *et al.*, 1982a). Initially, the affinity or adsorption of the crude or the partially purified [12- to 15-fold after $(NH_4)_2SO_4$ precipitation] mitotic factors to different chromatographic materials such as DEAE-Sephadex, CM-Sephadex, hydroxylapatite, DEAE-cellulose, arginine-agarose, pentylagarose, and phosphocellulose was tested. A combination of $(NH_4)_2SO_4$ precipitation and chromatography on arginine-agarose and pentylagarose was reported by Wu and Gerhart (1980) to result in a 23-fold enrichment of MPF from *Xenopus* oocytes. The behavior of crude or partially purified mitotic factors on molecular sieving columns, such as Ultrogel ACA-34 and Sephacryl S-200, was investigated, and precipitation with polyethylene glycol (PEG)-6000 was also tried. These preliminary experiments indicated that $(NH_4)_2SO_4$ precipitation and chromatography on hydroxylapatite, Sephacryl S-200, and DEAE-cellulose should be useful. However, a combination of these procedures resulted, at best, in only about a 50-fold enrichment of the mitotic factor activity.

Our observations that mitotic factors are preferentially localized on chromosomes and can be released by mild digestion with endonucleases suggest that they may be DNA-binding proteins (Adlakha *et al.*, 1982a,b). Therefore, affinity chromatography of the partially purified mitotic factors (enriched 50-fold, as mentioned previously) was performed on a DNA-cellulose column (0.8 × 3.0 cm). The column was eluted with a linear gradient of 0.01–0.6 M NaCl in a 10-mM sodium phosphate buffer containing protease and phosphatase inhibitors. The proteins with MPA eluted as a single peak with 0.15–0.20 M NaCl, and the specific activity of mitotic factors was about 200-fold higher than that of the crude mitotic extracts (Fig. 1 and Table III). By our definition, one unit of activity causes GVBD in 25–75% of the oocytes injected with a volume of 65–70 nl of the extracts or column fractions. This definition of a unit of activity was chosen over the one previously used by Wu and Gerhart (1980) and Gerhart *et al.* (1984) due to the nonlinearity of the assay. However, we subsequently observed that affinity chromatography on DNA-cellulose of crude mitotic cell extracts also resulted in a 200-fold enrichment of the mitotic factors in a single step (Table IV) (Adlakha *et al.*, 1985c). More recently we have improved this technique of DNA-cellulose affinity chromatography and have obtained about 500-fold purification of the mitotic factors in a single step (R. C. Adlakha and P. N. Rao, unpublished data). Although the elution profile was very similar to the one shown in Fig. 1, all the activity was recovered in a single fraction rather than spread over three to five fractions (Fig. 2). Furthermore, the recovery of the active mitotic factors was also significantly greater, using the single step purification procedure (63% compared to 19%; see Tables III an IV). Fast chromatographic methods that only minimally dilute the sample such as

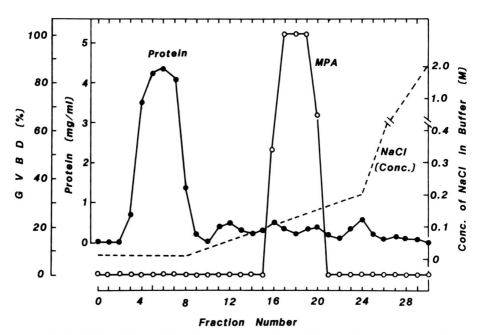

Fig. 1.   DNA-cellulose affinity chromatography of mitotic factors. An extract from N_2O blocked mitotic HeLa cells (~8mg proteins/ml) was concentrated about 5-fold by ultrafiltration on Amicon YM10 filters, and dialyzed extensively against 10 mM Na_2HPO_4, 1 mM PMSF, 1 mM ATP, 5 mM NaF, 5 mM sodium β-glycerolphosphate, 10 mM $MgCl_2$, 2 mM EGTA, 1 mM EDTA, and 5% glycerol, pH 6.5 (dialysis buffer) prior to loading on a DNA-cellulose column (0.8 x 3.0 cm) preequilibrated with the dialysis buffer. The column was eluted with a linear gradient of 0.01–0.6 M NaCl in the dialysis buffer and 0.5 ml fractions were collected and evaluated for MPA by microinjection into *Xenopus* oocytes. (From Adlakha *et al.*, 1985c.)

affinity chromatography and high performance liquid chromatography (HPLC), are considered essential if biologically active mitotic factors are to be purified.

Several investigators have been working for a number of years to purify the MPF from mature frog oocytes (Masui and Markert, 1971; Wasserman and Masui, 1976; Wasserman and Smith, 1978; Drury, 1978; Wu and Gerhart, 1980; Lewis and Potter, 1982). Yet the reports in the literature indicate only limited success. Even under the best conditions, no greater than 50-fold purification of MPF has been reported (Wu and Gerhart, 1980; Gerhart *et al.*, 1984). The purification of mitotic factors proved to be more difficult than anticipated because of the following problems: (1) the instability of the mitotic factors even at 4°C, especially if the protein

TABLE III

Purification of Mitotic Factors from Extracts of Mitotic HeLa Cells[a]

Purification step	Total volume (ml)	Total protein (mg)	Total activity (units $\times$ 10^{-3})	Specific activity (units/mg of protein $\times$ 10^{-3})	Recovery (%)	Purification (-fold)
100,000 g supernatant	20	189.00	1,715	9.074	100	1
Ultrafiltration through YM10	4	168.20	1,502	8.930	87.6	1
20–40% $(NH_4)_2$ SO_4 precipitation	2	33.74	1,115	33.04	65	3.7
Hydroxylapatite	1	2.76	841	304.51	49	34.1
DNA-cellulose	1	0.181	325	1,796.72	19	198.0

[a] A unit of activity is defined as the amount of activity that causes GVBD in 25–75% of the oocytes injected with a volume of 65–70 nl of the extracts. All the chromatographic procedures were carried out at 4°C, and all the elution buffers contained phosphatase and protease inhibitors. The active fractions from hydroxylapatite chromatography were pooled, dialyzed, and loaded onto the DNA-cellulose column. The elution conditions of DNA-cellulose column are given in the legend to Fig. 1. (From Adlakha et al., 1985c.)

TABLE IV

Purification of Mitotic Factors from Extracts of Mitotic HeLa Cells in a Single Step by Affinity Chromatography on DNA-Cellulose Column[a]

Step	Total volume (ml)	Total protein (mg)	Total activity (units $\times 10^{-3}$)	Specific activity (units/mg of protein $\times 10^{-3}$)	Recovery (%)	Purification (-fold)
100,000 g supernatant	5	40.60	363	8.73	100	1
Ultrafiltration through YM10	2	40.28	330	8.193	91	1
DNA-cellulose	1	0.141	228	1,617.30	63	185.3

[a] The elution conditions from the DNA-cellulose column were essentially the same as described in the legend to Fig. 1 and in the footnote to Table III. (From Adlakha et al., 1985c.)

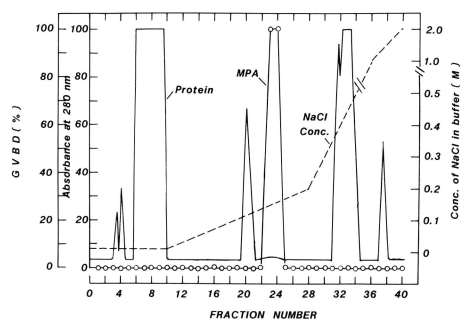

Fig. 2. DNA-cellulose affinity chromatography of crude mitotic extract. An extract from N$_2$O-blocked or Colcemid-arrested mitotic HeLa cells (~40 mg proteins) was extensively dialyzed against 10 mM sodium phosphate buffer, pH 6.5 containing 1 mM PMSF, 1 mM ATP, 5 mM NAF, 5 mM sodium β-glycerolphosphate, 10 mM MgCl$_2$, 2 mM EGTA, 1 mM EDTA, and 30% glycerol (dialysis buffer). After removing the insoluble material by centrifugation, the soluble proteins were loaded on a DNA-cellulose column (0.8 x 4.0 cm) preequilibrated with the dialysis buffer. The column was eluted with a stepwise gradient 0.01–2 M NaCl in the dialysis buffer containing 30% glycerol, and 0.5 ml fractions were collected and evaluated for MPA as described in Fig. 1.

concentration was ≤50μg/ml; (2) mitotic factors constituted only a small fraction of the total cellular proteins in crude extracts; (3) extensive dilution during fractionation; (4) low recovery; and (5) nonlinearity of the assay.

It is also possible that mitotic factors represent a complex mixture of proteins rather than a single protein or that the purification procedures used also enrich factors that specifically inactivate or block the action of mitotic factors. Similar problems have been encountered by other investigators. We have been successful, however, in surmounting most of these difficulties by using affinity chromatography on small columns. Attempts are now in progress in our laboratory to purify further the mitotic factors from the 500-fold enriched fraction, using various combinations of DEAE-cellulose, hydroxylapatite, DNA-cellulose, and HPLC.

C. Characterization of the Mitotic Factors

Some characteristics of the mitotic factors were studied by using the 200-fold purified fraction. The molecular weight of the crude or partially purified mitotic factors, as determined by molecular sieving chromatography on Sephacryl S-200, appeared to be about 100,000 (Fig. 3). It has recently been reported that the molecular weight of MPF from mature oocytes is also about 100,000 (Gerhart *et al.*, 1984).

On nondenaturing polyacrylamide gels run at 4°C, we have observed much streaking along the lanes, with no clear bands when stained with Coomassie blue (data not shown). Preliminary attempts to elute the proteins with MPA from the nondenaturing gels have been unsuccessful,

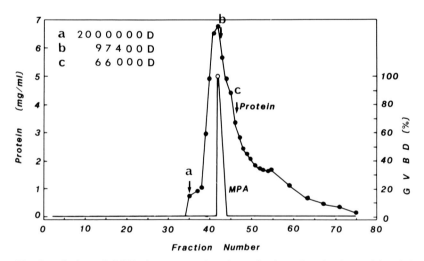

Fig. 3. Sephacryl S-200 chromatography; determination of molecular weight of the mitotic factors. The molecular weight of the mitotic factors was determined by molecular sieving chromatography on Sephacryl S-200 column. An extract from mitotic HeLa cells was concentrated and dialyzed as explained in legend to Fig. 1. About 95–100 mg of mitotic proteins were loaded on S-200 column (2.5 x 110 cm) preequilibrated with the dialysis buffer. The column was eluted with the dialysis buffer and 5-ml fractions were collected and evaluated for MPA in oocytes. To determine the molecular weight the column was separately calibrated by loading molecular weight standards such as (a) blue dextran (2,000,000); (b) phosphorylase *b* (97,400); and (c) bovine serum albumin (66,000).

probably because of extensive dilution and instability during isolation even at 4°C. Sodium dodecyl sulfate–polyacrylamide gel electrophoresis (SDS–PAGE) analysis of the partially purified mitotic factors (200-fold) indicated the presence of several polypeptides with a major band of 50,000 (Fig. 4). It is difficult to determine at this stage whether the active mitotic factor is a dimer of 50,000. Most of the polypeptides observed in the purified fraction were found to be phosphoproteins with little or no phosphorylation at the 50K band, as revealed by ^{32}P-labeling and autoradiography (Fig. 5, lane B). Interestingly, most of the phosphoproteins in the purified fraction were dephosphorylated when partially purified, prelabeled mitotic factors were mixed with G_1 cell extracts but not when mixed with late S-phase cell extracts (Fig. 5, lanes C and E). We have previously shown that mixing of mitotic cell extracts with G_1-cell extracts results in the inactivation of mitotic factors under the influence of the inhibitory factors present in G_1 cells (Adlakha et al., 1983). We have also shown that IMF can be induced both in G_0 phase and mitotic cells by UV irradiation (Adlakha et al., 1983, 1984a; for details, see Section III). Therefore, when extracts from UV-treated mitotic cells prelabeled with ^{32}P were purified on the DNA-cellulose column, and the fraction corresponding to the active fraction was separated by gel electrophoresis and then examined by autoradiography, most of the phosphoproteins normally found in the purified mitotic fraction were again absent (Fig. 5, lane D). These data suggest that UV irradiation may induce or activate some phosphatases in mitotic cells similar to phosphatases present in G_1 cells that may specifically dephosphorylate mitotic, nonhistone proteins.

The activity of the purified mitotic factors was also found to be stabilized by the presence of phosphatase inhibitors and by the addition of bovine serum albumin (BSA) as a carrier protein, as reported previously for crude mitotic factors (Adlakha et al., 1982a) or for MPF (Wu and Gerhart, 1980; Gerhart et al., 1984). Therefore, buffers containing these inhibitors were used in all chromatographic procedures.

Incubation of the partially purified mitotic factors with mitosis-specific monoclonal antibodies, MPM-1 and MPM-2, recently developed in our laboratory (Davis et al., 1983; see also Chapter 9, this volume) did not destroy the MPA of mitotic factors (Table V). However, when proteins from the purified fraction were separated on SDS–PAGE, transferred to nitrocellulose paper, and then allowed to react with MPM-1 or MPM-2, several polypeptides were recognized by the antibodies while the major peptide band at 50K was not recognized (Fig. 6, lanes 4 and 7). It is difficult to determine whether the 50K protein is responsible for the phosphorylation of the other proteins present in the purified fraction. How-

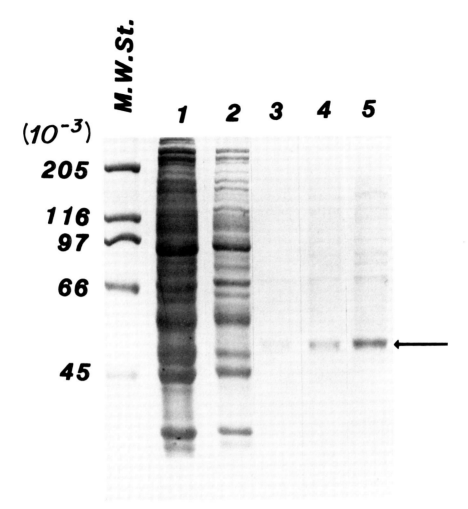

Fig. 4. Electrophoretic separation of the proteins from the crude and affinity-purified mitotic factors on 10% polyacrylamide gels containing 0.1% SDS. Mitotic factors were partially purified by DNA-cellulose affinity chromatography as described in legend to Fig. 1. Proteins were separated on SDS–PAGE and stained with Coomassie blue. Lanes: 1, crude mitotic extract; 2, unbound fractions from DNA cellulose column; 3, 200-fold purified mitotic factors; 4, same as in lane 3 (twice the amount); 5, same as in lane 3 (four times the amount). Arrow indicates the major polypeptide band of about 50 K. The migration of molecular weight standards (M.W.St.) is also shown. (From Adlakha *et al.*, 1985c.)

Autoradiogram

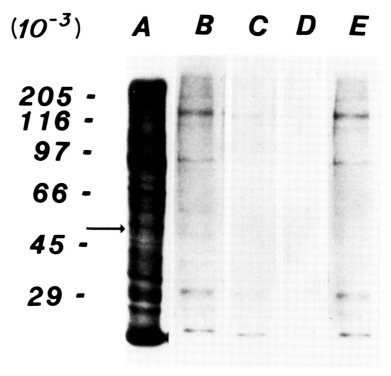

Fig. 5. Detection of phosphoproteins present in the purified mitotic factors fraction. HeLa cells synchronized in late S phase by double thymidine block method were labeled with ^{32}P (10 μCi/ml) and placed in N_2O. ^{32}P-labeled extracts were prepared from control and UV-irradiated mitotic HeLa cells and enriched (200-fold) for mitotic factor activity as described in Fig. 1, separated by electrophoresis on SDS–PAGE and gels autoradiographed. Lanes: A, crude mitotic extract; B. 200-fold purified mitotic factors fraction; C, mixture of ^{32}P-labeled enriched fraction (lane B) and unlabeled G_1 cell extract; D, fraction corresponding to the enriched fraction (lane B) from extracts of UV-irradiated mitotic cells; E, mixture of enriched fraction (lane B) and unlabeled S-phase extract. The migration of molecular weight standards is also known. (From Adlakha *et al.*, 1985c.)

Stained with

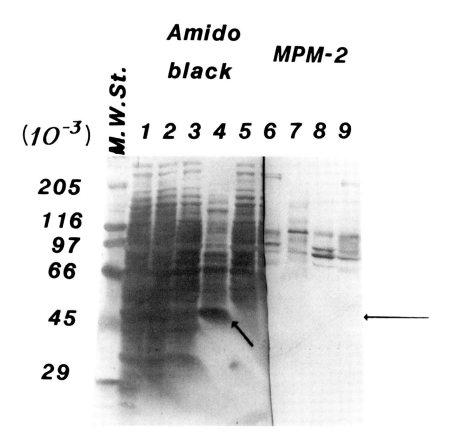

Fig. 6.   Reactivity of the mitosis-specific monoclonal antibody MPM-2 with the crude and partially purified mitotic factors. Mitotic factors were purified 200-fold by affinity chromatography on DNA-cellulose as described in Fig. 1. Polypeptides from the crude unbound and 200-fold purified mitotic fractions were separated on SDS-gradient (6–20%) polyacrylamide gels, electrophoretically transferred to nitrocellulose sheets, and stained either with amido black or with MPM-2 by an indirect immunoperoxidase procedure. Lanes: 1 and 9, crude mitotic extract; 2 and 8, 20–40% (NH_4)$_2$$SO_4$ precipitate fraction of crude mitotic extract; 3, 5, and 6, unbound fraction from DNA-cellulose column; 4 and 7, partially purified mitotic factors. Migration of the molecular weight standards (M.W.St.) is also shown. Arrows indicate the position of the major polypeptide of 50K present in partially purified mitotic factors. Note that the MPM-2 antibody does not recognize the major polypeptide (lane 7). (From Adlakha *et al.*, 1985c.)

TABLE V

Incubation of the Purified Mitotic Factors with Mitotic-Specific
Monoclonal Antibodies, MPM-1 or MPM-2, Does Not Destroy the MPA
of Mitotic Factors[a]

Oocytes injected with:	No. of oocytes injected	No. of oocytes showing GVBD (%)
Mitotic factors (control)	25	100
Mitotic factors + buffer (control)	25	92
Mitotic factors + MPM-1	25	92
Mitotic factors + MPM-2	25	88
Mitotic factors + IgG (control)	25	88

[a] Purified mitotic factor proteins (15–20 μg) were incubated with 100 μg of mouse immunoglobulin in 0.15 M NaCl for 0.5 hr at 37°C followed by 1.5 hr at 4°C. The mixture was then incubated for 1 hr at 4°C with sufficient amount of pelleted Pansorbin (Calbiochem) for MPM-2, IgG, and control mouse IgG or Con A-Sepharose (Pharmacia) for MPM-1, and IgM to bind all detectable mouse Ig. The supernatants after centrifugation were assayed for MPA, and the gel profile of the proteins remaining after adsorption are shown in Fig. 6. (From Adlakha *et al.*, 1985c.)

ever, these results do suggest that mitosis-specific monoclonal antibodies should be useful in removing some phosphoproteins recognized by these antibodies (Fig. 6, lanes 4 and 7), resulting in the further purification of the mitotic factors.

D. Association of Protein Kinase Activity with the Purified Mitotic Factors

We have also tested the partially purified mitotic factors for kinase activity. Using histones as the substrate, we have observed a 50- to 60-fold increase in the specific activity of a kinase in the 500-fold purified fraction compared with the crude extract. However, when casein or phos-vitin was used as a substrate, no significant increase in protein phosphorylation was observed. The mitotic factor-associated kinase is neither stimulated by cAMP nor inhibited by PK'I', an inhibitor of the cAMP-dependent protein kinase, indicating that this kinase is not cAMP-dependent. Similarly, Ca^{2+}, calmodulin, and spermine had no stimulatory effect on kinase activity. In the absence of histones, little phosphorylation was observed. When immature oocyte extract was used as a substrate, no

TABLE VI

Association of Protein Kinase Activity with the Purified Mitotic Factors[a]

Assay conditions	Activity (cpm)	Specific activity (pmol/mg/min)	Activity (-fold increase)
Crude extract	0.0	0.00	0.0
+Histones	2,383.0	0.60	1.0
Purified MF	0.0	0.00	0.0
+cAMP	0.0	0.00	0.0
+Ca²⁺	349.7	2.91	4.9
Purified MF with			
+oocyte extract	18.9	0.16	0.3
+cAMP	0.0	0.00	0.0
+Ca²⁺	133.9	1.12	1.9
Purified MF with			
+Histones	4,173.4	34.78	58.0
+cAMP	5,067.4	42.23	70.4
+Ca²⁺	4,669.8	38.92	64.9
+PK'I'	4,273.0	35.61	59.5
Purified MF with			
+Histones and			
+Oocyte extract	7,785.4	64.88	108.1
+cAMP	8,430.1	70.25	117.1
+Ca²⁺	8,595.8	71.67	119.7

[a] Data represent an average of duplicate experiments. Amounts of proteins used: crude extract, 100 μg; purified MF, 3 μg; oocyte extract, 50 μg; PK'I', 5 μg. The final concentrations of cAMP and Ca²⁺ were 10 mM and 3 mM, respectively.

significant phosphorylation was observed. However, when histones were used along with immature oocyte extract, a significant (2-fold) increase in kinase activity was observed (Table VI; Adlakha *et al.,* 1985b). It is possible that in the oocyte extract, the mitotic factors activate a kinase or set of kinases, that phosphorylate histones or other oocyte proteins, thereby augmenting the phosphorylation observed. Studies are now in progress using other substrates, such as ribosomal protein S_6 and extracts of HeLa cells in S or G_2 phase and separating on gels the proteins that are phosphorylated. Further purification of mitotic factors to homogeneity is necessary to establish conclusively whether the factors themselves act as a kinase or as activators of an *in vivo* kinase responsible for the GVBD and chromosome condensation.

III. INHIBITORS OF THE MITOTIC FACTORS

What happens to the mitotic factors at the end of mitosis? If the mitotic factors bind to chromatin to bring about its condensation, do they dissociate or become degraded or inactivated by yet undetermined factors at the mitosis–G_1 transition? We reasoned that if the mitotic factors become inactivated or overridden by other factors, then it should be possible to detect these factors in G_1 cells. Therefore, studies were carried out to determine the fate of mitotic factors as a cell completes mitosis and enters G_1. Extracts from mitotic HeLa cells were mixed with extracts from cells in G_1 phase in different proportions; the mixtures were then injected into *X. laevis* oocytes to determine the effects of G_1 cell extracts on the MPA of mitotic cell extracts (Adlakha *et al.*, 1983). Mitotic cell extracts diluted with the extraction buffer, with or without BSA up to 10 mg/ml in corresponding proportions, served as the controls. Mitotic cell extracts could be diluted with the extraction buffer down to a 20% concentration without any loss of MPA. In contrast, the mitotic cell extract mixed with G_1-cell extract did not exhibit any MPA, even at mitotic extract concentrations as high as 66.6% (Fig. 7). In general, when mitotic and G_1-cell extracts with equal protein content were mixed, a complete inactivation of the mitotic factors occurred as concentration of the mixture was reduced to 50%. These results suggest that G_1 cells contain certain factors (IMF) that can inactivate the MPA in mitotic cell extracts.

Extracts from HeLa cells in G_1 phase collected between 2 and 7 hr after reversal of N_2O blockade (Rao, 1968) exhibited strong IMF activity. However, cells collected at later times were less active (Fig. 7). Furthermore, IMF could be detected as early as 1.5 hr after reversal of N_2O blockade, when 10–15% of the cells had not yet completed mitosis. These results suggest that IMF are either activated or newly synthesized at the end of mitosis, i.e., at telophase. Kinetic studies using cycloheximide, an inhibitor of protein synthesis, indicated that IMF are activated rather than newly synthesized (Adlakha *et al.*, 1983). These studies also revealed that, during the cell cycle, the IMF activity fluctuates in a cyclical manner. Extracts from early S-phase cells were more effective in neutralizing the activity of mitotic factors than those from cells either in mid-S or late S phase (Fig. 8). Extracts from early, mid-, and late G_2-phase HeLa cells had no IMF activity. These results were not surprising because mitotic factors are known to accumulate during G_2-phase (Sunkara *et al.*, 1979a). These data indicated that the activity of IMF coincided well with the process of chromosome decondensation, which begins at telophase and continues until the beginning of S phase when the chromatin reaches its

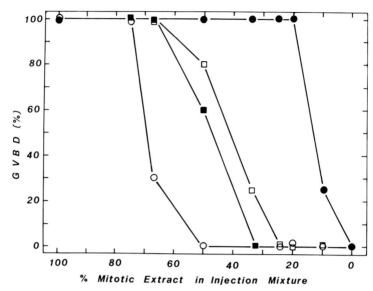

Fig. 7. Effect of G_1 cell extracts on the MPA of the mitotic cell extracts. HeLa cells in G_1 phase were collected at different times (i.e., at 1.5, 2, 3, 4, 5, 6, 7, and 7.5 hr) after reversal of the N_2O block. G_1 cell extracts were made using 6–8 x 10^7 cells/ml with a protein content ranging from 8 to 11 mg/ml. Similarly, mitotic cell extracts were made using 4 x 10^7 cells/ml containing approximately 8 mg/ml of protein. Whenever necessary, extracts were concentrated to give a protein content of 8 mg/ml by ultrafiltration using Amicon YM10 filters. Extracts of G_1 cells at different points during G_1 were separately mixed with mitotic extracts in various proportions so as to obtain a mitotic extract concentration of 100, 75, 66.6, 50, 33.3, 25, 20, 10, and 0% in the injection mixture. These mixtures were incubated for 1 hr at 4°C prior to injection into *Xenopus* oocytes. A minimum of 10 oocytes was injected for each dilution. A volume of 65 nl (containing approximately 500–550 ng of proteins) was actually injected. The percentage of GVBD in the injected oocytes was determined at 2–3 hr after injection by scoring the oocytes for the appearance of a white spot in the animal hemisphere. In doubtful cases oocytes were fixed in 7.5% TCA and dissected to check for the breakdown of the germinal vesicle. Mitotic cell extracts diluted with the extraction buffer (in the presence or absence 10 mg/ml of BSA) to give corresponding concentrations served as controls. These data represent an average of five experiments and the differences in % GVBD between the experiments did not exceed 10%. Typically, oocytes from the same female were used for a given experiment. Dilution of mitotic extract with extracts from G_1 cells at 1.5 hr, □—□; 2 hr, ○—○; 7.5 hr, ■—■ after reversal of the N_2O block. Dilution with buffer, ●—●. The data obtained with extracts from G_1 cells collected at 3,4,5,6, or 7 hr after reversal were identical to those of G_1 cells at 2 hr, and hence these data are not presented. (From Adlakha *et al.*, 1983.)

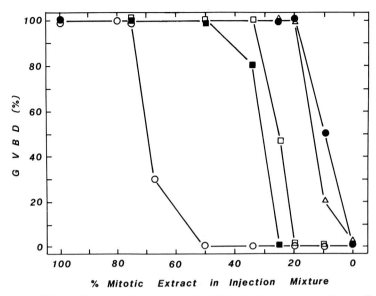

Fig. 8. Effect of S-phase cell extracts on the MPA of mitotic extracts. HeLa cells were synchronized in S phase by double thymidine block. Early, mid-, and late S-phase cells were collected at 2, 4, and 6 hr, respectively, after reversal of the second TdR block. Extracts from early, mid-, or late S-phase cells were prepared using $12-15 \times 10^7$ cells/ml, with a protein concentration of 22–27 mg/ml, whereas the mitotic extract was prepared by using 4×10^7 mitotic cells/ml, with approximately 8 mg/ml of protein, thus giving an S:M protein ratio of 3:1. Aliquots from each of these S-phase extracts were separately mixed with mitotic extracts in various proportions and mixtures were injected into oocytes to test for MPA as explained in the legend to Fig. 7. ●—●, Dilution with buffer (negative control); ○—○, dilution with mid-G_1 cell extract as in Fig. 7 (positive control); dilution with extracts from early (■—■), mid- (□—□), and late (△—△) S-phase cells. (From Adlakha *et al.*, 1983.)

most decondensed state (Pederson, 1972; Pederson and Robbins, 1972; Schor *et al.*, 1975; Hildebrand and Tobey, 1975; Nicolini *et al.*, 1975; Moser *et al.*, 1975, 1981; Rao *et al.*, 1977; Hittelman and Rao, 1976, 1978; Rao and Hanks, 1980; Hanks *et al.*, 1983a).

A. Stimulation or Activation of IMF in Quiescent (G_0) Human Diploid Fibroblasts (WI-38)

The morphology of PCCs reveals that chromatin is more condensed in noncycling G_0-phase human fibroblasts than in cycling G_1 cells (Hanks and Rao, 1980). To determine if noncycling G_0 cells contained any IMF activity, extracts were prepared from quiescent WI-38 cells 7 days after they became confluent. When the extracts were mixed with mitotic cell extracts and tested for MPA, they exhibited little or no IMF activity. However, IMF activity could be induced in WI-38 cells in G_0 by exposing them to UV irradiation, which has been shown to cause chromosome decondensation and unscheduled DNA synthesis (Schor *et al.*, 1975; Hittelman and Pollard, 1982, 1984; Mullinger and Johnson, 1985). Activity of IMF in UV-irradiated G_0 cells was further enhanced if DNA synthesis was prevented by hydroxyurea and arabinosylcytosine (Ara-C) after UV treatment (Fig. 9). The activation of IMF in G_0 cells by UV irradiation did not depend on the synthesis of new proteins but on activation of existing

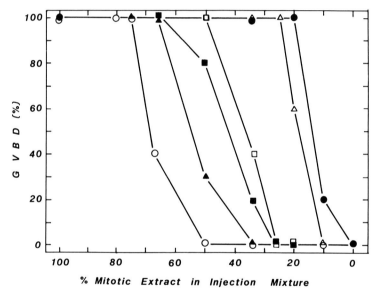

Fig. 9. Activation of IMF in quiescent (G_0) WI-38 human diploid fibroblasts by UV irradiation. WI-38 cells in G_0 phase were collected 7–10 days after they became confluent. G_0 cells were UV-irradiated for 10 sec at 90 ergs cm^{-2} sec^{-1} and incubated for varying times in the presence or absence of cycloheximide (25 μg/ml), Ara-C (10^{-4} M), or hydroxyurea (10^{-2} M). Extracts from the control and treated G_0 cells were prepared so that the protein concentration level would equal that of mitotic extracts (approximately 8 mg/ml). Extracts of G_0 cells from the different treatments were separately mixed with mitotic extracts and tested for MPA as in Fig. 7. ●—●, Dilution with buffer (negative control); ○—○, dilution with mid-G_1 cell extract (positive control); △—△, extracts from untreated G_0 cells pretreated with Ara-C and hydroxyurea; extracts from G_0 cells, UV-irradiated and incubated in the presence or absence of cycloheximide for 2 hr (□—□) and 4 hr (■—■); extracts from G_0 cells, UV-irradiated and incubated for 2 hr in the presence of Ara-C and hydroxyurea (▲—▲). The data presented here are the average of two experiments. (From Adlakha *et al.*, 1983.)

ones, because activation occurred even in the presence of cycloheximide (Adlakha *et al.*, 1983). More recently, we have observed that UV irradiation of mitotic HeLa cells resulted in the decondensation of chromosomes, inactivation of mitotic factors, and activation of IMF. These studies also indicated that IMF inactivate the mitotic factors by directly binding to them and forming an inert complex; Mg^{2+} or polyamines (putrescine, spermidine, and spermine), agents known to promote chromatin condensation (Rao and Johnson, 1971), partially restored the MPA of the extracts from UV-irradiated mitotic cells (Adlakha *et al.*, 1984b).

B. Presence of IMF in V79-8 Cells That Lack G_1 and G_2 Periods in Their Cell Cycle

The activity of IMF is maximum in G_1, minimum during S, and absent in G_2 phase in HeLa cells, which exhibit a G_1 period of about 10 hr. In which phase of the cell cycle would the activity of IMF be expressed in cells that lack a G_1 period? Because the Chinese hamster cell line V79-8 has no measurable G_1, we decided to study the regulation of IMF during the cell cycle of V79-8 cells. In this cell line, we observed that IMF activity was maximal in early S-phase cells and decreased rapidly as the cells traversed S phase. Little or no activity was present in the extracts of late S phase cells (Fig. 10). When V79-8 cells were grown to confluency, they became arrested in G_1 phase; the extracts of these cells possessed as much IMF activity as early S-phase cells. These results indicated that IMF, which are usually present during G_1-phase in other cell types, were manifest during early S phase in the V79-8 cells that lacked a G_1 period. These studies also suggested that IMF did not exhibit species specificity and are probably present in all eukaryotic cells as is the case with mitotic factors (Johnson *et al.*, 1971; Kishimoto *et al.*, 1982; Rao and Adlakha, 1985).

C. Characterization of the IMF

Our recent studies (Adlakha *et al.*, 1983) have indicated that IMF are nondialyzable, nonhistone proteins sensitive to inactivation by proteases, but not by RNase, ethylene glycol bis(β-aminoethyl ether) N, N'-tetraacetic acid (EGTA), or ethylenediaminetetraacetic acid (EDTA). The IMF activity could not have been due to a protease because three different protease inhibitors, phenylmethyl sulfonyl fluoride (PMSF), antipain, and trypsin inhibitor were present in the extraction buffer. Unlike the mitotic factors, IMF are heat stable (at 65°C for 15 min). They are also

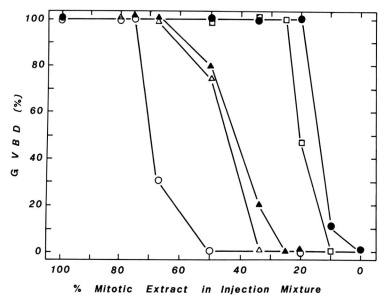

Fig. 10.   Presence of IMF in V79-8 cells that lack G_1 and G_2 phases in their cell cycle. V79-8 cells were synchronized in mitosis by selective detachment after a 4-hr Colcemid block. Early, mid-, and late S-phase cells were obtained by collecting cells at 2, 5, and 8 hr, respectively, after the reversal of the Colcemid block. Extracts from confluent cultures of V79-8 cells and from those in early, mid-, and late S phase were prepared so that the protein concentration level would equal that of the mitotic extracts. These extracts were separately mixed with extracts from mitotic HeLa cells in various proportions and tested for MPA as described in Fig. 7. ●—●, Buffer (negative control); ○—○, extracts of HeLa G_1 cells (positive control); extracts from V79-8 cells in early S (△—△), mid-S (□—□), and confluent cultures (▲—▲). Data from experiments involving extracts from late S-phase V79-8 cells were identical to those of the buffer control. (From Rao and Adlakha, 1985.)

stable over a broad pH range (6.0–10.0) but are extremely sensitive to low pH (below 5.0). The apparent molecular weight is greater than 12,000. These properties of IMF are summarized in Table VII. Some of these characteristics of IMF distinguish them clearly from the mitotic factors (Table VIII). The pH dependency of these factors is also in agreement with the results of the early cell fusion experiments of Obara *et al.* (1973), which suggested that high pH favored telophasing and low pH favored prophasing and also PCC. It appears that at low pH the mitotic factors are active, whereas the IMF are either inactive or less active. These studies support the hypothesis that the mitotic factors and the IMF play a role in the regulation of nuclear events associated with mitosis.

TABLE VII

Preliminary Characterization of Inhibitors of Mitotic Factors

Treatment of G_1 cell extract	Relative IMF activity $(\%)^a$
None	100
Papain[b]	0
RNase[c]	100
Temperature[d]	
0°C for 2 days	80
−70°C for 2 months	100
25–65°C for 15 min	100
75°C for 15 min	75
100°C for 15 min	0
pH[e]	
4.0	20
5.0–5.5	75
6.0–8.0	100
8.5–10.0	75–80

[a] A freshly prepared mid-G_1 cell extract from 8×10^7 cells/ml was mixed with mitotic extract from 4×10^7 cells/ml to give a M:G_1 protein ratio of 1:1, and the mixture was injected into *X. laevis* oocytes for MPA determination. A mixture of mitotic extract with extraction buffer served as control. If the G_1-cell extract completely inactivated the MPA of the mitotic extract at a 50% dilution, i.e., a protein ratio of 1:1, then the activity of the G_1 cell extract was considered to be 100%. An average of 10 oocytes was injected for each sample.

[b] G_1 cell extract was treated with the protease papain (500 μg/ml) in extraction buffer containing cysteine (5 mM) and 2-mercaptoethanol (0.05 mM) for 1 hr at 25°C. Antipain (80 μg/ml) was then added for 15 min at 25°C to neutralize the excess papain before being mixed with mitotic extract in various proportions and eventually being injected into oocytes for MPA determination. In these experiments, we made certain that papain activity was completely neutralized by antipain so that it could not inactivate the mitotic factors, which are known to be proteins.

[c] G_1 cell extracts were incubated with RNase (1.5 units/ml) at 25°C for 1 hr. At the end of incubation, G_1 extracts were mixed with mitotic extracts and tested for MPA.

[d] G_1 cell extracts were incubated at different temperatures for 15 min and at the end of the incubation, extracts were centrifuged at 10,000 g for 15 min at 4°C to remove any precipitate before being mixed with mitotic extract for injection into oocytes.

[e] G_1 cell extracts were dialyzed overnight against buffers of different pH with three changes each and redialyzed for 12 hr against the extraction buffer (pH 6.5) before the mixing experiments were carried out to test the activity.

TABLE VIII

Comparison of the Mitotic Factors with IMF[a]

Property	Mitotic factors	IMF
Induces GVBD in amphibian oocytes	Yes	No
Induces amphibian meiotic chromosome condensation	Yes	No
Induction of premature condensation in interphase nuclei	Yes	Induces chromosome decondensation or "telophasing"
Affinity to chromatin	Very high	None
Dependent on new protein synthesis	Yes	No
Inactivated by protease	Yes	Yes
Inactivated by RNase	No	No
Inactivated by EGTA or EDTA	Yes, at concentrations higher than 5 mM	No
Activity sensitive to Ca^{2+}	Yes	No
Activity dependent on Mg^{2+}	Probably yes	No
Heat sensitivity	Over 37°C	Over 75°C
Effect of pH	Inactivated at pH 7.5 and above	Inactivated at pH 5.0 and below but active up to pH 10.0
Activity stabilized by the presence of phosphatase inhibitors (ATP, NaF, and sodium β-glycerolphosphate)	Yes, greatly	No effect
Molecular size	Approximately 100,000	>12,000
Sedimentation value	4–5 S	NT[b]

[a] Data drawn from Sunkara et al. (1979a,b, 1982) and Adlakha et al. (1982a, b, 1983).
[b] NT, not tested.

D. Role of IMF in the Regulation of Chromosome Decondensation

The data in the preceding section indicate that the IMF are activated at telophase and are present throughout the G$_1$ period, thus coinciding with the process of chromosome decondensation, which is known to begin at telophase and continue throughout G$_1$ phase. It is tempting to speculate that the IMF, which are antagonistic to mitotic factors, may serve the reverse function of the mitotic factors, i.e., the regulation of chromosome decondensation. Our results suggest that activation of IMF at telophase may lead to a rapid inactivation of the mitotic factors and consequently may result in the decondensation of chromosomes.

This proposition is further strengthened by our observations of the activation of IMF in quiescent (G_0 phase) and mitotic cells by UV irradiation and also by some of our earlier observations:

1. As mentioned previously, in cell fusion experiments, Rao and Johnson (1970) and Rao *et al.*(1975) observed that the entry of G_2 cells into mitosis was delayed after fusion with G_1- or S-phase cells. The G_2 nucleus in the heterophasic binucleate cell would wait until the G_1- or S-phase nucleus had completed DNA synthesis; then both nuclei synchronously entered mitosis. It was speculated at that time that G_1- and S-phase components were causing decondensation of chromatin in G_2 nuclei, thus blocking them from entering mitosis. Our present results seem to provide the experimental evidence to support that assumption.

2. Rao and Smith (1981) noted that in binucleate cells formed by the fusion of HeLa cells in G_2 phase with G_0 human diploid fibroblasts (HDF), the G_0 component had no effect on the progression of the G_2 nucleus in mitosis. In contrast, fusion of G_2 cells with a UV-treated G_0 HDF delayed the entry of the G_2 nucleus into mitosis. Knowing now that UV irradiation causes chromatin decondensation and also activates IMF, we can readily explain the G_2 delay in the G_0-UV/G_2-binucleate cells by assuming the presence of IMF in UV-irradiated G_0 cells, which would neutralize the mitotic factors progressively accumulating in the G_2 cells. Hence, the G_2 component needs extra time to compensate for the loss and then to build up the factors to the critical concentration required for entry into mitosis. In light of these data, it appears that IMF may play an important role in the regulation of chromosome decondensation. However, whether the activation of IMF is the cause or the result of chromosome decondensation remains to be elucidated.

E. IMF-Like Activities in Other Cell Systems

Gerhart *et al.* (1984) recently described factors similar to the IMF that inactivate MPF at the end of meiosis in oocytes. Lohka and Masui (1983, 1984) reported that a cytoplasmic preparation from activated *Rana pipiens* eggs can induce the transformation of demembranated *X. laevis* sperm nuclei into pronuclei and then mitotic chromosomes. Lokha and

Maller (1985) developed an assay in which addition of either MPF or crude fractions of unfertilized eggs could cause nuclear envelope breakdown, chromosome condensation, and spindle formation *in vitro* with nuclei from a variety of sources. They also observed that the mere addition of calcium to an *in vitro*-assembled preparation of chromosomes and spindles will lead to chromosome decondensation and nuclear envelope formation, indicating an important role for Ca^{2+} in these events. The development of these *in vitro* assays may facilitate purification of MPF and the IMF-like factors (for details see Chapter 3, this volume).

Kirschner and colleagues, (Miake-Lye *et al.*, 1983; Karsenti *et al.*, 1984; Gerhart *et al.*, 1984; Newport and Kirschner, 1984; Miake-Lye and Kirschner, 1985) through an elegant series of *in vivo* and *in vitro* studies using partially purified preparations of MPF, developed a model to explain how the frog egg (embryonic) cell cycle is regulated. According to this model, M phase results from the presence of MPF and S phase results from the absence of MPF (or presence of anti-MPF). When MPF activity is high, the cytoplasm is in the M-phase state, and when MPF activity is low, the cytoplasm is in the S-phase state. Addition of cytostatic factor (CSF) (Meyerhof and Masui, 1979a,b) maintains MPF at high levels, presumably by protecting it from anti-MPF or stopping the appearance of anti-MPF, thus arresting the cell cycle at mitosis. The initiation of M is determined by the onset of MPF activation or synthesis (or the inactivation of the anti-MPF), and the duration of M is determined by the stability of MPF. It is not known at this time what molecular mechanism actually drives the MPF cycle. Kirschner and colleagues also observed that phosphorylation of lamins A and C, the major structural proteins of the nuclear lamina, is one of the first changes that occur in response to MPF (Newport and Kirschner, 1984; Miake-Lye and Kirschner, 1985). We are currently repeating these studies in frog embryos using the 500-fold purified mitotic factors and the IMF. Our preliminary results are very similar to the observations of M. W. Kirschner, R. Miake-Lye, and J. W. Newport (R. C. Adlakha and P. N. Rao, unpublished data).

In addition to inducing inactivation of the mitotic factors, IMF upon incubation with mitotic extracts have been shown to induce dephosphorylation of mitotic nonhistone proteins (Adlakha *et al.*, 1984a) and mitosis-specific phosphoprotein antigens (Adlakha *et al.*, 1985a; for details, see Chapter 9, this volume). They specifically decrease the activity of a mitosis-specific kinase (Halleck *et al.*, 1984c; for details, see Chapter 8, this volume). IMF appear to inactivate the mitotic factors by binding to them and forming an inert complex (Adlakha *et al.*, 1984b). It is not yet clear whether the IMF also possess a phosphatase activity that is specific to the proteins phosphorylated by the mitosis-related kinases; more than one mechanism may be involved.

IV. ROLE OF PROTEIN PHOSPHORYLATION AND DEPHOSPHORYLATION IN THE REGULATION OF NUCLEAR EVENTS ASSOCIATED WITH MITOSIS AND MEIOSIS

The covalent modification of proteins via reversible phosphorylation–dephosphorylation has been suggested as an important mechanism for the regulation of numerous intracellular events, including those of mitosis (for reviews, see Krebs and Beavo, 1979; Cohen, 1982; Fischer, 1983; Ingebritsen and Cohen, 1983; Nestler and Greengard, 1983; Maller, 1983, 1985; Adlakha *et al.*, 1985a). Histone phosphorylation has been shown to correlate very closely with the entry of cells into mitosis (for reviews, see Gurley *et al.*, 1978; and Chapter 11, this volume). Induction of PCC in interphase cells also resulted in the increased phosphorylation of their histones H1 and H3 (Krystal and Poccia, 1981; Ajiro *et al.*, 1983; Hanks *et al.*, 1983b). However, it is generally accepted that superphosphorylation of histone H1 by itself is not sufficient for chromosome condensation (Gorovsky and Keevert, 1975; Tanphaichitr *et al.*, 1976; Krystal and Poccia, 1981; Allis and Gorovsky, 1981; Hanks *et al.*, 1983b).

In recent years, evidence has been accumulating which suggests that the phosphorylation and dephosphorylation of nonhistone proteins (NHP) may also be an important control mechanism in the events associated with mitosis and meiotic maturation of oocytes. During meiotic maturation of *Xenopus* oocytes, the incorporation of ^{32}P into proteins is maximal prior to GVBD and chromosome condensation (Maller *et al.*, 1977; Wu and Gerhart, 1980; Nielsen *et al.*, 1982). Increases in the phosphorylation of nuclear matrix (Henry and Hodge, 1983), nuclear lamina (Gerace and Blobel, 1980; Jost and Johnson, 1981; Ottoviano and Gerace, 1985), intermediate filaments (Robinson *et al.*, 1981; Evans and Fink, 1982; Bravo *et al.*, 1982), nucleolar proteins (Shibayama *et al.*, 1982), and high mobility group (HMG) proteins (Bhorjee, 1981; Paulson and Taylor, 1982) have been observed in cells during G_2 and mitosis. Results of our latest studies, discussed below, suggest that phosphorylation of NHP may be causally related to the entry of cells into mitosis, while dephosphorylation of NHP is related to the exit of cells from mitosis (Sahasrabuddhe *et al.*, 1984; Adlakha *et al.*, 1985a).

A. Phosphorylation of Nonhistone Proteins Associated with Mitosis in HeLa Cells

Our studies on the mitotic factors referred to in Section II indicate that certain NHP extractable with 0.2 M NaCl from mitotic HeLa cells induce GVBD and chromosome condensation in *Xenopus* oocytes. Since the MPA of the mitotic factors was stabilized by phosphatase inhibitors, we examined whether phosphorylation of NHP plays a role in the condensation of chromosomes during mitosis. HeLa cells synchronized in late S phase were incubated with ortho[^{32}P]phosphate, and the cells were subsequently collected while in G_2, mitosis, G_1 or S. Cytoplasmic, nuclear, or chromosomal proteins were extracted, precipitated by trichloroacetic acid (TCA), and counted or separated by SDS–PAGE and labeled proteins were then detected by autoradiography. The results indicated an 8- to 10-fold increase in the phosphorylation of NHP from mid-G_2 to mitosis (Fig. 11), followed by a corresponding size decrease as the cells divided and entered G_1 (Fig. 12). The rate of NHP phosphorylation increased progressively during G_2 traverse and reached a peak in mitosis. It was extremely low during G_1 and decreased further in S phase to about 3–5% of that in mitotic cells (Table IX). Autoradiography of the NHP separated on SDS–PAGE revealed eight prominent, extensively phosphorylated

TABLE IX

Rate of Phosphorylation of Nonhistone Proteins During G_1 and S Phase in HeLa Cells[a]

Hours after reversal of N$_2$O block	Mitotic index (%)	Incorporation of ^{32}P into NHP (cmp/10^6 cells)	
		Cytoplasmic	Nuclear
0	18	309	721
1	10	220	578
2	5	183	472
3	3	155	398
10[b]	2	141	269

[a] Mitotic HeLa cells after reversal of the N$_2$O block were incubated with ^{32}P at different times for 1 hr. Cells were collected at the end of 1 hr of labeling; then, extracts were prepared and processed for incorporation of label into nonhistone proteins. (From Sahasrabuddhe *et al.*, 1984.)

[b] At 10 hr after the reversal of N$_2$O, most of the cells were in S phase.

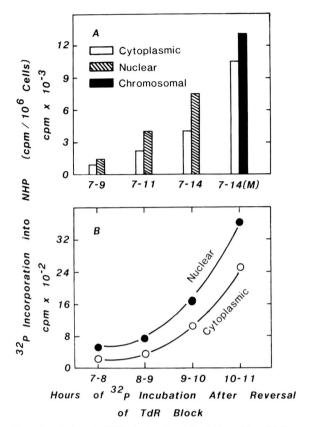

Fig. 11. Phosphorylation of NHP during G_2–mitosis transition. HeLa cells were synchronized in S phase by the excess thymidine (2.5 mM) double-block method. At the end of S phase (i.e., 7 hr after reversal), the cells were incubated with ^{32}P (A) continuously or (B) for 1 hr. Cells were collected and extracts prepared and processed for scintillation counting. 7-14(M) represents mitotic cells, whereas the other time points refer to G_2 cells. (From Sahasrabuddhe *et al.*, 1984.)

protein bands with molecular masses ranging from 27.5 to 100 kDa in mitotic cell extracts. These NHP were rapidly dephosphorylated during the M–G_1 transition (Fig. 13). Westwood *et al.* (1985) reported similar changes in protein phosphorylation during the cell cycle of CHO cells. They also observed eight proteins that show increased phosphorylation during late G_2–M transition. However, Song and Adolph (1983), working with HeLa cells, reported that NHP from isolated metaphase chromosomes were strikingly dephosphorylated in comparison to those of S-phase chromatin. The differences between these studies most likely stem

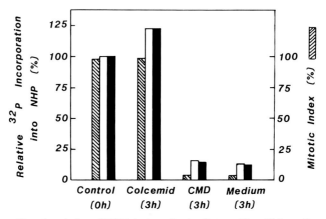

Fig. 12. Phosphorylation of NHP during mitosis–G_1 transition. HeLa cells were labeled with ^{32}P during G_2 and were synchronized in mitosis by the N_2O block method. Mitotic cells were allowed to divide in the presence or absence of cycloheximide (CMD). Three hours after reversal of the N_2O block, cells were collected and extracts prepared and processed for scintillation counting. □, Cytoplasmic extracts; ■, nuclear or chromosomal extracts; ▨, mitotic index. (From Sahasrabuddhe *et al.*, 1984.)

from the different experimental protocols used for extraction of NHP (Sahasrabuddhe *et al.*, 1984).

To determine whether the degree of phosphorylation of NHP changes when cells are held in mitosis for a prolonged period, unlabeled mitotic HeLa cells were obtained by reversal of the N_2O block. The cells were then incubated with ortho[^{32}P]phosphate while they were held in mitosis for another 8–10 hr by N_2O, Colcemid, or Taxol. Cell samples were taken at different times during incubation, and extracts were prepared and counted for incorporation of ^{32}P into NHP. Radioactivity increased as a function of time (Table X). Similar results were obtained with each of the three agents used to block mitosis. Separation of proteins labeled with ^{32}P while cells were held in mitosis by SDS–PAGE and subsequent autoradiography of the gels revealed that the same eight major proteins identified earlier were phosphorylated. The amount of label in each protein band increased with time (Fig. 14).

To determine whether this increased uptake of ^{32}P was due to superphosphorylation of NHP or turnover of the existing phosphate, ^{32}P prelabeled mitotic HeLa cells were incubated for 9 hr in medium containing Colcemid and unlabeled pyrophosphate. Cell samples were taken at 3-hr intervals and extracts were prepared and counted for incorporation of radioactivity into NHP. In this set of experiments, radioactivity incorporated into NHP decreased linearly with time, reaching the 50% mark in

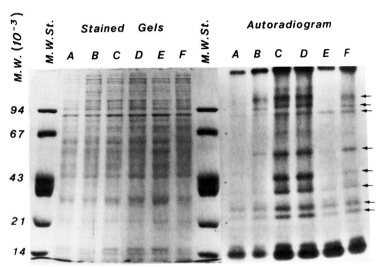

Fig. 13. Identification of the proteins phosphorylated during G_2–mitosis and dephosphorylated during mitosis–G_1 transition. Extracts from labeled G_2, mitosis, and G_1 cells were prepared as described in the captions to Figs. 11 and 12. Proteins were separated by SDS–PAGE and the gels were stained with Coomassie blue. Approximately 35 μg of protein per slot was loaded. For autoradiography, Kodak XAR-5 film was exposed to dried gels. Lanes: A, G_2-cytoplasmic; B, G_2-nuclear; C, mitosis-cytoplasmic; D, mitosis-chromosomal; E, G_1-cytoplasmic; F, G_1-nuclear, M.W. St.; molecular weight standards. Arrows indicate the eight major protein bands phosphorylated during mitosis. Note the decrease in the intensity of labeling of these eight bands in early G_1 (lanes E, F) as compared with mitotic extracts (lanes C, D). (From Sahasrabuddhe *et al.*, 1984.)

about 6 hr (data not shown). These results indicated that, rather than superphosphorylation, there is a turnover of existing phosphates associated with these proteins.

Kinetic studies on the entry of cells into mitosis in the presence of cycloheximide revealed that the commitment of a cell to enter mitosis was not only dependent on continued protein synthesis but also on immediate NHP phosphorylation (Fig. 15). X-ray-induced mitotic delay resulted in a corresponding delay in NHP phosphorylation (Fig. 16).

When HeLa cells were irreversibly arrested in G_2 by the alkylating agent cis-acid [*cis*-4-([[(2-chloroethyl)nitrosoamino]carbonyl]amino)cyclohexanecarboxylic acid] (Al-Bader *et al.*, 1978), nonhistone protein phosphorylation was completely blocked (Table XI). Cells blocked in this manner also lack certain G_2-specific proteins (Al-Bader *et al.*, 1978). These data further support our conclusion that NHP phosphorylation is closely associated with the entry of cells into mitosis. From the results

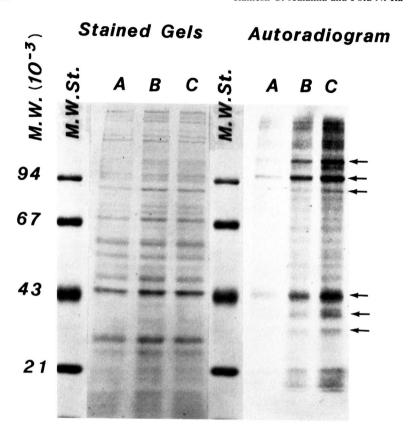

Fig. 14.   Identification of the proteins phosphorylated during prolonged mitotic arrest. HeLa cells synchronized in mitosis by N₂O block were labeled with ³²P and further incubated in the presence of Colcemid. Cells were collected at 2, 4, and 6 hr after the addition of Colcemid, and extracts were prepared as described in the captions to Figs. 11 and 12. Proteins were separated by SDS–PAGE and gels stained. Approximately 25 µg/ml proteins per slot were loaded. For autoradiography Kodak XAR-5 film was exposed to dried gels. Lanes A, B, C, chromosomal NHP from mitotic cells collected at 2, 4, and 6 hr, respectively. M.W. St., molecular weight standards. Arrows indicate the major protein bands phosphorylated during prolonged mitotic arrest. Note the increase in the intensity of labeling in these protein bands as a function of time (compare lane A with lanes B and C in autoradiogram). (From Sahasrabuddhe *et al.,* 1984.)

TABLE X

**Increase in Uptake of ³²P by Chromosome-Bound
NHP during Prolonged Mitotic Arrest**[a]

Hours held in mitosis	Incorporation of ³²P into NHP (cpm/10⁶ cells)		
	N_2O	Colcemid	Taxol
1	572	511	459
3	1,893	1,564	1,427
6	3,478	3,316	3,197
8	4,394	3,948	3,913
10	5,405	4,826	4,709

[a] HeLa cells were synchronized in mitosis by the N_2O block method; ortho[³²P]phosphate was added, cells were divided into three groups, exposed to three different mitotic-blocking agents, and incubation was continued. Cell samples taken at various times of incubation were processed for the incorporation of label into NHP. (From Sahasrabuddhe *et al.*, 1984.)

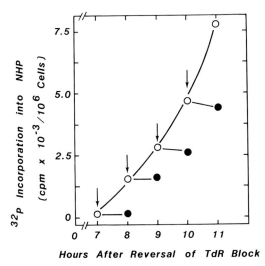

Hours After Reversal of TdR Block

Fig. 15. Effect of cycloheximide on the phosphorylation of NHP during G_2–mitosis transition. HeLa cells were synchronized in S phase as monolayer cultures. ³²P was added to all the dishes when the cells were entering G_2 (i.e., 7 hr after reversal of second TdR block) and incubation continued. Dishes were grouped into two sets. In the first set, one of the dishes was taken every hour and pulse-treated with cycloheximide for 1 hr. The other set served as a control. Cell samples taken at 1-hr intervals from the control and treated culture were processed for determination of the incorporation of ³²P into NHP. ○—○, Control; ●—●, cycloheximide-treated. Arrows indicate the time of addition of cycloheximide. (From Sahasrabuddhe *et al.*, 1984.)

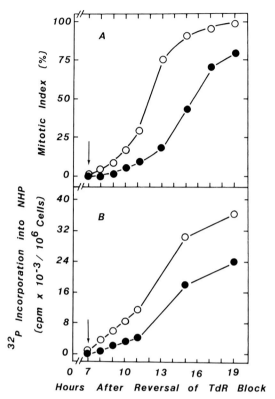

Fig. 16.   Effect of X-irradiation on mitotic accumulation and NHP phosphorylation. HeLa cells were synchronized in S phase by double TdR block method. At 7 hr after reversal of the second TdR block, a batch of cells was exposed to 500 rads of X rays (indicated by arrow), while another batch of unirradiated cells served as control. ^{32}P was added to both the control and treated cultures, and incubation was continued. Cell samples were taken at different times for mitotic index and the preparation of extracts to determine the incorporation of ^{32}P into NHP. (A) Mitotic index; (B) ^{32}P incorporation into NHP. ○—○ Control; ●—●, X irradiation. (From Sahasrabuddhe *et al.*, 1984.)

summarized in this section, it appears that the phosphorylation–dephosphorylation of NHP is a dynamic process, with the equilibrium shifting to phosphorylation during G_2–M and dephosphorylation during M–G_1 transition (for details, see Sahasrabuddhe *et al.*, 1984).

Chromosome decondensation induced by UV irradiation, which is associated with the inactivation of mitotic factor activity (Adlakha *et al.*, 1984a, and Section III, A), resulted in a significant dephosphorylation of NHP from prelabeled mitotic cells (Table XII). Furthermore, the rate of phosphorylation of NHP in UV-irradiated mitotic HeLa cells was signifi-

TABLE XI

Incorporation of ^{32}P into NHP in HeLa Cells Arrested in G_2 and Mitosis[a]

| Incubation with ^{32}P (hr) | Incorporation of ^{32}P into NHP of cells arrested in | | | | Incorporation of ^{3}H-labeled amino acids into G_2-arrested cells (cmp/10^6 cells) |
| | G_2 | | Mitosis | | |
	cpm/10^6 cells	Relative increase cpm (%)	cpm/10^6 cells	Relative increase in cpm (%)	
2	1,179	100	1,630	100	616
4	1,408	119	3,389	208	872
6	1,783	151	4,646	285	1,263
8	1,835	156	7,239	444	1,483

[a] HeLa cells were synchronized in S phase by the double thymidine block method. Immediately after reversal of the second thymidine block, cis-acid (75 μg/ml) was added. After 1 hr incubation with cis-acid cells were washed and incubated in fresh medium containing Colcemid. At 20 hr after the drug reversal, loosely attached mitotic cells were removed and replated in different dishes. Fresh medium was added to the G_2-arrested cells that remained attached to the dishes. Both the G_2-arrested and the mitotic cell populations were separately incubated with ^{32}P for a period of 8 hr. Cell samples were taken at 2-hr intervals; extracts were then prepared and processed for incorporation of label into NHP. Protein synthesis in the G_2-arrested cells was also monitored separately by incubating cells with a mixture of ^{3}H-labeled L-amino acids (1 μCi/ml). (From Sahasrabuddhe *et al.*, 1984.)

TABLE XII

Level of Dephosphorylation of NHP in UV-irradiated Mitotic HeLa Cells Prelabeled with ^{32}P during Prolonged Mitotic Arrest[a]

| Hours held in mitosis | Incorporation of ^{32}P into NHP (cpm/10^6 cells) | | | | | |
| | NHP in cytoplasmic fraction | | | NHP in chromosomal fraction | | |
	Control	UV-treated	Relative loss (%)	Control	UV-treated (%)	Relative loss
0	189,083	189,083	0	163,493	163,493	0
1	175,750	141,200	19.6	150,533	137,450	8.7
2	157,614	100,160	36.45	143,054	101,854	28.8
4	135,187	72,187	46.60	127,183	75,038	41.0

[a] From Adlakha *et al.*, (1984a).

TABLE XIII

Rate of Phosphorylation of Control and UV-Irradiated Mitotic HeLa Cells during Prolonged Mitotic Arrest[a]

	Incorporation of ^{32}P into NHP (cpm/10^6 cells)					
	NHP in cytoplasmic fraction			NHP in chromosomal fraction		
Hours held in mitosis	Control cells	UV-treated cells	Control (%)	Control cells	UV-treated cells	Control (%)
1	739	589	79.7	580	359	61.9
2	1,240	696	56.1	1,110	505	45.5
3	3,090	1,451	46.9	2,322	1,106	47.6
4	6,282	3,044	48.4	4,777	2,533	53.0

[a] From Adlakha et al., (1984a).

cantly lower than the NHP phosphorylation rate in untreated mitotic cells; the rate decreased further with time (Table XIII and Fig. 5, lane D). These results suggest that UV irradiation may induce or activate phosphatases that specifically dephosphorylate these mitotic NHP. These findings strongly suggest that phosphorylation of this subset of NHP, in addition to phosphorylation of other proteins, represents a crucial mechanism for the regulation of mitosis.

B. Role of Phosphorylation in the Activity of the Mitotic Factors

To establish more definitely the role of protein phosphorylation in the activity associated with mitotic factors, several experiments were performed using purified mitotic factors. Injection of alkaline phosphatase (APase) into oocytes completely inhibited the meiotic maturation induced either by injection of partially purified mitotic factors or stimulation by progesterone in a dose- and time-dependent manner (Table XIV). Cytological examination of the APase-injected oocytes did not reveal any toxic effects, and the oocytes that did not mature had a normal germinal vesicle. No inhibition was observed when APase was injected within 1 hr prior to maturation (Adlakha et al., 1985b). These data indicate that the in vivo action of the APase is time- and dose-dependent. Similarly, while this work was in progress, Hermann et al. (1984) reported that microinjection

TABLE XIV

Effect of Microinjection of APase into *X. laevis* **Oocyte on the MPA Induced by Stimulation with Progesterone or Injection of Partially Purified Mitotic Factors**[a]

Meiotic maturation of oocytes induced by:	Oocytes injected with APase or buffer (control) after stimulation	Time of injection of APase or buffer after stimulation (hr)	GVBD induction (%)
Stimulation with progesterone (10 µg/ml for 15 min)	None	—	100
Progesterone	Buffer	1, 2, 3, 4, 5, or 6	100
Progesterone	APase (1000 units/ml)	2, 3, 4, or 5	0
Progesterone	APase (200 units/ml)	2, 3, 4, or 5	10–20
Progesterone	APase (50 units/ml)	2, 3, 4, or 5	100
Injection of mitotic factors (100 µg protein/ml)	None	—	100
Mitotic factors	Buffer	0.25, 0.50, 0.75, 1.0, or 1.50	100
Mitotic factors	APase (1000 units/ml)	0.25 or 0.50	0
Mitotic factors	APase (1000 units/ml)	0.75, 1.0, or 1.50	90
Mitotic factors	APase (200 units/ml)	0.25 or 0.50	25
Mitotic factors	APase (200 units/ml)	0.75, 1.0, or 1.50	100

[a] Data represent an average of three separate injection experiments. GVBD was scored at 8–9 hr after stimulation with progesterone or 2–3 hr after injection of mitotic factors. A volume of 65–70 nl of buffer, APase, or mitotic factors was injected into each oocyte. An average of 10 oocytes was injected with each treatment. In each experiment, mitotic factors were freshly purified 200-fold for each experiment as described in Fig. 1 and Table III. (From Adlakha *et al.*, 1985c.)

of APase totally blocked the maturation of *Xenopus* oocytes induced by progesterone or MPF. However, we have observed that pretreatment of the mitotic factors with APase, attached to beaded agarose, has no effect on the MPA (Table XV). These results suggest that mitotic factors need not necessarily be in a phosphorylated form in order to induce maturation. Injection of acid phosphatase proved to be toxic to the oocytes under our experimental conditions. The finding that pretreatment of the mitotic factors with immobilized phosphatase had no effect on GVBD was surprising because the activity of the mitotic factors has been shown to be stabilized by the presence of phosphatase inhibitors. However, it is possible that phosphatase inhibitors could be stabilizing the activity of mitotic factors by blocking or inhibiting phosphatases present in the crude or partially purified fractions or in immature oocytes. To test this hypothe-

TABLE XV

Effect of Microinjection of Mitotic Factors Preincubated with APase Attached to Beaded Agarose on the Induction of GVBD in *X. laevis* Oocytes[a]

Oocytes injected with:	Concentration of extract (%)	GVBD induction (%)
Crude extract from mitotic	100, 50, 33, or 25	100
HeLa cells (control)	20	25
Crude mitotic cell extract	100, 50, 33, or 25	100
+ APase (200 μg/ml)	20	10
200-fold purified mitotic	100, 50, 33, or 25	100
factors (control)	20	0
200-fold purified mitotic	100, 50, 33, or 25	100
factors + APase (200 μg/m)	20	0

[a] Extracts from mitotic HeLa cells (4×10^7 cells/ml) or 200-fold purified mitotic factor fraction were incubated at 0°C or 25°C for 2 hr, then centrifuged and injected into *X. laevis* oocytes to determine the MPA.

sis, various dilutions of mitotic extracts, freshly prepared in either the presence or absence of phosphatase inhibitors, were injected either into untreated control oocytes or into oocytes preinjected with phosphatase inhibitors. Mitotic extracts, prepared in the absence of phosphatase inhibitors but injected into oocytes preinjected with phosphatase inhibitors, were found to be active at a higher dilution (20–25%) than when injected into untreated oocytes (33%). Mitotic extracts prepared in the presence of phosphatase inhibitors were also active at 20–25% dilution when injected into either untreated oocytes or into oocytes preinjected with phosphatase inhibitors (Table XVI). Furthermore, mitotic extracts that were prepared in the absence of phosphatase inhibitors showed a negligible amount of MPA after the extracts were stored for 30 days at −70°C, and injected into control oocytes. However, these extracts were active up to a 25% dilution when injected into oocytes that had been preinjected with phosphatase inhibitors. These results support our hypothesis that phosphatase inhibitors stabilize the activity of mitotic factors—not by preventing the dephosphorylation of the mitotic factors themselves—but by counteracting the action of phosphatases that act in the oocyte on other proteins that must be phosphorylated for GVBD to occur.

The suggestion that phosphorylation and dephosphorylation of NHP may play an important role in mitosis is further supported by studies in

TABLE XVI

Effect of Microinjection of Extracts from Mitotic HeLa Cells Made in the Absence or Presence or Phosphatase Inhibitors on the Induction of GVBD in *X. laevis* Oocytes Preinjected with or without Phosphatase Inhibitors

Oocytes injected with:	Extract concentration	Oocytes preinjected with phosphatase inhibitors	GVBD induction (%)	
			Fresh extract	1-month-old extract
Extract from mitotic	100, 50, 33	−	100	0
HeLa cells made in		+	100	100
the absence of	25	−	0	0
phosphatase inhibi-		+	50	25
tors				
Extract from mitotic	100, 50, 33, 25	−	100	100
HeLa cells made in		+	100	100
the presence of	20	−	30	25
phosphatase inhibi-		+	50	25
tors				

our laboratory with monoclonal antibodies specific to mitotic phospho-proteins (Davis *et al.*, 1983; Vandre *et al.*, 1984; for details see Chapter 9, this volume).

V. CONCLUSIONS

In this chapter, we have presented a comprehensive review of our studies on mitotic factors and their inhibitors (IMF) which seem to play an important role in the regulation of nuclear events associated with mitosis and meiosis, i.e., nuclear envelope breakdown and re-formation and chromosome condensation and decondensation. We have been able to achieve for the first time a 500-fold purification of the mitotic factors. A protein kinase activity has been found to be associated with the purified mitotic factors. We have also demonstrated the existence of IMF in the G_1 period that are antagonistic to the action of mitotic factors. Several other laboratories have recently shown the presence of factors similar to IMF in other cell types. Because of the complexity of the bioassay, the progress made in the purification of IMF has been limited. However, some *in vitro* assays have recently been developed to facilitate the purification and characterization of these factors.

Our studies with mitotic factors reviewed here and other studies of MPF (Maller *et al.*, 1977; Drury, 1978; Wu and Gerhart, 1980; Lewis and Potter, 1982; Maller, 1983, 1985; Gerhart *et al.*, 1984; Newport and Kirschner, 1984; Miake-Lye and Kirschner, 1985) suggest that meiotic maturation of *Xenopus* oocytes is not only correlated with but also induced by posttranslational modification of proteins, especially phosphorylation, since maturation occurs very rapidly and in the absence of protein synthesis. The involvement of protein phosphorylation in this activity is shown directly by our results reported here in which the injection of alkaline phosphatase completely inhibited the meiotic maturation of oocytes induced either by injection of purified mitotic factors or by stimulation with progesterone. Recently, Halleck *et al.*, (1984a,b) identified certain protein kinases in mitotic cell extracts that are absent from G_1 cell extracts by using an *in situ* kinase assay in nondenaturing polyacrylamide gels. Some of these kinases were observed even when no exogenous substrate was added, suggesting either the autophosphorylation of kinases themselves or the co-migration of the endogenous substrates.

In addition to inducing GVBD and chromosome condensation, both the MPF and the mitotic factors stimulate autoamplification of the factors, as judged by serial injections (Adlakha *et al.*, 1982a; Schorderet-Slatkine, 1972; Drury and Schorderet-Slatkine, 1975). This autoamplification also occurs in the absence of protein synthesis (Adlakha *et al.*, 1982a; Wu and Gerhart, 1980; Drury and Schorderet-Slatkine, 1975). These results suggest that immature oocytes may contain MPF in an inactive form that can be activated when MPF or a small amount of the mitotic factors are injected into oocytes. Autophosphorylation is a possible mechanism by which the autoamplification could occur.

In light of these observations, we speculate that a protein kinase cascade mechanism is involved in the breakdown of the germinal vesicle or nuclear envelope and the condensation of chromosomes at meiosis or mitosis. As yet there is no conclusive evidence that the protein factors from mitotic cells or mature oocytes that induce meiotic maturation have kinase activity or that they activate a kinase preexisting in an inactive form, because most of the studies have been done using crude or partially purified preparations. The results presented here suggest the latter possibility. Another related possibility that has been raised in this study is whether more than one protein is required for MPA. It is possible that there is a common activation mechanism, such as phosphorylation, but that different proteins are required to be phosphorylated to promote the different events during mitosis. For example, one protein may induce chromosome condensation, another the breakdown of the nuclear lamina, and yet another the organization of the spindle. Moreover, the relationship between the factors needed to induce PCC and those needed to induce oocyte maturation has yet to be elucidated. For example, if a kinase cascade mechanism is involved in oocyte maturation, only an initial trigger may be needed because the oocyte, as a specialized cell, may contain large quantities of the other effectors needed, albeit in an inactive form. A similar cascade mechanism in mitotic cells might require not only the synthesis or activation of the "trigger" but also that of the other proteins involved in the cascade mechanism.

ACKNOWLEDGMENTS

We thank William F. Lindsey, Hélène Bigo, and Monica Rasmus for their excellent technical assistance during the course of these studies, and Josephine Neicheril for her superb secretarial assistance in the preparation of this manuscript. We are also grateful to David A. Wright, Chintaman G. Sahasrabuddhe, Frances M. Davis, Walter N. Hittelman, and Nagindra Prashad for useful discussions during the course of these investigations. This work was supported in part by research grants CA-27544 and CA-34783 from the National Cancer Institute, Department of Health and Human Services.

REFERENCES

Adlakha, R. C., Sahasrabuddhe, C. G., Wright, D. A., Lindsey, W. F., and Rao, P. N. (1982a). Localization of mitotic factors on metaphase chromosomes. *J. Cell Sci.* **54,** 193–206.

Adlakha, R. C., Sahasrabuddhe, C. G., Wright, D. A., Lindsey, W. F., Smith, M. L., and Rao, P. N. (1982b). Chromosome bound mitotic factors: Release by endonucleases. *Nucleic Acids Res.* **10,** 4107–4117.

Adlakha, R. C., Sahasrabuddhe, C. G., Wright, D. A., and Rao, P. N. (1983). Evidence for the presence of inhibitors of mitotic factors during G_1 period in mammalian cells. *J. Cell Biol.* **97,** 1707–1713.

Adlakha, R. C., Sahasrabuddhe, C. G., Wright, D. A., Bigo, H., and Rao, P. N. (1984a). Role of nonhistone protein phosphorylation in the regulation of mitosis in mammalian cells. *In* "Growth, Cancer, and the Cell Cycle", (P. Skehan and S. J. Friedman, eds.), pp. 59–69. Humana Press, Clifton, New Jersey.

Adlakha, R. C., Wang, Y. C., Wright, D. A., Sahasrabuddhe, C. G., Bigo, H., and Rao, P. N. (1984b). Inactivation of mitotic factors by ultraviolet irradiation of HeLa cells in mitosis. *J. Cell Sci.* **65,** 279–295.

Adlakha, R. C., Davis, F. M., and Rao, P. N. (1985a). Role of phosphorylation of nonhistone proteins in the regulation of mitosis. *In* "Control of Animal Cell Proliferation" (A. L. Boynton and H. L. Leffert, eds.), Vol. 1, pp. 485–513. Academic Press, New York.

Adlakha, R. C., Prashad, N., Wright, D. A., Rasmus, M., and Rao, P. N. (1985b). Protein kinase activity associated with purified mitotic factors from HeLa cells. *J. Cell Biol.* **101,** 142a.

Adlakha, R. C., Wright, D. A., Sahasrabuddhe, C. G., Davis, F. M., Prashad, N., Bigo, H., and Rao, P. N. (1985c). Partial purification and characterization of mitotic factors from HeLa cells. *Exp. Cell Res.* **160,** 471–482.

Ajiro, K., Nishimoto, T., and Takhashi, T. (1983). Histone H1 and H3 phosphorylation during premature chromosome condensation in a temperature-sensitive mutant (tsBN₂) of baby hamster kidney cells. *J. Biol. Chem.* **258,** 4534–4538.

Al-Bader, A. A., Orengo, A., and Rao, P. N. (1978). G₂-phase-specific proteins of HeLa cells. *Proc. Natl. Acad. Sci. U.S.A.* **75,** 6064–6068.

Allis, C. D., and Gorovsky, M. A. (1981). Histone Phosphorylation in macro- and micronuclei of *Tetrahymena thermophila. Biochemistry* **20,** 3828–3833.

Baserga, R. (1976). "Multiplication and Division in Mammalian Cells." Dekker, New York.

Baserga, R. (1981). The cell cycle. *N. Engl. J. Med.* **304,** 453–459.

Bhorjee, J. S. (1981). Differential phosphorylation of nuclear nonhistone high mobility group proteins HMG 14 and HMG 17 during the cell cycle. *Proc. Natl. Acad. Sci. U.S.A.* **78,** 6944–6948.

Bravo, R., Fey, S. J., MoseLarsen, P., and Celis, J. E. (1982). Modification of vimentin polypeptides during mitosis. *Cold Spring Harbor Symp. Quant. Biol.* **46,** 379–385.

Cohen, P. (1982). The role of protein phosphorylation in neural and hormonal control of cellular activity. *Nature (London)* **296,** 613–620.

Davis, F. M., and Rao, P. N. (1982). Antibodies specific for mitotic human chromosomes. *Exp. Cell Res.* **137,** 381–386.

Davis, F. M., Tsao, T. Y., Fowler, S. K., and Rao, P. N. (1983). Monoclonal antibodies to mitotic cells. *Proc. Natl. Acad. Sci. U.S.A.* **80,** 2926–2930.

Drury, K. C. (1978). Method for the preparation of active maturation promotional factor (MPF) from *in vitro* maturing oocytes of *Xenopus laevis. Differentiation* **10,** 181–186.

Drury, K. C., and Schorderet-Slatkine, S. (1975). Effects of cycloheximide on the "autocatalytic" nature of the maturation promoting factor (MPF) in oocytes of *Xenopus laevis. Cell* (Cambridge, Mass.) **4,** 269–274.

Evans, R. M., and Fink, L. M. (1982). An alteration in the phosphorylation of vimentin-type intermediate filaments is associated with mitosis in cultured mammalian cells. *Cell* (Cambridge, Mass.) **29,** 43–52.

Fischer, E. H. (1983). Cellular regulation by protein phosphorylation. *Bull. Inst. Pasteur (Paris)* **81,** 7–31.

Gerace, L., and Blobel, G. (1980). The nuclear envelope lamina is reversibly depolymerized during mitosis. *Cell* (Cambridge, Mass) **19,** 277–287.

Gerhart, J. C., Wu, M., and Kirschner, M. W. (1984). Cell cycle dynamics of an M-phase-specific cytoplasmic factor in *Xenopus laevis* oocytes and eggs. *J. Cell Biol.* **98,** 1247–1255.

Gorovsky, M. A., and Keevert, J. B. (1975). Absence of histone F₁ in a mitotically dividing genetically inactive nucleus. *Proc. Natl. Acad. Sci. U.S.A.* **72,** 2672–2676.

Gurley, L. R., Tobey, R. A., Walters, R. A., Hildebrand, C. E., Hohmann, P. D., D'Anna, J. A., Barham, S. S., and Deaven, L. L. (1978). Histone phosphorylation and chromatin structure in synchronized mammalian cells. *In* "Cell Cycle Regulation" (J. R. Jeter, I. L. Cameron, G. M. Padilla, and A. M. Zimmerman, eds.), pp. 37–60. Academic Press, New York.

Halleck, M. S., Lumley-Sapanski, K., Reed, J. A., Iyer, A. P., Mastro, A. M., and Schlegel, R. A. (1984a). Characterization of protein kinases in mitotic and meiotic cell extracts. *FEBS Lett.* **167,** 193–198.

Halleck, M. S., Reed, J. A., Lumley-Sapanski, K., and Schlegel, R. A. (1984b). Injected mitotic extracts induce condensation of interphase chromatin. *Exp. Cell Res.* **153,** 561–569.

Halleck, M. S., Reed, J. A., Lumley-Sapanski, K., and Schlegel, R. A. (1984c). Involvement of protein kinases in mitotic-specific events. *In* "Growth, Cancer, and the Cell Cycle" (P. Skehan and S. J. Friedman, eds.), pp. 143–150. Humana Press, Clifton, New Jersey.

Hanks, S. K., and Rao, P. N. (1980). Initiation of DNA synthesis in the prematurely condensed chromosomes of G_1 cells. *J. Cell Biol.* **87,** 285–291.

Hanks, S. K., Gollin, S. M., Rao, P. N., Wray, W., and Hittelman, W. N. (1983a). Cell cycle-specific changes in the ultrastructural organization of prematurely condensed chromosomes. *Chromosoma* **88,** 333–342.

Hanks, S. K., Rodriguez, L. V., and Rao, P. N. (1983b). Relationship between histone phosphorylation and premature chromosome condensation. *Exp. Cell Res.* **148,** 293–302.

Henry, S. M., and Hodge, L. D. (1983). Nuclear matrix: A cell cycle-dependent site of increased intranuclear protein phosphorylation. *Eur. J. Biochem.* **133,** 23–29.

Hermann, J., Mulner, O., Belle, R., Marot, J., Tso, J., and Ozon, R. (1984). *In vivo* effects of microinjected alkaline phosphatase and its low molecular weight substrates on the first meiotic cell division in *Xenopus laevis* oocytes. *Proc. Natl. Acad. Sci. U.S.A.* **81,** 5150–5154.

Hildebrand, C. E., and Tobey, R. A. (1975). Cell cycle-specific changes in chromatin organization. *Biochem. Biophys. Res. Commun.* **63,** 134–139.

Hittelman, W. N., and Pollard, M. (1982). A comparison of the DNA damage and chromosome repair kinetics after γ-irradiation. *Radiat. Res.* **92,** 497–509.

Hittelman, W. N., and Pollard, M. (1984). Visualization of chromatin events associated with repair of ultraviolet light-induced damage by premature chromosome condensation. *Carcinogenesis (London)* **5,** 1277–1285.

Hittelman, W. N., and Rao, P. N. (1976). Premature chromosome condensation: Conformational changes of chromatin associated with phytohemagglutinin-stimulation of peripheral lymphocytes. *Exp. Cell Res.* **100,** 219–222.

Hittelman, W. N., and Rao, P. N. (1978). Mapping G_1 phase by the structural morphology of the prematurely condensed chromosomes. *J. Cell. Physiol.* **95,** 333–341.

Ingebritsen, J. S., and Cohen, P. (1983). Protein phosphatases: Properties and role in cellular regulation. *Science* **221,** 331–338.

Johnson, R. T., and Rao, P. N. (1970). Mammalian cell fusion: Induction of premature chromosome condensation in interphase nuclei. *Nature (London)* **226,** 717–722.

Johnson, R. T., Rao, P. N., and Hughes, S. D. (1971). Mammalian cell fusion. III. A HeLa cell inducer of premature chromosome condensation active in cells from a variety of species. *J. Cell. Physiol.* **77,** 151–158.

Jost, E., and Johnson, R. T. (1981). Nuclear lamina assembly synthesis and disaggregation during the cell cycle in synchronized HeLa cells. *J. Cell Sci.* **47,** 25–53.

Karsenti, E., Newport, J., Hubble, R., and Kirschner, M. W. (1984). The interconversion of metaphase and interphase microtubule arrays, as studied by the injection of centrosomes and nuclei into *Xenopus* eggs. *J. Cell Biol.* **98,** 1730–1745.

Kishimoto, T., Kuriyama, R., Kondo, H., and Kanatani, H. (1982). Generality of the action of various maturation-promoting factors. *Exp. Cell Res.* **137,** 121–126.

Krebs, E. G., and Beavo, J. A. (1979). Phosphorylation-dephosphorylation of enzymes. *Annu. Rev. Biochem.* **48**, 923–959.

Krystal, G. W., and Poccia, D. L. (1981). Phosphorylation of cleavage stage histone H1 in mitotic and prematurely condensed chromosomes. *Exp. Cell Res.* **134**, 41–48.

Kuehl, L. (1974). Nuclear protein synthesis. *In* "The Cell Nucleus" (H. Busch, ed.), Vol. 3, pp. 345–375. Academic Press, New York.

Kuehl, L., Barton, D. J., and Dixon, G. H. (1980). Binding of the high mobility group protein, H6, to trout testis chromatin. *J. Biol. Chem.* **255**, 10671–10675.

Lewis, R. V., and Potter, R. L. (1982). Anion exchange chromatography of proteins. *Int. Symp. HPLC Proteins, Pept. Polynucleotides, 2nd, 1982,* p. 13a (abstr.).

Lohka, M. J., and Maller, J. L. (1985). Induction of nuclear envelope breakdown, chromosome condensation, and spindle formation in cell-free extracts. *J. Cell Biol.* **101**, 518–523.

Lohka, M. J., and Masui, Y. (1983). Formation *in vitro* of sperm pronuclei and mitotic chromosomes by amphibian ooplasmic components. *Science* **220**, 719–721.

Lohka, M. J., and Masui, Y. (1984). Roles of cytosol and cytoplasmic particles in nuclear envelope assembly and sperm pronuclear formation in cell-free preparations from amphibian eggs. *J. Cell Biol.* **98**, 1222–1230.

Maller, J. L. (1983). Interaction of steriods with the cyclic nucleotide system in amphibian oocytes. *Adv. Cyclic Nucleotide Res.* **15**, 295–336.

Maller, J. L. (1985). Regulation of amphibian oocyte maturation. *Cell Differ.* **16**, 211–221.

Maller, J. L., and Krebs, E. G. (1980). Regulation of oocyte maturation. *Curr. Top. Cell. Regul.* **16**, 271–311.

Maller, J. L., Wu, M., and Gerhart, J. C. (1977). Changes in protein phosphorylation accompanying maturation of *Xenopus laevis* oocytes. *Dev. Biol.* **58**, 295–312.

Masui, Y., and Clarke, H. (1979). Oocyte maturation. *Int. Rev. Cytol* **57**, 185–223.

Masui, Y., and Markert, C. L. (1971). Cytoplasmic control of nuclear behavior during meiotic maturation of frog oocytes. *J. Exp. Zool.* **177**, 129–146.

Mazia, D. (1974). The cell cycle. *Sci. Am.* **1**, 55–64.

Meyerhof, P. G., and Masui, Y. (1979a). Chromosome condensation activity in *Rana pipiens* eggs matured *in vivo* and in blastomeres arrested by cytostatic factor (CSF). *Exp. Cell Res.* **123**, 345–353.

Meyerhof, P. G., and Masui, Y. (1979b). Properties of a cytostatic factor from *Xenopus* eggs. *Dev. Biol.* **72**, 182–187.

Miake-Lye, R., and Kirschner, M. W. (1985). Induction of early mitotic events in a cell-free system. *Cell* (Cambridge Mass.) **41**, 165–175.

Miake-Lye, R., Newport, J. W., and Kirschner, M. W. (1983). Maturation promoting factor induces nuclear envelope breakdown in cycloheximide-arrested embryos of *Xenopus laevis*. *J. Cell Biol.* **97**, 81–91.

Moser, G. C., Muller, H., and Robbins, E. (1975). Differential nuclear fluorescence during the cell cycle. *Exp. Cell Res.* **91**, 73–78.

Moser, G. C., Fallon, R. J., and Meiss, H. K. (1981). Fluorometric measurements and chromatin condensation patterns of nuclei from 3T3 cells throughout G_1. *J. Cell. Physiol.* **106**, 293–301.

Mullinger, A. M., and Johnson, R. T. (1985). Manipulating chromosome structure and metaphase status with ultraviolet light and repair synthesis inhibitors. *J. Cell Sci.* **73**, 159–186.

Nelkin, B., Nicholas, C., and Vogelstein, B. (1980). Protein factor(s) from mitotic CHO cells induce meiotic maturation in *Xenopus laevis* oocytes. *FEBS Lett.* **109**, 233–238.

Nestler, E. J., and Greengard, P. (1983). Protein phosphorylation in the brain. *Nature (London)* **305**, 583–588.

Newport, J. W., and Kirschner, M. W. (1984). Regulation of the cell cycle during early *Xenopus* development. *Cell* (Cambridge, Mass.) **37**, 731–745.

Nicolini, C., Ajiro, K., Borun, T. W., and Baserga, R. (1975). Chromatin changes during the cell cycle of HeLa cells. *J. Biol. Chem.* **250**, 3381–3385.

Nielsen, P. J., Thomas, G., and Maller, J. L. (1982). Increased phosphorylation of ribosomal protein S6 during meiotic maturation of *Xenopus*. *Proc. Natl. Acad. Sci. U.S.A.* **79**, 2937–2941.

Obara, Y., Yoshida, H., Chai, L. S., Weinfeld, H., and Sandberg, A. A. (1973). Contrast between the environmental pH dependencies of prophasing and nuclear membrane formation in interphase-metaphase cells. *J. Cell Biol.* **58**, 608–627.

Obara, Y., Chai, L. S., Weinfeld, H., and Sandberg, A. A. (1974a). Prophasing of interphase nuclei and induction of nuclear envelopes around metaphase chromosomes in HeLa and Chinese Hamster homo- and heterokaryons. *J. Cell Biol.* **62**, 104–113.

Obara, Y., Chai, L. S., Weinfeld, H., and Sandberg, A. A. (1974b). Synchronization of events in fused interphase-metaphase binucleate cells: Progression of the telophase like nucleus. *J. Natl. Cancer Inst. (U.S.)* **53**, 247–249.

Ottaviano, Y., and Gerace, L. (1985). Phosphorylation of the nuclear lamins during interphase and mitosis. *J. Biol. Chem.* **260**, 624–632.

Pardee, A. B., Dubrow, R., Hamlin, J. L., and Kletzien, R. F. (1978). Animal cell cycle. *Annu. Rev. Biochem.* **47**, 715–750.

Paulson, J. R., and Taylor, S. S. (1982). Phosphorylation of histones 1 and 3 and nonhistone high mobility group 14 by an endogenous kinase in HeLa metaphase chromosomes. *J. Biol. Chem.* **257**, 6064–6072.

Pederson, T. (1972). Chromatin structure and the cell cycle. *Proc. Natl. Acad. Sci. U.S.A.* **69**, 2224–2228.

Pederson, T., and Robbins, E. (1972). Chromatin structure and the cell division cycle. Actinomycin binding in synchronized HeLa cells. *J. Cell Biol.* **55**, 322–327.

Prescott, D. M. (1976). "Reproduction of Eukaryotic Cells." Academic Press, New York.

Rao, P. N. (1968). Mitotic synchrony in mammalian cells treated with nitrous oxide at high pressure. *Science* **160**, 774–776.

Rao, P. N. (1982). The phenomenon of premature chromosome condensation. *In* "Premature Chromosome Condensation: Application in Basic, Clinical and Mutation Research" (P. N. Rao, R. T. Johnson, and K. Sperling, eds.), pp. 1–41. Academic Press, New York.

Rao, P. N., and Adlakha, R. C. (1985). Chromosome condensation and decondensation factors in the life cycle of eukaryotic cells. *In* "Mediators in Cell Growth and Differentiation" (R. J. Ford and A. L. Maizel, eds.), pp. 45–69. Raven Press, New York.

Rao, P. N., and Hanks, S. K. (1980). Chromatin structure during the prereplicative phases in the life cycle of mammalian cells. *Cell Biophys.* **2**, 327–337.

Rao, P. N., and Johnson, R. T. (1970). Mammalian cell fusion: Studies on the regulation of DNA synthesis and mitosis. *Nature (London)* **225**, 159–164.

Rao, P. N., and Johnson, R. T. (1971). Mammalian cell fusion. IV. Regulation of chromosome formation in interphase nuclei by various chemical compounds. *J. Cell. Physiol.* **78**, 217–224.

Rao, P. N., and Johnson, R. T. (1974). Regulation of cell cycle in hybrid cells. *Cold Spring Harbor Conf. Cell Proliferation* **1**, 785–800.

Rao, P. N., and Smith, M. L. (1981). Differential response of cycling and noncycling cells to inducers of DNA synthesis and mitosis. *J. Cell Biol.* **88**, 649–653.

Rao, P. N., Hittelman, W. N. and Wilson, B. A. (1975). Mammalian cell fusion. VI. Regulation of mitosis in binucleate HeLa cells. *Exp. Cell Res.* **90**, 40–46.

Rao, P. N., Wilson, B. A., and Puck, T. T. (1977). Premature chromosome condensation and cell cycle analysis. *J. Cell. Physiol.* **91**, 131–142.

Robinson, S. I., Nelkin, B., Kaufmann, S., and Vogelstein, B. (1981). Increased phosphorylation rate of intermediate filaments during mitotic arrest. *Exp. Cell Res.* **133**, 445–449.

Sahasrabuddhe, C. G., Adlakha, R. C., and Rao, P. N. (1984). Phosphorylation of nonhistone proteins associated with mitosis in HeLa cells. *Exp. Cell Res.* **133**, 439–450.

Schor, S. L., Johnson, R. T., and Waldren, C. A. (1975). Changes in the organization of chromosomes during the cell cycle: Response to ultraviolet light. *J. Cell Sci.* **17**, 539–565.

Schorderet-Slatkine, S. (1972). Action of progesterone and related steroids on oocyte maturation in *Xenopus laevis*. An *in vitro* study. *Cell Differ.* **1**, 179–189.

Shibayama, T., Nakaya, K., Matsumoto, S., and Nakamura, Y. (1982). Cell-cycle-dependent change in the phosphorylation of the nucleolar proteins of *Physarum polycephalum in vivo*. *FEBS Lett.* **139**, 214–216.

Smith, L. D., and Ecker, R. E. (1971). The interaction of steroids with *Rana pipien* oocytes in the induction of maturation. *Dev. Biol.* **25**, 233–247.

Song, M. K. H., and Adolph, K. W. (1983). Phosphorylation of nonhistone proteins during the HeLa cell cycle. Relationship to DNA synthesis and mitotic chromosome condensation. *J. Biol. Chem.* **258**, 3309–3318.

Sunkara, P. S., Wright, D. A., and Rao, P. N. (1979a). Mitotic factors from mammalian cells induce germinal vesicle breakdown and chromosome condensation in amphibian oocytes. *Proc. Natl. Acad. Sci. U.S.A.* **76**, 2799–2802.

Sunkara, P. S., Wright, D. A., and Rao, P. N. (1979b). Mitotic factors from mammalian cells: A preliminary characterization. *J. Supramol. Struct.* **11**, 189–195.

Sunkara, P. S., Wright, D. A., Adlakha, R. C., Sahasrabuddhe, C. G., and Rao, P. N. (1982). Characterization of chromosome condensation factors of mammalian cells. *In* "Premature Chromosome Condensation: Application in Basic, Clinical and Mutation Research" (P. N. Rao, R. T. Johnson, and K. Sperling, eds.), pp. 233–251. Academic Press, New York.

Tanphaichitr, N., Moore, K. C., Granner, D. K., and Chalkley, R. (1976). Relationship between chromosome condensation and metaphase lysine-rich histone phosphorylation *J. Cell Biol.* **69**, 43–50.

Vandre, D. D., Davis, F. M., Rao, P. N., and Borisy, G. G. (1984). Phosphoproteins are components of microtubule organizing centers. *Proc. Natl. Acad. Sci. U.S.A.* **81**, 4439–4443.

Wasserman, W. J., and Masui, Y. (1976). A cytoplasmic factor promoting oocyte maturation: Its extraction and preliminary characterization. *Science* **191**, 1266–1268.

Wasserman, W. J., and Smith, L. D. (1978). The cyclic behavior of a cytoplasmic factor controlling nuclear membrane breakdown. *J. Cell Biol.* **78**, R15–R22.

Weintraub, H., Buscaglia, M., Ferrez, M., Weiller, S., Boulet, A., Fabre, F., and Baulieu, E. E. (1982). "MPF" activity in *Saccharomyces cerevisiae*. *C. R. Seances Acad. Sci., Ser. 3* **295**, 787–790.

Westwood, J. T., Church, R. B., and Wagenaar, E. B. (1985). Changes in protein phosphorylation during the cell cycle of Chinese hamster ovary cells. *J. Biol. Chem.* **260**, 10308–10313.

Wu, M., and Gerhart, J. C. (1980). Partial purification and characterization of the maturation-promoting factor from eggs of *Xenopus laevis*. *Dev. Biol.* **79**, 465–477.

8

Mitosis-Specific Cytoplasmic Protein Kinases

MARGARET S. HALLECK,[1]
KATHERINE LUMLEY-SAPANSKI,
AND ROBERT A. SCHLEGEL

Department of Molecular and Cell Biology
The Pennsylvania State University
University Park, Pennsylvania 16802

I. INTRODUCTION

Transit of cells through mitosis is closely associated with the reversible phosphorylation of numerous nuclear components (Davis *et al.*, 1983; Sahasrabuddhe *et al.*, 1984). Although the identity and cellular function of the proteins modified in this manner are for the most part unknown, phosphorylation of several defined molecular species has been correlated with changes in the organization of structures in which these species are integral components. For example, as chromosomes condense at mitosis, histones H1 and H3 become highly phosphorylated (Gurley *et al.*, 1978). Similarly, phosphorylation of nuclear lamins accompanies dissolution of the nuclear membrane (Gerace and Blobel, 1980). Such findings have prompted the search for protein kinases present (or activated) only at (or just prior to) mitosis.

Cells in mitosis contain cytoplasmically transmissable factors which, when introduced into interphase cells by fusion (Rao and Johnson, 1970) or injection (Halleck *et al.*, 1984c), are able to induce nuclear events simulating those seen at mitosis. Circumstantial evidence suggests that

[1] Present address: Department of Pharmacology, The University of Texas Medical School, P.O. Box 20708, Houston, Texas 77225.

these "mitotic factors" may be protein kinases (Nelkin *et al.*, 1980; Adlakha *et al.*, 1982a; Laskey, 1983). Should this be the case, mitotic events such as chromosome condensation and nuclear membrane breakdown might be directly induced by these factors through phosphorylation of molecular species such as histones and nuclear lamins. Or, alternatively, mitotic factors might initiate a phosphorylation cascade leading to activation of the protein kinase(s) ultimately responsible for the phosphorylation of the relevant structural components. Whichever the case, it might be expected that mitosis-specific protein kinases should be identifiable in cellular cytoplasmic extracts. We present data herein which demonstrate that in fact there are protein kinase activities that are expressed predominantly at mitosis and, further, that they have properties in common with mitotic factors.

II. IDENTIFICATION OF PROTEIN KINASES
IN CELLULAR EXTRACTS

In order to examine quickly and easily the variety of protein kinases present in different populations of cells, extracts were first electrophoresed in nondenaturing polyacrylamide gels, then phosphorylation reactions were performed directly in the gels in the presence of [^{32}P]ATP as phosphate donor. Figure 1 presents an autoradiogram of a gel in which cytoplasmic extracts prepared from mammalian culture cells were examined in this manner. Several discrete bands of kinase activity were readily visible, whether phosphorylations were performed in the presence or absence of exogenous substrate.

When extracts prepared from a population of cells arrested in mitosis were compared with those prepared from a population in interphase (G$_1$, S, and G$_2$), very few, if any, differences were seen in the pattern of protein bands revealed by staining with Coomassie blue (Halleck *et al.*, 1984a). This result was not unexpected since mitotic and interphase protein patterns were essentially identical when examined by SDS–polyacrylamide gel electrophoresis (Sahasrabuddhe *et al.*, 1984). However, when kinase activities were displayed in the nondenaturing system, notable differences were seen between mitotic and interphase extracts. When exogenous histone substrate was used, the kinase activities of interphase extracts were found primarily in the C region of gels (Fig. 1a). Bands were also seen in this region in mitotic extracts, but in different positions. Even more striking, and in further contrast to interphase extracts, mitotic extracts contained considerable activity in the A and B regions as well. The

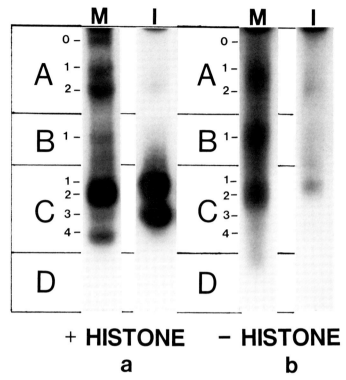

Fig. 1. Protein kinase activities present in mitotic vs. interphase cell extracts. Mitotic (M) and interphase (I) extracts were prepared from HeLa cells, as described in the legend to Fig. 2, applied at 50 μg/lane and separated on a polyacrylamide gel (3% stacking; 5% resolving) as described by Laemmli (1970) except that SDS was excluded from all solutions and glycerol was added to the gel (6% stacking; 10% resolving) which was cast to Gel Bond PAG backing (FMC Corp., Rockland, ME). Following electrophoresis (50 V stacking; 200 V resolving), the gel was rinsed for 1 hr in a total of 1 liter of 50 mM Tris-HCl, pH 7.0, at 4°C with three changes of buffer, cut in half and each piece placed in 25 ml of phosphorylation buffer (150 mM NaCl, 4 mM magnesium acetate, 30 mM Tris acetate, 3 mM EGTA, 10 mM NaF, 0.1 mM sodium phosphate, pH 7.0) at 37°C. The phosphorylation solution in (a) contained 4 mg/ml histone (IIA, Sigma). Following incubation at 37°C with gentle agitation for 30 min, phosphorylation was initiated by adding [γ-^{32}P]ATP (New England Nuclear) diluted to 2–3 Ci/mmole, and at a final concentration of 2 μCi/ml. Incubation was continued for 45 min after which the reaction was terminated by removal of the phosphorylation solution and addition of 200 ml of an ice-cold solution of 10% TCA, 1% H$_3$PO$_4$ (TCA-fix). The gel was refrigerated overnight, then rinsed with gentle agitation for 8–12 hr with two to three changes of cold TCA-fix which was finally replaced with 25% methanol, 7.5% acetic acid, containing 0.01% Coomassie Blue (R250). After overnight staining followed by destaining, the gel was dried in an oven at 60°C. For finest detail (shown here) autoradiography was performed using Kodak X-Omat XRP film and a Fast Detail Screen (Picker International) with exposure times of from 1 to 3 days.

net result of this comparison was that five bands of kinase activity (A0, A1, B1, C2, and C4) identifiable in mitotic extracts were not detectable in interphase extracts.

In the absence of exogenous substrate, fewer and generally less intense bands were seen. A1, B1, and C2 were relatively intense in mitotic extracts. None of these bands was seen in interphase extracts, just as was the case when histone was added; only A2 and C1 were faintly visible.

The protein kinases found after mitotic cells are lysed and mitotic chromosomes removed by centrifugation are not strictly speaking cytoplasmic kinases, since at mitosis nuclear and cytoplasmic contents intermix upon dissolution of the nuclear membrane. It is therefore possible that the kinase activities identified as mitosis-specific may be present in the nucleus during interphase and liberated at mitosis. This possibility has not yet been examined.

III. CELL CYCLE SPECIFICITY OF KINASE ACTIVITIES

Having in hand a detection system that could resolve multiple protein kinase species, populations of cells synchronized in various phases of the cell cycle were examined for the presence of each of the bands of activity identified in mitotic and interphase extracts. G_1-phase extracts were prepared from cells arrested in mitosis, then released into G_1 for 4–5 hr. As seen in Fig. 2, the activity profile of these extracts was quite dissimilar from that of the extracts of the mitotic cells from which the G_1 cells were derived. Only band A1 of mitotic extracts was visible in G_1 extracts. Bands C1 and C3, characteristic of interphase extracts, were intense; only band A2 of interphase extracts was not found in G_1. Band A2 was, however, seen faintly in extracts prepared from cells synchronized in S phase by a double thymidine block. In fact, with the presence of this band, S-phase extracts took on the appearance of interphase extracts. This similarity held true with extracts prepared from a population of G_2 cells, derived from the S-phase cells. However, the G_2 pattern also began to resemble that of mitotic extracts with bands B1 and sometimes C4 becoming just barely visible.

Because populations of cells that have been similarly synchronized may differ to some degree in their exact position within the cell cycle, minor variations in kinase patterns between different preparations have sometimes been observed. The data in Fig. 2 are, therefore, representative of only the particular samples presented. In order to convey a more accurate overall picture of the cell cycle distribution of kinase activities, Fig. 3

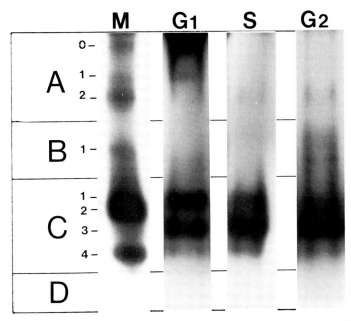

Fig. 2. Protein kinase activities as a function of cell cycle position. Human D98/AH2 cells (a HeLa derivative) were grown as monolayer cultures in media specified by Sunkara *et al.* (1979a), and mitotic cells were collected by shake-off from cultures treated with Colcemid, as previously described (Halleck *et al.,* 1984a). Interphase cells (examined in Fig. 1) were harvested by trypsinization of untreated, exponentially growing cultures from which mitotic cells had been removed by shake-off. To obtain cells synchronized in G_1, mitotic cells, collected by shake-off at 10-min intervals for 3 hr from cultures growing exponentially (without Colcemid) were pooled on ice (Tobey *et al.,* 1967), plated, and incubated at 37°C for 4–5 hr before harvesting by trypsinization. Cells synchronized in S phase or G_2 phase were prepared using a double thymidine block as described by Sunkara *et al.* (1979a) except that 0.02 μg/ml of Colcemid was added immediately after the reversal of the second block to prevent the earliest cells from exiting mitosis. Any cells in mitosis were then removed by shake-off at the time of harvest. Cytoplasmic extracts were prepared as previously described (Halleck *et al.,* 1984a) except that 1 mM PMSF and 1–2 μg/ml of soybean trypsin inhibitor were added to buffer A. Electrophoresis and autoradiography were performed as described in the legend to Fig. 1.

presents a composite of data compiled from the examination of several preparations of each type of synchronized population. Flow cytometry was used to verify the cell cycle distribution of each population and representative DNA profiles are included. Mitotic extracts always contained the six bands seen in Figs. 1 and 2, but sometimes also contained a faint band at C3, characteristic of neighboring G_1- and G_2-phase cells. G_1-phase cells, in addition to reproducibly containing bands C1 and C3,

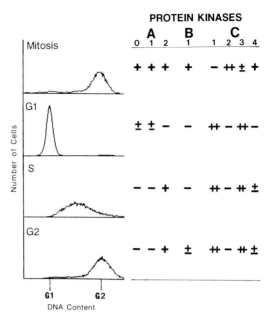

Fig. 3. Cell cycle distribution of protein kinase activities. Samples were fixed and stained by the procedure of Crissman and Steinkamp (1982) and analyzed by an Epics V flow cytometer. $(-)$, Absent, never seen; $(\pm)$, sometimes seen, faint when present; $(+)$, always present, seen as faint or moderately heavy band; $(++)$, always present, seen as heavy band.

sometimes contained faint A0 and A1 bands characteristic of mitotic cells. S-phase cells did not have either of these "mitosis-specific" activities, but sometimes had a faint C4 band. This band was also seen in G_2-phase extracts as was, sometimes, another "mitosis-specific" band, B1. Thus it appears that as cells progress through the cell cycle, kinase activities appear and disappear in an orderly fashion. However, the profile of one cell cycle compartment may overlap to a limited extent into an adjacent compartment, tending to make differences between cell cycle phases less distinct.

It is quite possible that some of the kinase species identified may share common catalytic subunits and be converted from one form to another, either within or between cell cycle phases, by binding of other, possibly regulatory, subunits. If so, it might be possible to produce conditions *in vitro* that would favor the stability of some species over others. Initial attempts using this approach have met with some success as described in Sections V,B and C.

IV. COMPARISON OF KINASE ACTIVITIES WITH PROTEIN KINASES IMPLICATED IN MITOTIC EVENTS

Since some degree of cell cycle specificity has been ascribed to several previously identified protein kinases (reviewed by Laskey, 1983; Matthews and Huebner, 1984), it is incumbent to ask whether the kinase activities identified on nondenaturing gels resemble any of them. A kinase that first appears in late G_2, persists throughout mitosis, and prefers histone H1 as substrate has been variously described by Lake and Salzman (1972), Hardie *et al.* (1976), and Langan (1976, 1978). Another candidate whose activity is related to events specific to mitosis is a nuclear membrane-associated protein kinase that specifically phosphorylates a lamin-sized polypeptide (Lam and Kasper, 1979). Because both of these kinases have been partially characterized, the properties of kinases identified in nondenaturing gels were compared with them.

A. cAMP Dependence

Both the histone kinase termed "mitosis-associated" by Matthews and Huebner (1984) and the putative lamin-phosphorylating kinase are cAMP-independent enzymes. Accordingly, the cAMP dependence of the kinases identified on gels was examined.

A universal characteristic of cAMP-dependent protein kinases is their dissociation into regulatory and catalytic subunits in the presence of cAMP, thus activating the catalytic subunit. Phosphorylation reactions were therefore performed in the presence or absence of 10 μM cAMP and the extent of phosphorylation compared. As seen in Fig. 4a, the profiles produced by mitotic extracts under either set of conditions were indistinguishable. Interphase extracts, however, exhibited somewhat enhanced activity in the B region in the presence of cAMP. As a further test, extracts were incubated with cAMP *prior* to electrophoresis. Because the pI of the dissociated catalytic subunit of cAMP-dependent protein kinases precludes its entry into the gel (McClung and Kletzien, 1981a), the disappearance of a band following preincubation with cAMP identifies it as being cAMP-dependent. As can be seen in Fig. 4b, when mitotic extracts were preincubated with 10 μM cAMP (conditions effective in preventing detection of cAMP-dependent protein kinase prepared from beef heart; Sigma), no alteration in the pattern of activities was readily apparent. (The absence of band C4 in this extract is probably due to the presence of small amounts of Ca^{2+}, which was only later discovered to inhibit this

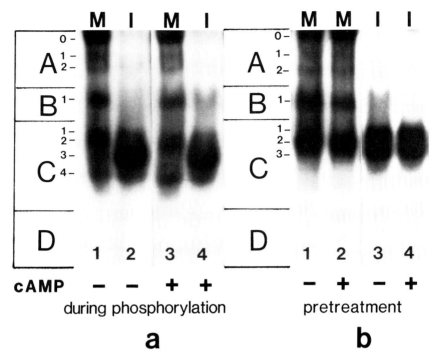

Fig. 4.   cAMP dependence of protein kinase activities. Mitotic and interphase extracts were analyzed as described in the legend to Fig. 1. In (a), cAMP was omitted from lanes 1 and 2 and present at 10 μM in lanes 3 and 4 during phosphorylation with histone as substrate. In (b), samples in lanes 1 and 3 were incubated without, and samples in lanes 2 and 4 with, 10 μM cAMP for 5 min at 37°C prior to electrophoresis. (From Halleck *et al.*, 1984b.)

activity, as discussed below.) In contrast, the faint activity sometimes seen in the B region in samples from interphase cells was completely eliminated by the cAMP pretreatment. Accordingly, in all of the experiments described in this chapter (those already described as well as those that follow) extracts were routinely preincubated with cAMP in order to distinguish (by elimination) this minor cAMP-dependent activity from the major cAMP-independent kinase activity in this region.

It is perhaps a bit curious that since numerous cAMP-dependent protein kinases exist in the cytoplasm, only one minor cAMP-dependent kinase is detected in this assay. One possible explanation for this result might be that many of these kinases are not truly soluble but rather are associated with structures that are removed by high-speed centrifugation. Another explanation for the apparent absence of these species might be that the assay is internally biased against identification of cAMP-dependent protein kinases as such, since histone can cause dissociation of the holoenzyme into its constituent subunits (Miyamoto *et al.*, 1971; Johnson *et al.*, 1975; Corbin *et al.*, 1975). Were cAMP-dependent protein kinases present in extracts and were they affected in this manner, addition of cAMP during phosphorylation would not identify them as cAMP-dependent through enhancement of their activity since dissociation of subunits and activation would be accomplished by histone whether cAMP were present or not. Preincubation with cAMP eliminates the possibility of such a bias, however, since exposure to cAMP and the possibility of dissociation precede exposure to histone, so that bands of cAMP-dependent activity would still be detected by their disappearance upon pretreatment. A final possibility, that some ingredient in buffers or some mechanical manipulation might cause dissociation of catalytic and regulatory subunits during isolation or electrophoresis, cannot be excluded, although other investigators have detected cAMP-dependent protein kinases in similar gel systems with extracts prepared using protocols similar to those used here (McClung and Kletzien, 1981a,b; Iyer *et al.*, 1984a,b).

The virtual absence of kinase band C4 in panel 4b in Figure 4 should also be noted. As will be discussed below, this kinase is particularly sensitive to the presence of calcium ions. Panels 4a and 4b represent separate experiments carried out before the effects this divalent cation were studied in our system; thus Ca^{2+} levels may have been higher in 4b, inhibiting C4 in the entire gel.

B. Histone Substrate Specificity

Evidence has accumulated implicating the mitosis-associated histone kinase in the superphosphorylation of histone H1 observed in mitotic chromosomes (Langan, 1976; Hohmann *et al.*, 1975, 1976). Not only is H1 a preferred substrate of the enzyme, but the sites on H1 phosphorylated *in vitro* by this kinase are the same as those which at mitosis become highly phosphorylated *in vivo* (Langan *et al.*, 1980). It was therefore asked whether H1 might be a preferred substrate of any of the kinases identified in nondenaturing gels.

Phosphorylation reactions were performed in the presence of exogenously added histone substrates that differed in their content of H1. SDS and acid/urea gel electrophoretic profiles of the four preparations of histone used are shown in Fig. 5. Unfractionated (whole) calf thymus histone purchased from a commercial vendor contained very little H1. In contrast, whole histone prepared by us from calf thymus contained substantial amounts of H1. Fractions containing predominantly H1 or core his-

a
b

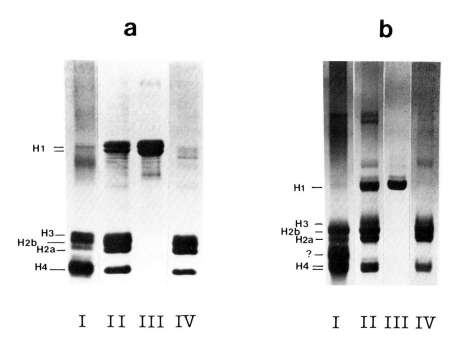

I II III IV   I   II III IV

Fig. 5.   Analysis of histones used as substrates. Histone preparations were analyzed on SDS gels (a) as described by Laemmli (1970) and on acetic acid–2.5 *M* urea gels (b) as described by Panyim and Chalkley (1969). Substrate I, commercially prepared unfractionated calf thymus whole histone (Sigma, IIAS); substrate II, unfractionated histone, prepared by 0.4 *N* H$_2$SO$_4$ extraction of calf thymus; substrate III, histone H1, prepared by 5% HClO$_4$ extraction of calf thymus; substrate IV, core histones, prepared by 0.4 *N* H$_2$SO$_4$ extraction of calf thymus following extraction of H1 by 5% HClO$_4$ (Johns and Butler, 1962; Bonner *et al.,* 1968).

tone were prepared from calf thymus by their differential solubilities in perchloric acid.

Each of these samples was examined for its ability to be phosphorylated by the kinases of mitotic extracts; the results are presented in Fig. 6. A comparison of the profiles produced in the presence of the two whole histone preparations indicates that all kinases were active in the presence of the sample containing H1, but that none of the kinases in the A and B regions phosphorylated to any significant extent the sample deficient in H1. When H1 was the only substrate available, it might be expected that those kinases which phosphorylated only whole histone containing H1 (bands in the A and B regions) would be active, plus perhaps the kinase at band C2 which was more active when H1 was present in whole histone. Such was seen to be the case. In contrast, when core histones were used

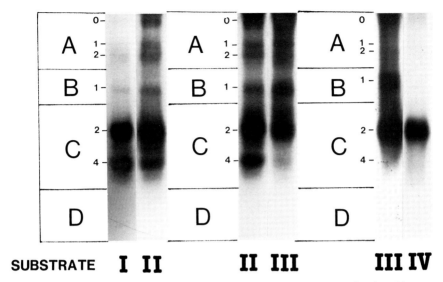

SUBSTRATE   **I II**   **II III**   **III IV**

Fig. 6.   Activity of protein kinases in mitotic extracts in the presence of various histone substrates. Three mitotic extracts were used, each pair of lanes containing the same extract in the same amount. Assays were performed in the presence of the histone preparations indicated, which were analyzed in Fig. 5 and described in its legend.

as substrate, band C2 was usually the only kinase active, though a band at C4 was occasionally seen. It is not clear why C4 was not always active in the presence of core histones, since it phosphorylated whole histone deficient in H1. One possible explanation is that H1 is required for the activation of this kinase but does not itself serve as a substrate, as has been reported for at least one other kinase (Abdel-Ghany et al., 1984). Or, alternatively, core histones preparations may carry with them variable amounts of calcium which, as described below, is inhibitory to the activity of C4. The results of this analysis therefore lead to the conclusions that kinases in the A and B regions show a strong preference for H1 as substrate, C2 can phosphorylate both H1 and core histones, and C4 phosphorylates best a mixture of H1 and core histones.

These conclusions should be verifiable by identifying which histones are phosphorylated by each kinase. To perform this identification, bands were cut from a gel similar to the one in Fig. 6 where whole histone (II) was the substrate, using the autoradiogram as a template. Each band was then applied to a separate well of an SDS-polyacrylamide gel. The results of such a second-dimension electrophoretic analysis are shown in Fig. 7. Examination of the SDS gel stained for protein indicates that some his-

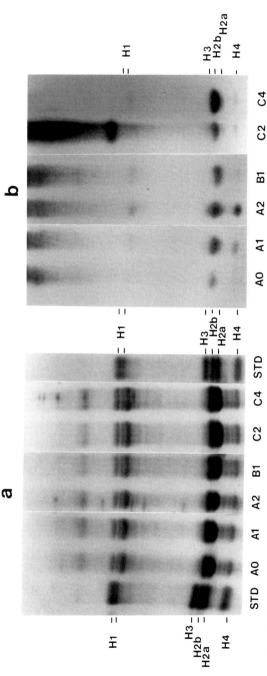

Fig. 7. Second-dimension analysis of histones phosphorylated by mitotic protein kinases. Mitotic extract was electrophoresed into a nondenaturing gel in the first dimension and assayed for protein kinase activity in the presence of unfractionated calf thymus histone (substrate II, see Figs. 5 and 6). Using the autoradiogram as a template, bands were cut from the fixed and dried gel, rehydrated, and electrophoresed in a lane of a second-dimension SDS-polyacrylamide gel (5% stacking; 15% resolving), as described by Poccia and Green (1986). STD = calf thymus histone reference (second dimension only): (a) Coomassie staining for protein; (b) autoradiogram.

tone species were proportionately underrepresented when compared to the original histone sample used as substrate. Histones H2a and H2b were recovered and present in the amounts expected. However, the recoveries of H1, H3, and H4 were not proportionately as large. This result was consistent, though somewhat more or less pronounced, from gel to gel. Whether this effect was due to differential precipitation of histones into the first-dimension gel following the kinase assay, or to differential solubilization by SDS when running into the second dimension is unknown. However, proportionately more H1 was recovered in the second dimension when 20% rather than 10% TCA was used to precipitate the substrate after the first-dimension separation.

It was still possible to identify the histone species phosphorylated, however, since each species was distinctly visible, and differential recovery was consistent across the gel. As concluded from examination of the first dimension only, kinases in the A and B regions do to some extent phosphorylate H1 (or an H1 degradation product; see below). However, the conclusion that these kinases prefer H1 appears to be erroneous, at least when they are presented with a mixture of H1 and core histones. Such a result might be explained if H1 were able to activate these kinases, as noted earlier. The conclusion from the first-dimension analysis that C2 phosphorylates both H1 and core histones appears to be quite correct. Also correct is the conclusion that C4 is able to phosphorylate a mixture of H1 and core histones.

Beyond these general observations, several other specific points need to be addressed. First, the origin of the phosphorylated higher molecular weight material seen in the second-dimension gel is not known. It could have been produced by crosslinking of histones during fixation, phosphorylation of endogenous substrates (examined in detail below), or phosphorylation of contaminants within the histone preparation, although these contaminants are not detectable by Coomassie staining. Second, the species in the H1 region of the gel that was phosphorylated by each of the kinases runs slightly faster than the two main H1 subtypes. The fact that this band is much more prominent by Coomassie staining after incubation than before (all other lanes vs. STD, Fig. 7), and that recovery of this species was variable from gel to gel, argues strongly that it is a degradation product of a major H1 subtype, generated during the phosphorylation incubation period. If that is the case, the degradation product must serve as a better substrate than the parent from which it was derived since the two major H1 subtypes are not phosphorylated. However, the species in this region that is the most heavily phosphorylated of any, and phosphorylated only by C2, runs more slowly than the two main subtypes. Although there is disagreement as to whether the electrophoretic mobility of H1 in SDS gels is altered by phosphorylation (Hnilica, 1975; Hardison and Chalkey, 1978; Paulson, 1980; Yonemoto *et al.*, 1985), a retardation of very highly phosphorylated H1 would explain the result observed. Finally, there is the consistent observation that histone H2b is the most preferred of the core histones as substrate. Even though the proportionately poor recovery of H3 and H4 might explain their relatively light phosphorylation, differential phosphorylation of these two histones is still observed since band A2 is relatively effective at phosphorylating both H3 and H4, and band A1 at phosphorylating H4, while the C-region kinases are totally ineffective in modifying H3. On the other hand, H2a was recovered in good quantity but did not appear to be phosphorylated to any appreciable extent by any of the kinases. Taylor (1982) has noted that a kinase associated with isolated mitotic chromosomes phosphorylates H1 and H3 of the endogenous chromatin, but when free histones are added to the same system, H2b is the preferred substrate. The results presented here may reflect that bias when free histone vs. chromatin is the substrate.

If one conclusion emerges from both the one- and two-dimensional analyses, it is that of all the kinases in mitotic extracts, only C2 prefers H1 as substrate when presented with a mixture of all histones, although each kinase is able to phosphorylate what is probably an H1 degradation product along with several core histones. Whether C2 is related to or identical with the mitosis-associated kinase described by others cannot be answered by experiments such as these. It would clearly be of interest to determine whether a purified preparation of that enzyme would result in a band of kinase activity in nondenaturing gels which migrated to a position corresponding to one of the kinase bands seen in mitotic extracts.

V. COMPARISON OF KINASE ACTIVITIES WITH MITOTIC FACTORS

Fusions between mitotic and interphase cells demonstrate that cells in mitosis contain cytoplasmically transmissible factors which are able to induce both breakdown of the nuclear membrane and condensation of chromatin in interphase cells (Rao and Johnson, 1970). These distinctive cytological events seen in somatic cells are also observed in germ cells when cytoplasm from a mature oocyte is injected into an immature recipient (Masui and Markert, 1971; Smith and Ecker, 1971; Schorderet-Slatkin, 1972; Reynhout and Smith, 1974), thus clearly implicating cytoplasmic involvement in the induction of the nuclear processes implicit to meiotic maturation as well. Although these maturation-promoting factors (MPF) only persist in the cytoplasm of oocytes until maturation is complete, they reappear after eggs are fertilized, since cytoplasm from cleaving embryos is also able to promote maturation following injection into immature oocytes. Since this ability fluctuates relative to the mitotic cell cycle (Wasserman and Smith, 1978), the cytoplasmic factors responsible for inducing nuclear events during mitosis in somatic cells appear to be similar to those generated in oocytes undergoing meiotic maturation. Indeed, when cytoplasmic extracts prepared from culture cells arrested in mitosis were injected into immature amphibian oocytes, germinal vesicle (nucleus) breakdown (GVBD) and chromosome condensation were observed (Sunkara *et al.*, 1979a; Nelkin *et al.*, 1980).

Based on these findings, the ability of injected somatic extracts to induce oocyte maturation was adopted as the criterion by which cells from a variety of different organisms were tested for the presence of mitotic factors (Adlakha *et al.*, 1982a, 1983, 1984b; Kishimoto *et al.*, 1982; Weintraub *et al.*, 1982). It might be questioned, however, whether it is valid to use a germ cell recipient in assays for activity of somatic cell extracts without evidence that somatic extracts contain factors capable of promoting mitotic events in other somatic cells. Attempts to demonstrate such a capacity, by injecting extracts into somatic culture cells using microneedles, have not been successful (Adlakha *et al.*, 1982a). However, exposure of nuclei in a syncytial embryo to somatic extracts has provided proof of the competency of mitotic extracts toward somatic nuclei (Halleck *et al.*, 1984c).

In these experiments, advantage was taken of the rapid divisions which take place in early cleavage stage embryos, resulting in a large number of somatic nuclei within a volume equivalent to that of the unfertilized egg; thus these embryos are readily amenable to injection. Ordinarily, each nucleus is within its own cytoplasmic compartment within the embryo. If, however, just after fertilization, embryos are placed in cytochalasin B, which prevents cleavage but allows nuclear division (Hammer *et al.*, 1971), by 4–5 hr of development 75–100 somatic nuclei are accumulated in a common cytoplasm, accessible by a single injection. Such syncytial embryos are still developing, however, undergoing many rapid mitotic divisions. But if they are treated with cycloheximide, inhibition of protein synthesis arrests nuclei in interphase after they have completed one more round of DNA synthesis (Miake-Lye *et al.*, 1983; Forbes *et al.*, 1983).

When cycloheximide-arrested syncytial embryos were fixed and stained with a DNA-specific fluorescent Hoescht dye, many nuclei exhibiting a stippled, networklike pattern of fluorescence were observed in squash preparations (Fig. 8a). When embryos were injected with extracts

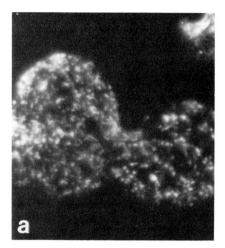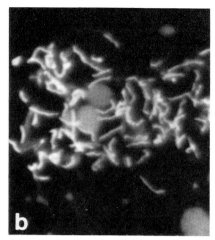

Fig. 8.   Induction of chromosome condensation in somatic nuclei by somatic extracts. Cycloheximide-arrested syncytial frog embryos uninjected (a) or injected with mitotic extract (b) were stained for DNA, squashed, and viewed by fluorescence microscopy (Halleck *et al.*, 1984c).

prepared from mitotic cells, a dramatic change in the pattern of fluorescent staining of some nuclei was observed. The diffuse, patchy pattern of interphase nuclei was replaced by long, thin, fibrous strands resembling partially condensed chromosomes (Fig. 8b). The extent of condensation increased with time such that by 2 hr after injection some spreads took on the appearance of mitotic chromosomes. When embryos were injected with interphase extracts at protein concentrations comparable to those of mitotic extracts, the transformation of chromatin from the interphase to the mitotic configuration was never observed.

It is of interest to note that the condensed chromosomes observed in embryos persisted for at least 2–3 hr after injection of mitotic extract. In contrast to these results, condensed chromosomes produced in response to purified oocyte MPF were reported to decondense within 1 hr after exposure both in injected embryos (Newport and Kirschner, 1984) and in somatic nuclei suspended in embryonic extracts (Miake-Lye and Kirschner, 1985). This reversibility of chromosome condensation was suggested to result from inactivation of MPF, since activity of this factor (and thus chromosome condensation) could be preserved by the addition of cytostatic factor (CSF), which is normally present in the cytoplasm of unfertilized frog eggs and is thought to maintain metaphase arrest through stabilization of MPF (Newport and Kirschner, 1984). Purified preparations of

MPF apparently do not contain CSF activity. If an analogy can be made to extracts from somatic cells, the persistence of chromosomes seen in embryos injected with mitotic extracts signals the presence of both MPF- and CSF-like activities in these preparations.

The foregoing studies suggest, but by no means prove, that the factors in mitotic extracts responsible for inducing meiotic maturation are the same as those that induce chromosome condensation in syncytial nuclei; it is still quite possible that separate molecular species are active in the two different assays. At present, much greater use has been made of the older oocyte system in providing a preliminary characterization of the factors responsible for the activity of mitotic extracts in this assay (Sunkara *et al.*, 1979b; Nelkin *et al.*, 1980; Adlakha *et al.*, 1982a, 1983). The following sections present information and studies that address the question of whether any of the kinases identified in mitotic extracts share the properties of mitotic factors so characterized.

A. General Characteristics

Some information has been provided using the oocyte assay as to the location of mitotic factors within cells. Although extracts are generally collected from mitotic cells lysed in hypotonic buffer, a second sequential extraction of the insoluble fraction using high salt (0.2 M NaCl) yields additional material that demonstrates an even greater specific activity in the oocyte assay. These results suggest that mitotic factors are located in both the cytoplasmic and chromosomal compartments of the mitotic cells but may be preferentially associated with chromosomes (Adlakha *et al.*, 1982a,b). When the material recovered from chromosomes by the high salt was examined on nondenaturing gels, the kinase activity profile generated was equivalent to that of the cytoplasmic extracts presented in Figs. 1 and 2 (data not shown). In some cases, the specific activities of the chromosomal fractions appear to be greater. Therefore, mitosis-specific protein kinases also appear to be associated with chromosomes, as do mitotic factors.

The conditions under which the biological activity of mitotic extracts is most stable also demonstrate similarities between mitotic factors and protein kinases. Nelkin *et al.* (1980) found that inclusion of sodium fluoride, β-glycerophosphate, and MgATP in buffers used in the preparation and storage of mitotic extracts was necessary to preserve factor activity. Since these substances are, respectively, a phosphatase inhibitor, an alternate phosphatase substrate, and a protein kinase stabilizer, routinely included in buffers used for preparation and assay of protein kinases, modification of mitotic factors by phosphorylation appears to be important to their activation.

When mitotic cells are harvested after being held in Colcemid for various lengths of time and then are fused to interphase cells to test for factor efficacy, exposure to the drug for 8–12 hr is optimal for maximal accumulation of the ability to induce nuclear membrane breakdown and chromosome condensation in interphase cells (K. L. S., unpublished results). Concomitantly, a particular mitosis-specific protein kinase, band C4, is enhanced in extracts prepared from cells blocked for a similar length of time (data not shown).

Ammonium sulfate fractionation has been used to enrich mitotic extracts threefold with respect to their ability to induce meiotic maturation in oocytes (Nelkin *et al.*, 1980). When extracts fractionated in the identical manner were examined on nondenaturing gels, the activity of all kinases was enriched as judged by a considerable increase in band intensities relative to unfractionated extracts (Halleck *et al.*, 1984a). Similar extracts were also found to be biologically active when assayed in syncytial embryos (Halleck *et al.*, 1984b).

This comparison of mitotic factors and mitosis-specific kinases with respect to their location, stability, accumulation, and partial purification involved only rather general characteristics. In addition, however, several much more specific properties of mitotic factors have been identified. In the remainder of this section, the relationship between mitosis-specific protein kinases and mitotic factors with respect to these properties is addressed.

B. Ca^{2+} Sensitivity

The biological activity of factors promoting chromosome condensation and nuclear membrane breakdown appears to be highly sensitive to calcium ions. When extracts are prepared from mitotic cells either in the absence of ethylene glycol bis(β-aminoethyl ether)N,N^1-tetraacetic acid (EGTA) to chelate Ca^{2+}, or in the presence of 1 mM Ca^{2+}, mitotic factor activity is attenuated, as assayed by injection into oocytes (Sunkara *et al.*, 1979b; Adlakha *et al.*, 1982a). A similar conclusion has been reached using an *in vitro* system. When demembranated sperm are added to a cell-free preparation of unfertilized frog eggs, a nuclear membrane forms around the newly dispersed chromatin and a sperm pronucleus is produced. If, and only if, precautions have been taken to exclude Ca^{2+}, breakdown of the pronuclear membrane then commences, chromosomes condense (Lohka and Masui, 1984), and spindles are formed (Lohka and Maller, 1985). Addition of Ca^{2+} to the system at this point reverses this process and nuclei re-form. These results suggest that Ca^{2+} has an inhibitory effect on factor activity, although it has also been proposed that Ca^{2+} may actually operate by inactivating CSF (Newport and Kirschner, 1984). Whichever mode of inactivation is correct, it was of interest to examine the sensitivity of protein kinases to Ca^{2+}.

There are several ways in which Ca^{2+} could affect protein kinases. It could serve as either an inhibitor or an activator of the enzyme, acting directly on the enzyme molecule or indirectly, through the action of a Ca^{2+}-binding regulator. In order to help distinguish among these possible modes of action, Ca^{2+} was added to extracts either prior to electrophoresis or later, during the phosphorylation reaction, since predictions concerning the effects of Ca^{2+} differ depending on the point of addition. As seen in Fig. 9, when 5 mM Ca^{2+} was included during phosphorylation, the

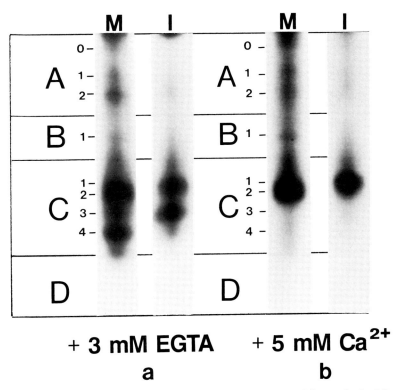

Fig. 9. Ca^{2+} sensitivity of kinase activities. Extracts prepared from mitotic (M) and interphase (I) cells in the presence of 3 mM EGTA were electrophoresed into nondenaturing gels and assayed for protein kinase activity in the presence of histone substrate and either no added Ca^{2+} (a) or 5 mM added Ca^{2+} (b).

most evident change was the disappearance of bands in the C region in both mitotic and interphase extracts. Band C3 was eliminated from interphase extracts in the presence of Ca^{2+}, as was the small amount of this activity sometimes seen in mitotic extracts. In mitotic extracts, band C4 was completely eliminated by Ca^{2+}, and the activity of band B1 appeared to be slightly enhanced. Similar results were obtained with 10, 3, and 1 mM Ca^{2+}. When Ca^{2+} was added to extracts before electrophoresis, the same alterations were seen as when it was added later, during the phosphorylation reaction (data not shown).

At a minimum these results indicate that Ca^{2+} is inhibitory to some protein kinases in the C region. Because the point at which Ca^{2+} was added did not affect the outcome, the possibility that Ca^{2+} promotes the binding of a dissociable inhibitory protein can be eliminated, since in the absence of Ca^{2+} during the electrophoresis the dissociated inhibitor and protein kinase would most likely not migrate similarly and, therefore, the inhibitor would not be available to bind when Ca^{2+} was added during phosphorylation. This conclusion leaves two other possibilities: Ca^{2+} may act directly on the kinases (or on permanently bound regulatory subunits), or it may promote the dissociation of an activator protein (consistent with insensitivity to time of addition). These two possible modes of action cannot at present be distinguished. However, if inhibition is reversible, removal of Ca^{2+} added prior to electrophoresis, using EGTA prior to phosphorylation, would restore a band of activity only if Ca^{2+} acted directly, since a dissociable activator would presumably be separated from the kinase during electrophoresis.

On the other hand, Ca^{2+} appears to be slightly stimulatory to the kinase activity of band B1. Again, the possibility that Ca^{2+} promotes the binding of a dissociable activator can be eliminated, since its addition to the phosphorylation buffer enhanced activity equally as well as its addition to the extract before electrophoresis. But, whether the enhancement of B1 by Ca^{2+} is due to direct activation by the cation or indirectly through dissociation of an inhibitor cannot be determined by these experiments.

Finally, since Ca^{2+} appears to inhibit the activity of C4 while stimulating B1 in mitotic extracts, it might be asked whether the two kinases share the same catalytic subunit and are simply interconverted from one form to the other through interaction with the cation. This cannot be the case, however, because kinases would have to be able to translocate physically within the gel to explain the effects of Ca^{2+} when added following electrophoresis.

These results are consistent with the suggestion that a Ca^{2+}-regulated protein kinase is responsible for the sensitivity to Ca^{2+} observed when mitotic extracts are tested for biological activity. In addition, they emphasize the need to carefully regulate Ca^{2+} levels during preparation of extracts as well as during assay in order to prevent selective inhibition or enhancement of kinase activities. Apparently, the inclusion of 3 mM EGTA in the isolation buffers used to prepare extracts is normally sufficient to prevent inhibition of the Ca^{2+}-sensitive kinases. However, occasionally an extract is found to have little or no activity in the C4 region, if mitotic, (see Fig. 4b, lanes 1 and 2) or C3 region, if interphase, suggesting that inhibitory levels of Ca^{2+} may have been inadvertently introduced during isolation. Although addition of EGTA to levels higher than 3 mM during isolation would seem to be a solution to the problem, this action might also serve to reduce the activity of band B1 and perhaps other kinases whose activity is dependent on the presence of minimal amounts of Ca^{2+}.

C. Inhibition by Interphase Extracts

Evidence has been marshaled in support of the hypothesis that as cells exit mitosis antagonistic substances inactivate mitotic factors. When extracts from somatic cells synchronized in G_1, S, or G_2 phase were mixed with mitotic extract, incubated, and then injected into immature oocytes, the biological activity of the mitotic extract was found to have been neutralized, specifically and exclusively by extracts from G_1 cells (Adlakha *et al.*, 1983). It was suggested that so-called inhibitors of mitotic factors (IMF) might serve the reverse function of mitotic factors, since they appeared to be present (or most active) during the period when chromatin was actively decondensing. If, interphase extracts do, in fact, contain such attenuating factors, the question can be asked whether addition of a G_1-phase extract to a mitotic extract results in the reduction or elimination of any mitosis-specific protein kinase band(s).

Mitotic and G_1-phase extracts were adjusted to equal protein concentrations and then a series of mixtures was prepared. Decreasing amounts of interphase extract were added to a constant amount of mitotic extract such that the former represented 50, 33, and 20% of the total protein content of the mixture. Following incubation at room temperature, the entire contents of each mixture was loaded onto and electrophoresed through a nondenaturing gel, then assayed for protein kinase activity. As can be seen in Fig. 10, when equal amounts of G_1 and mitotic extract were mixed, the result was what might be expected if all of the protein kinase activities of the two extracts were simply added, although the B and C regions were somewhat obscured due to the combined activities (lane 2). But, when one-half or one-quarter as much G_1 extract was mixed, band

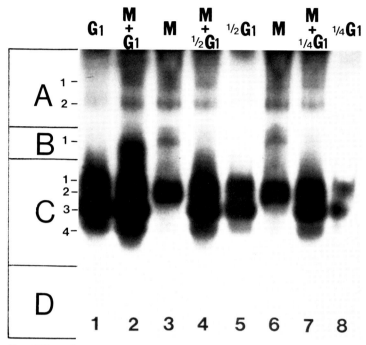

Fig. 10.   Effect of mixing mitotic and interphase extracts. Mitotic and G_1-phase extracts were each adjusted to 6.8 mg/ml of protein, mixed, incubated at room temperature for 15 min, separated on nondenaturing gels, and assayed for protein kinase activity in the presence of histone substrate. Lanes 3 and 6 contained 54.4 μg of mitotic extract; lanes 1, 5, and 8 contained 54.4, 27.2, and 13.6 μg of G_1-phase extract, respectively; lanes 2, 4, and 7 contained 54.4 μg of mitotic extract plus 54.4, 27.2, or 13.6 μg of G_1-phase extract, respectively, i.e., a mixture of the extracts in adjacent lanes. (From Halleck *et al.*, 1984b.)

B1, usually prominent in mitotic extracts, could no longer be seen (lanes 4 and 7). Even though the same amount of mitotic extract was applied to the gel as in lanes 3 and 6 (mitotic extract alone), mixing of just one-half or even one-quarter as much G_1 extract completely eliminated this mitosis-specific protein kinase. In striking contrast, all of the other bands in the mixtures appeared to be of the intensities expected by a simple summation of the activities of the two different extracts (shown in adjacent lanes on either side of each mixture). Since these experiments were completed before the effects of calcium ions on kinase activity had been systematically investigated and controlled for, the absence of kinase activity at band C4 in the mitotic extract used in this and other similar experiments was not recognized as abnormal. It would be of obvious interest to repeat this experiment using mitotic extracts in which C4 is present, since the Ca^{2+} sensitivity shared by mitotic factors and band C4 mark this kinase as one of some interest.

These results demonstrate that at least one mitosis-specific protein kinase shares with mitotic factors the property of inactivation by G_1-phase extracts. They also help to explain the difficulty occasionally encountered in preparing mitotic extracts with significant amounts of B1 activity since contamination of a population of mitotic cells by only a small fraction of G_1-phase cells could so effectively abrogate detection of this kinase. Further, this system may provide an effective biochemical assay with which to isolate and further characterize IMF, if in fact they correspond to the inhibitor of kinase B1. For example, it has been suggested that IMF inactivate mitotic factors by specifically binding to their active site, thus forming an inert complex (Adlakha *et al.*, 1984b). If this were the case, it might be expected that such a complex would migrate to a location on gels different from that of the active protein kinase. IMF appear to be active over a broad range of pH (6–10) but are inactivated when exposed to buffers with a pH of less than 5 (Adlakha *et al.*, 1983). If mitosis-specific protein kinases were stable under these conditions, exposure of a gel containing an interphase–mitotic mixture to a pH of less than 5 might selectively neutralize or dissociate IMF from the complex, resulting in a band of protein kinase activity at a previously silent location in the gel.

D. Endogenous Substrates

In addition to inducing GVBD and chromosome condensation, mitotic extracts, when injected into frog oocytes, apparently stimulate an amplification of the factors responsible for these events, since activity can be serially propagated from oocyte to oocyte by transfer of only a small amount of cytoplasm, after just the initial injection of mitotic extract (Adlakha *et al.*, 1982a). Should mitotic factors and MPF be identical, this response could correctly be termed autoamplification. One mechanism by which (auto)amplification might occur is (auto)phosphorylation. If *auto*phosphorylation is, in fact, the mechanism, it is noteworthy that when phosphorylations by mitotic extracts were performed in the absence of exogenous substrate, bands of activity were seen at the positions of some mitosis-specific protein kinases (Fig. 1b), suggesting either autophosphorylation or phosphorylation of an endogenous substrate that comigrated with the kinases during electrophoresis. Should this activity represent autophosphorylation, examination of the phosphorylated products would reveal the molecular weights of at least some of the polypeptides which compose the kinases.

To perform such an analysis, a lane was cut from a nondenaturing gel phosphorylated in the absence of added substrate and applied horizontally across the top of a second-dimension denaturing gel. Since the most intensely phosphorylated band was B1, the most intense and most well-resolved polypeptides originated from that region in the second-dimension gel. Of the five species identified, with relative molecular weights of 105, 70, 58, 47, and 38K (Fig. 11), only two (105 and 58K) appeared to be

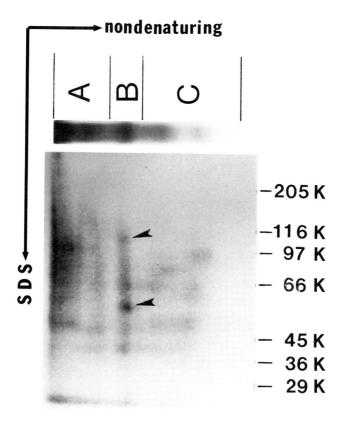

Fig. 11. Molecular weight of proteins phosphorylated in the absence of exogenous substrate. A lane containing mitotic extract was cut from a nondenaturing gel phosphorylated in the absence of endogenous substrate, equilibrated overnight in sample buffer containing 0.1% SDS, and applied to a second dimension SDS gel (5% stacking; 10% resolving). An autoradiogram of the first-dimension analysis of the exised lane has been mounted above the second-dimension autoradiogram to aid in the identification of the various regions of the gel. Arrows indicate polypeptides unique to the B region. Positions of molecular weight standards are shown on the right-hand portion of the gel. (From Halleck *et al.*, 1984b.)

unique to the B region, the other species being seen, although less intensely, in the A and C regions of the gel as well. The phosphorylated peptides common to all regions may conceivably represent ubiquitous endogenous substrates or, alternatively, subunits shared by each of the protein kinases. The species unique to the B region might then represent additional subunits which are able to be autophosphorylated by this particular mitosis-specific protein kinase.

Several other approaches have been used to gather information concerning mitosis-specific phosphoproteins. Using monoclonal antibodies, Davis et al. (1983) found that phosphoproteins of 182, 118, and 70K were present in mitotic cell extracts, but not in extracts from interphase cells. By labeling synchronized, cycling cells with ortho[^{32}P]phosphate, Adlakha et al. (1984a) identified nonhistone proteins of 100, 92, 70, 55, 43, 36, and 27.5K which were specifically phosphorylated at mitosis. Whether any of these phosphorylated polypeptides identified by three independent methods are the same, or related, is not clear at this time.

It should be emphasized that the ability of mitosis-specific protein kinases to autophosphorylate may not in any way be related to the ability of mitotic extracts to amplify MPF. If mitotic factors and MPF are identical or homologous, autophosphorylation might indeed be responsible for autoamplification. If, however, they are not, MPF might merely provide a fortuitous substrate for protein kinases in mitotic extracts, which once activated, activates other stored MPF (autoamplification). Since autoamplification following injection of mitotic factors into somatic cells has not been examined, it is unknown whether autophosphorylation is an important mechanism by which the factors function in their normal milieu. Thus, when kinases are assayed for their ability to autophosphorylate, a negative result (e.g., band C4) may have limited interpretation with respect to physiological significance. At a minimum, however, this method provides information on some of the components that make up those mitosis-specific protein kinases that do autophosphorylate.

E. Presence in Mitoplasts

Although there can be no doubt that cytoplasmic extracts of ruptured mitotic cells contain factors capable of inducing oocyte maturation, it cannot be stated with certainty from this conclusion that such factors are actually present in the cytoplasm of intact cells. It is quite possible that factors are found only in association with chromosomes *in vivo,* but that during hypotonic lysis and extraction of cells, they are leached from the chromosomes and thus artifactually appear in cytoplasmic extracts. Evidence against such an explanation was provided by the study of Sunkara *et al.* (1980) in which cytochalasin B was used to induce extrusion of chromosomes from mitotic cells, and the resulting "mitoplasts" were fused with interphase cells. Induction of nuclear membrane breakdown and chromosome condensation within the interphase cells conclusively demonstrated that factor activity was located in the soluble compartment of mitotic cells.

Even though extracts prepared from mitoplasts have never been tested for their ability to induce oocyte maturation, thus possibly providing further circumstantial support for the identity of mitotic factors and MPF, the kinase activity of such extracts has been examined. As seen in Fig. 12, extracts from mitoplasts produced the same bands of kinase activity that were found in extracts of the mitotic cells from which the mitoplasts were derived. However, mitoplast extracts were less active on a per microgram of protein basis, and in each of three experiments, band C3 was somewhat enhanced relative to the other activities in the C region. These results indicate that the kinases identified are, most likely, normally located in

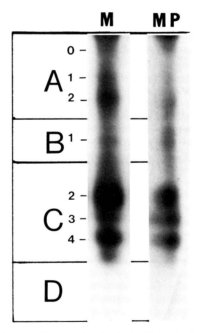

Fig. 12. Protein kinase activities of mitoplasts. Extracts from mitotic cells (M) and mitoplasts (MP; prepared according to Rao *et al.*, 1982) were electrophoresed into nondenaturing gels and assayed for protein kinase activity in the presence of histone substrate.

the cytoplasm of mitotic cells, consistent with their possible identity with mitotic factors. Whether or not the association of mitotic factors with isolated chromosomes might be artifactual was not addressed by these experiments.

VI. CONCLUSIONS AND PERSPECTIVES

A number of protein kinases that are found in the cytoplasm of mitotic cells are not found in the corresponding compartment of interphase cells. Although these species can also be recovered from isolated mitotic chromosomes, their association with interphase chromatin, or their presence in the soluble nucleoplasm of interphase cells, has yet to be investigated.

A comparison of these mitosis-specific kinase activities with mitotic factors has revealed some interesting similarities. Perhaps the most distinguishing feature of mitotic factors is their inactivation by Ca^{2+}; the mitosis-specific protein kinase C4 was found to be calcium-sensitive. The biological activity of mitotic factors can also be attenuated by mixing with cytoplasmic extracts from G_1-phase cells; the activity of the mitosis-specific protein kinase B1 was abolished by this operation. Should autophosphorylation be required for activation of mitotic factors, kinase B1 has been shown to possess this ability. However, the activity of B1 is not sensitive to Ca^{2+}, and, therefore, this kinase does not possess all of the characteristics of mitotic factors. Whether kinase C4 meets these criteria must await experiments to examine its response to G_1-phase inhibitory factors.

A third mitosis-specific protein kinase, C2, was the only kinase found to prefer histone H1 as substrate; it was neither sensitive to Ca^{2+} nor was its activity affected by G_1-phase extracts. One possible explanation for the disparate properties of the various kinases might be that C4, individually or in combination with other kinases such as B1, activates the kinases directly responsible for the modification of structural elements, such as phosphorylation of H1 in chromatin by C2. But, no matter what correlations can be demonstrated between specific protein kinases and mitotic factors, nor what scenarios of interactions can be devised, it is clear that individual kinases must be further purified and biochemically characterized in order to substantiate a causal role for these kinases in mitotic events.

Immunological approaches should provide a fruitful avenue for further investigations. Antisera generated against mitosis-specific protein kinases might be used to study the expression and subcellular localization of kinases as a function of position in the cell cycle using indirect immunofluorescence, to characterize the kinases biochemically following immunoprecipitation, and to aid in the purification of these kinases using immunoaffinity techniques. In addition, antibodies could be used to ask whether inactivation of a protein kinase *in situ* would perturb mitotic events. Since antibodies introduced into culture cells by red cell-mediated microinjection (for reviews, see Schlegel and Rechsteiner, 1986; Rechsteiner and Schlegel, 1986) remain active intracellularly and are able to neutralize the activity of the molecules against which they are directed (Schlegel, 1985), antibodies to protein kinases may be able to prevent injected interphase cells from entering mitosis. Such immunological approaches are now underway.

ACKNOWLEDGMENTS

The authors thank Jon A. Reed for establishing in our laboratory the methodology for needle microinjections of *Xenopus* oocytes and embryos. Our investigations were supported by a Nellie and Alvin Reid Memorial Grant, CD111, from the American Cancer Society. R.A.S. is an Established Investigator of the American Heart Association.

REFERENCES

Abdel-Ghany, M., Riegler, C., and Racker, E. (1984). A placental polypeptide activator of a membranous protein kinase and its relation to histone 1. *Proc. Natl. Acad. Sci. U.S.A.* **81,** 7388–7391.

Adlakha, R. C., Sahasrabuddhe, C. G., Wright, D. A., Lindsey, W. F., and Rao, P. N. (1982a). Localization of mitotic factors on metaphase chromosomes. *J. Cell Sci.* **54,** 193–206.

Adlakha, R. C., Sahasrabuddhe, C. G., Wright, D. A., Lindsey, W. F., Smith, M. L., and Rao, P. N. (1982b). Chromosome bound mitotic factors: Release by endonucleases. *Nucleic Acids Res.* **10,** 4107–4117.

Adlakha, R. C., Sahasrabuddhe, C. G., Wright, D. A., and Rao, P. N. (1983). Evidence for the presence of inhibitors of mitotic factors during G1 period in mammalian cells. *J. Cell Biol.* **97,** 1707–1713.

Adlakha, R. C., Sahasrabuddhe, C. G., Wright, D. A., Bigo, H., and Rao, P. N. (1984a). Role of nonhistone protein phosphorylation in the regulation of mitosis in mammalian cells. *In* "Growth, Cancer, and the Cell Cycle" (P. Skehan and S. J. Friedman, eds.), pp. 59–69. Humana Press, Clifton, New Jersey.

Adlakha, R. C., Wang, Y. C., Wright, D. A., Sahasrabuddhe, C. G., Bigo, H., and Rao, P. N. (1984b). Inactivation of mitotic factors by ultraviolet irradiation of HeLa cells in mitosis. *J. Cell Sci.* **65,** 279–295.

Bonner, J., Chalkley, R., Dahmus, M., Fambrough, D., Fujimura, F., Huang, R. C. C.,

Huberman, J., Jensen, R., Marushige, K., Ohlenbusch, H., Olivera, B. M., and Widholm, J. (1968). Isolation and characterization of nucleoprotein. *In* "Methods in Enzymology" (L. Grossman and K. Moldave, eds.), Vol. 12B, pp. 3–65. Academic Press, New York.

Corbin, J. D., Keeley, S. L., and Park, C. R. (1975). The distribution and dissociation of cyclic adenosine 3':5'-monophosphate-dependent protein kinases in adipose, cardiac, and other tissues. *J. Biol. Chem.* **250,** 218–225.

Crissman, H. A., and Steinkamp, J. A. (1982). Rapid, one step staining procedures for analysis of cellular DNA and protein by single and dual laser flow cytometry. *Cytometry* **3,** 84–90.

Davis, F. M., Tsao, T. Y., Fowler, S. K., and Rao, P. N. (1983). Monoclonal antibodies to mitotic cells. *Proc. Natl. Acad. Sci. U.S.A.* **80,** 2926–2930.

Forbes, D. J., Kirscher, M. W., and Newport, J. W. (1983). Spontaneous formation of nucleus-like structures around bacteriophage DNA microinjected into *Xenopus* eggs. *Cell (Cambridge, Mass.)* **34,** 13–23.

Gerace, L., and Blobel, G. (1980). The nuclear envelope lamina is reversibly depolymerized during mitosis. *Cell (Cambridge, Mass.)* **19,** 277–287.

Gurley, L. R., D'Anna, J. A., Barham, S. S., Deaven, L. L., and Tobey, R. A. (1978). Histone phosphorylation and chromatin structure during mitosis in Chinese hamster cells. *Eur. J. Biochem.* **84,** 1–15.

Halleck, M. S., Lumley-Sapanski, K., Reed, J. A., Iyer, A. P., Mastro, A. M., and Schlegel, R. A. (1984a). Characterization of protein kinases in mitotic and meiotic cell extracts. *FEBS Lett.* **167,** 193–198.

Halleck, M. S., Lumley-Sapanski, K., Reed, J. A., and Schlegel, R. A. (1984b). Involvement of protein kinases in mitotic-specific events. *In* "Growth, Cancer and the Cell Cycle" (P. Skehan and S. Friedman, eds.), pp. 143–150. Humana Press, Clifton, New Jersey.

Halleck, M. S., Reed, J. A., Lumley-Sapanski, K., and Schlegel, R. A. (1984c). Injected mitotic extracts induce condensation of interphase chromatin. *Exp. Cell Res.* **153,** 561–569.

Hammer, M. G., Sheridan, J. D., and Estersen, R. D. (1971). Cytochalasin B. II. Selective inhibition of cytokinesis in *Xenopus laevis* eggs. *Proc. Soc. Exp. Biol. Med.* **136,** 1158–1162.

Hardie, D. G., Matthews, H. R., and Bradbury, E. M. (1976). Cell cycle dependence of two nuclear histone kinase activities. *Eur. J. Biochem.* **66,** 37–42.

Hardison, R., and Chalkley, R. (1978). Polyacrylamide gel electrophoretic fractionation of histones. *Methods Cell Biol.* **17,** 235–252.

Hnilica, L. S. (1975). Methods for analysis of histones. *In* "Methods in Enzymology" (B. W. O'Malley and J. G. Hardman, eds.), Vol. 40, pp. 102–138. Academic Press, New York.

Hohmann, P., Tobey, R. A., and Gurley, L. R. (1975). Cell cycle-dependent phosphorylation of serine and threonine in Chinese hamster cell fl histones. *Biochem. Biophys. Res. Commun.* **63,** 126–133.

Hohmann, P., Tobey, R. A., and Gurley, L. R. (1976). Phosphorylation of distinct regions of fl histone. Relationship to the cell cycle. *J. Biol. Chem.* **251,** 3685–3692.

Iyer, A. P., Pishak, S. A., Sniezek, M. J., and Mastro, A. M. (1984a). Visualization of protein kinases in lymphocytes stimulated to proliferate with Concanvalin A or inhibited with a phorbol ester. *Biochem. Biophys. Res. Commun.* **121,** 392–399.

Iyer, A. P., Pishak, S. A., Sniezek, M. J., and Mastro, A. M. (1984b). Protein kinase activity in lymphocytes: Effects of Concanavalin A and a phorbol ester. *In* "Growth, Cancer and the Cell Cycle'' (P. Skehan and S. J. Friedman, eds.), pp. 151–157. Humana Press, Clifton, New Jersey.

Johns, E. W., and Butler, J. A. V. (1962). Further fractionation of histones from calf thymus. *Biochem. J.* **82**, 15–18.

Johnson, E. M., Hadden, J. W., Inoue, A., and Allfrey, V. G. (1975). DNA binding by cyclic adenosine 3′,5′-monophosphate dependent protein kinase from calf thymus nuclei. *Biochemistry* **14**, 3873–3884.

Kishimoto, T., Kuriyama, R., Kondo, H., and Kanatani, H. (1982). Generality of the action of various maturation-promoting factors. *Exp. Cell Res.* **137**, 121–126.

Laemmli, V. K. (1970). Cleavage of structural proteins during the assembly of the head of bacteriophage T4. *Nature (London)* **227**, 680–685.

Lake, R. S., and Salzman, N. P. (1972). Occurrence and properties of a chromatin-associated Fl-histone phosphokinase in mitotic Chinese hamster cells. *Biochemistry* **11**, 4817–4826.

Lam, K. S., and Kasper, C. B. (1979). Selective phosphorylation of a nuclear envelope polypeptide by an endogenous protein kinase. *Biochemistry* **18**, 307–311.

Langan, T. A. (1976). Characterization of growth-associated phosphorylation sites in calf-thymus lysine-rich (H1) histone. *Fed. Proc., Fed. Am. Soc. Exp. Biol.* **35**, 1623.

Langan, T. A. (1978). Methods for the assessment of site-specific histone phosphorylation. *Methods Cell Biol.* **19**, 127–142.

Langan, T. A., Zeilig, C. E., and Leichtling, B. (1980). Analysis of multiple site phosphorylation of H1 histone. *In* "Protein Phosphorylation and Bio-regulation" (G. Thomas, E. J. Podesta, and J. Gordon, eds.), pp. 70–82. Karger, Basel.

Laskey, R. A. (1983). Phosphorylation of nuclear proteins. *Philos. Trans. R. Soc. London, Ser. B* **302**, 143–150.

Lohka, M. J., and Maller, J. L. (1985). Induction of nuclear envelope breakdown, chromosome condensation, and spindle formation in cell-free extracts. *J. Cell Biol.* **101**, 518–523.

Lohka, M. J., and Masui, Y. (1984). Effects of Ca^{2+} ions on the formation of metaphase chromosomes and sperm pronuclei in cell-free preparations from unactivated *Rana pipiens* eggs. *Dev. Biol.* **103**, 434–442.

McClung, J. K., and Kletzien, R. F. (1981a). The effect of nutritional status and of glucocorticoid treatment on the protein kinase isozyme pattern of liver parenchymal cells. *Biochim. Biophys. Acta* **676**, 300–306.

McClung, J. K., and Kletzien, R. F. (1981b). An increase in type II cyclic AMP-dependent protein kinase is correlated with growth arrest in G1 phase. *Biochim. Biophys. Acta* **678**, 106–114.

Masui, Y., and Markert, C. L. (1971). Cytoplasmic control of nuclear behavior during meiotic maturation of frog oocytes. *J. Exp. Zool.* **177**, 129–146.

Matthews, H. R., and Huebner, V. D. (1984). Nuclear protein kinases. *Mol. Cell. Biochem.* **59**, 81–99.

Miake-Lye, R., and Kirschner, M. W. (1985). Induction of early events in a cell-free system. *Cell (Cambridge, Mass.)* **41**, 165–175.

Miake-Lye, R., Newport, J., and Kirschner, M. (1983). Maturation-promoting factor induces nuclear envelope breakdown in cycloheximide-arrested embryos of *Xenopus laevis. J. Cell Biol.* **97**, 81–91.

Miyamoto, E., Petzgold, G. L., Harris, J. S., and Greengard, P. (1971). Dissociation and concomitant activation of adenosine 3′,5′-monophosphate-dependent protein kinase by histone. *Biochem. Biophys. Res. Commun.* **44**, 305–312.

Nelkin, B., Nichols, C., and Vogelstein, B. (1980). Protein factor(s) from mitotic CHO cells induce meiotic maturation in *Xenopus laevis* oocytes. *FEBS Lett.* **109,** 233–238.

Newport, J. W., and Kirschner, M. W. (1984). Regulation of the cell cycle during early *Xenopus* development. *Cell (Cambridge, Mass.)* **37,** 731–745.

Panyim, S., and Chalkley, R. (1969). High resolution acrylamide gel electrophoresis of histones. *Arch. Biochem. Biophys.* **130,** 337–346.

Paulson, J. R. (1980). Sulfhydryl reagents prevent dephosphorylation and proteolysis of histones in isolated HeLa metaphase chromosomes. *Eur. J. Biochem.* **111,** 189–197.

Poccia, D. L., and Green, G. R. (1986). Nuclei and chromosomal proteins. *Methods Cell Biol.* **27,** 153–174.

Rao, P. N., and Johnson, R. T. (1970). Mammalian cell fusion: Studies on the regulation of DNA synthesis and mitosis. *Nature (London)* **225,** 159–164.

Rao, P. N., Sunkara, P. S., and Al-Bader, A. A. (1982). Isolation and characterization of mitoplasts. *In* "Techniques in Somatic Cell Genetics" (J. W. Shay, ed.), pp. 245–254. Plenum, New York.

Rechsteiner, M. C., and Schlegel, R. A. (1986). Erythrocyte-mediated transfer: Applications. *In* "Microinjection and Organelle Transplanation Techniques: Methods and Applications" (J. E. Celis, A. Graessmann, and A. Loyter, eds.), pp. 89–116. Academic Press, New York.

Reynhout, J. K., and Smith, L. D. (1974). Studies on the appearance and nature of a maturation-inducing factor in the cytoplasm of amphibian oocytes exposed to progesterone. *Dev. Biol.* **38,** 394–400.

Sahasrabuddhe, C. G., Adlakha, R. C., and Rao, P. N. (1984). Phosphorylation of nonhistone proteins associated with mitosis in HeLa cells. *Exp. Cell Res.* **153,** 439–450.

Schlegel, R. A. (1985). Red cell-mediated microinjection of antibodies. *Bibl. Haematol.* **51,** 134–141.

Schlegel, R. A., and Rechsteiner, M. C. (1986). Erythrocyte-mediated transfer: Methods. *In* "Microinjection and Organelle Transplantation Techniques: Methods and Applications" (J. E. Celis, A. Graessmann, and A. Loyter, eds), pp. 67–87. Academic Press, New York.

Schorderet-Slatkine, S. (1972). Action of progesterone and related steroids on oocyte maturation in *Xenopus laevis*. An in *vitro study. Cell Differ.* **1,** 179–189.

Smith, L. D., and Ecker, R. E. (1971). The interaction of steroids with Rana pipiens oocytes in the induction of maturation. *Dev. Biol.* **25,** 232–247.

Sunkara, P. S., Wright, D. A., and Rao, P. N. (1979a). Mitotic factors from mammalian cells induce germinal vesicle breakdown and chromosome condensation in amphibian oocytes. *Proc. Natl. Acad. Sci. U.S.A.* **76,** 2799–2802.

Sunkara, P. S., Wright, D. A., and Rao, P. N. (1979b). Mitotic factors from mammalian cells: A preliminary characterization. *J. Supramol. Struct.* **11,** 189–195.

Sunkara, P. S., Al-Bader, A. A., Riker, M. A., and Rao, P. N. (1980). Induction of prematurely condensed chromosomes by mitoplasts. *Cell Biol. Int. Rep.* **4,** 1025–1029.

Taylor, S. (1982). The in *vitro* phosphorylation of chromatin by the catalytic subunit of cAMP-dependent protein kinase. *J. Biol. Chem.* **257,** 6056–6063.

Tobey, R. A., Anderson, E. C., and Petersen, D. F. (1967). Properties of mitotic cells prepared by mechanically shaking monolayer cultures of Chinese hamster cells. *J. Cell. Physiol.* **70,** 63–68.

Wasserman, W. J., and Smith, L. D. (1978). The cyclic behavior of a cytoplasmic factor controlling nuclear membrane breakdown. *J. Cell Biol.* **78,** R15–R22.

Weintraub, H., Buscaglia, M., Ferrez, M., Weiller, S., Boulet, A., Fabre, F., and Bauleiu,

E. E. (1982). "MPF"activity in *Saccharomyces cerevisiae. C. R. Seances Acad. Sci., Ser. 3* **295,** 787–790.

Yonemoto, W., Jarvis-Morar, M., Brugge, J. S., Bolen, J. B., and Israel, M. A. (1985). Tyrosine phosphorylation within the amino-terminal domain of pp60[c-src] molecules associated with polyoma virus middle-sized tumor antigen. *Proc. Natl. Acad. Sci. U.S.A.* **82,** 4568–4572.

9

Antibodies to Mitosis-Specific Phosphoproteins

FRANCES M. DAVIS AND POTU N. RAO

Department of Medical Oncology
The University of Texas System Cancer Center
M.D. Anderson Hospital and Tumor Institute
Houston, Texas 77030

I. INTRODUCTION

The total cellular architecture of the mammalian cell undergoes pro-
found rearrangement during mitosis in order to accomplish the orderly
distribution of the genetic material to the two daughter cells. This reorganization involves both the nucleus and the cytoskeleton. The chromatin
in the nucleus is rearranged and condensed into discrete chromosomes,
and the nuclear architecture, including the nuclear membrane and the
nuclear matrix, is dispersed. The interphase cytoskeleton and, particularly, the cytoplasmic microtubules are reorganized into the mitotic spindle. What are the means and mechanisms by which the cell accomplishes
this coordinated reorganization?

One possibility is that there are new proteins or enzymatic activities
present in mitotic cells that are absent from cells during interphase. In
fact, earlier studies have shown that RNA synthesis is necessary until 2 hr
before mitosis (Tobey *et al.*, 1966) and protein synthesis until 1 hr before
mitosis (Petersen *et al.*, 1969). Using two-dimensional gel electrophoresis
of proteins from synchronized HeLa cells, Al-Bader *et al.*, (1978) found
10 protein species present in cells in G_2 or mitosis that were not detectable
in cells in S phase. However, other investigators have failed to detect any
differences in protein species between mitotic cells and S-phase cells of

259

Chinese hamster origin (Bravo and Celis, 1980). An alternative possibility is that preexisting proteins are subjected to posttranslational modifications. Such modifications could affect the function and/or association of these modified molecules with other elements within the cell. Nuclear lamins (Gerace and Blobel, 1980; Jost and Johnson, 1981) and cytoplasmic intermediate filament proteins (Robinson *et al.*, 1981; Evans and Fink; 1982; Bravo *et al.*, 1982) are examples of presynthesized proteins that are phosphorylated as the cell enters mitosis. It is likely that both new proteins and modifications of preexisting proteins will be found to be necessary for mitosis.

One approach to studying these putative mitosis-specific proteins or mitosis-specific protein modifications is immunological. Antibodies can be used to identify the proteins that are new or newly modified. Antibodies can be used to localize their cognate antigens within mitotic cells and thus to suggest their function. In addition, antibodies can be used to perturb the activity of their cognate antigens when the antibodies are introduced into cells by microinjection. We have taken such an immunological approach and raised antibodies to various extracts of HeLa cells synchronized in mitosis. In 1982, we reported the preparation of a rabbit antiserum that reacted specifically with mitotic chromosomes of human origin (Davis and Rao, 1982). In this study, the antigen was identified as a DNA–protein complex. Since then, we have isolated a variety of mouse monoclonal antibodies. These antibodies exhibit apparent specificity for cells in mitosis. The antigens recognized seem to be related to the changes in chromatin condensation and cytoskeletal architecture accomplished by mitotic cells.

II. MONOCLONAL ANTIBODIES TO CELLS IN MITOSIS

We have selected monoclonal antibodies exhibiting greater reactivity with mitotic cells than with interphase cells on the basis of one of two assays: indirect immunofluorescence or ELISA. Monoclonal antibodies to mitotic cells can be easily detected by indirect immunofluorescence, since the morphology of the cells can be visualized first by phase contrast microscopy to identify those cells in mitosis and then by fluorescence microscopy. Monoclonal antibodies to cells in mitosis can also be detected by ELISA by comparing the intensity of the signal produced by a monoclonal antibody permitted to react with corresponding extracts of mitotic cells and synchronized S-phase cells. Table I summarizes some of the characteristics of monoclonal antibodies to proteins of mitotic cells (MPM) that we have detected in either of these two ways.

TABLE I

Monoclonal Antibodies to Mitotic Cells[a]

Antibody (class)	Immunofluorescence pattern	Immunoblot polypeptides (MW × 10⁻³)	Species specific	Remarks
MPM-1 (IgM, κ)	Mitotic cells, some nuclei	Mammals: 182, 118, 70, and family of others Nematode: 127, 97, 77 *Physarum:* 127 Yeast: 110	No	Antibodies react only when proteins are phosphorylated Antibodies delay exit of cells from mitosis Antibodies recognize active MTOC
MPM-2 (IgG₁, κ)	Very similar to MPM-1; from same fusion			
MPM-3 (IgM)	Mitotic cells	190, and family of others	No	Antigens are phosphorylated proteins
MPM-4 (IgG)	Mitotic chromosomes	125	Yes	Chromosome spreads suggest banding
MPM-5 (ND)	Centrosomal area	(ND)	Yes	
MPM-6 (ND)	Centrosomal area	(ND)	Yes	
MPM-7 (TgG₁)	Chromosomes	81	No	
MPM-8 (ND)	Mitotic spindles	(ND)	(ND)	
MPM-9 (ND)	Centrosomal area	(ND)	(ND)	
MPM-10 (ND)	Mitotic cells	147, and family of others	No	

[a] ND, Not determined; MTOC, microtubule organizing center.

Monoclonal antibodies MPM-1 and MPM-2 react strongly with cells in mitosis, recognizing epitopes present throughout the mitotic cell. These epitopes reside on the chromosomes, in the cytoplasm, and on spindle-associated structures (Fig. 1). As we will show, these antibodies react with an extensive family of phosphorylated polypeptides. MPM-3 and MPM-10 also react with epitopes found throughout cells in mitosis; however, these two antibodies react with single major polypeptide bands. The antigen recognized by MPM-3 is also a phosphorylated polypeptide, whereas the phosphorylation state of the antigen recognized by MPM-10

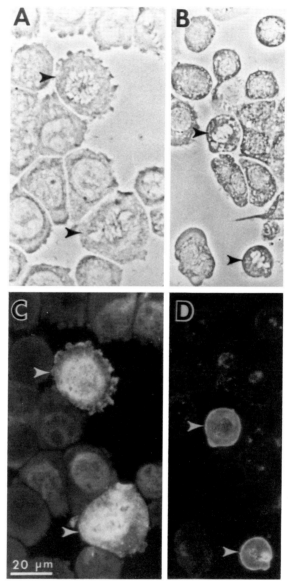

Fig. 1. Specificity of the MPM-1 antibodies to mitotic cells. Random populations of HeLa cells (A and C) and mosquito cells (B and D) were stained by indirect immunofluorescence with monoclonal antibody MPM-1. (A and B) phase-contrast photomicrographs; (C and D) fluorescent photomicrographs. Fluorescence is present both on the chromosomes and in the cytoplasm of the mitotic cells, which are indicated by arrows. (From Davis *et al.* 1983).

has not yet been determined. MPM-4 and MPM-7 react specifically with chromosomes in mitotic cells (Fig. 2). MPM-4 has been shown to be specific for cells of primate origin and to react with a single polypeptide with a molecular weight of 125,000, while MPM-7 reacts with a polypeptide of about 81,000. Like the antigens recognized by MPM-1, MPM-2, and MPM-3, the antigen recognized by MPM-7 cannot be detected when proteins of interphase cells are separated on gels. However, the antigen recognized by MPM-4 can be detected among the proteins of interphase cells. Therefore, the determinant recognized by MPM-4 may be masked during interphase, as has been reported for some determinants on vimentin by Franke and co-workers (1984), for a cytosine-5′-methyl transferase by Drahovsky and Kaul (1984), and for some epitopes of interchromatin granules as described by Clevenger and Epstein (1984). MPM-5, MPM-6, and MPM-9 react very strongly with epitopes localized in the area of the spindle poles in mitotic cells and only weakly, if at all, with nuclei in interphase cells (Fig. 3). This localization is similar to that reported by Pettijohn and co-workers (1984) for the NuMa antigen, a protein with a molecular weight of 250,000. Polypeptide species recognized by MPM-5, MPM-6, and MPM-9 have not yet been identified. MPM-8 seems to recog-

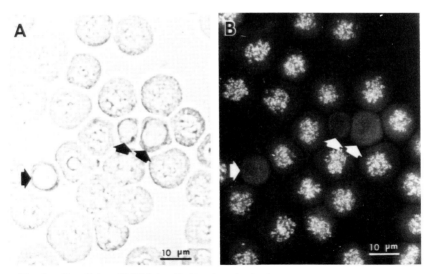

Fig. 2. Specificity of MPM-4 antiboides for mitotic chromosomes. HeLa cells arrested in mitosis by nitrous oxide were deposited on slides using a cytocentrifuge and were stained by indirect immunofluorescence with MPM-4 antibodies. (A) Phase contrast photomicrograph; (B) Fluorescent photomicrograph. Arrows indicate interphase cells *not* stained by the antibody.

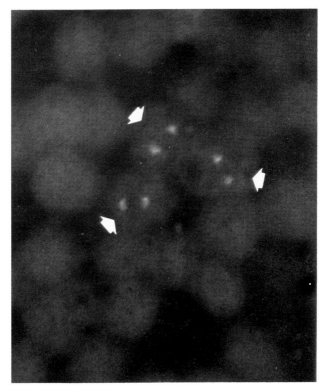

Fig. 3. Specificity of MPM-5 antibodies for spindle poles in mitotic cells. HeLa cells deposited on slides using a cytocentrifuge were stained by indirect immunofluorescence. The three metaphase cells (arrows) in the center have bright fluorescence in the centrosomal area.

nize specifically a component of the mitotic spindle that is not detectable in the cytoskeleton of interphase cells (Fig. 4)

III. CHARACTERIZATION OF ANTIGENS REACTIVE WITH MPM-1 AND MPM-2

The antigens reactive with the MPM-1 and MPM-2 monoclonal antibodies have been characterized by the electroimmunoblot methodology. Proteins were separated according to their molecular weights in one-dimensional sodium dodecyl sulfate (SDS)–polyacrylamide gels, transferred to

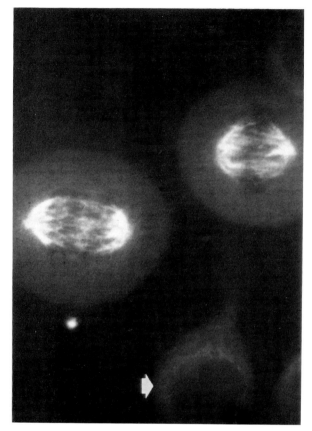

Fig. 4. Specificity of MPM-8 antibodies for spindle fibers in mitotic cells. Normal human fibroblasts (CRL 1508) were grown on slides, fixed in methanol, and stained by indirect immunofluorescence with the monoclonal antibody MPM-8. Spindle fibers in a metaphase cell and an anaphase cell are brightly fluorescent. The interphase cell on the bottom (arrow) shows only very weak perinuclear fluorescence.

nitrocellulose paper, and stained with antibodies by an indirect immuno-peroxidase technique (Davis *et al.,* 1983.).

A. A Family of Polypeptide Antigens

Surprisingly, a family of polypeptide bands of widely varying molecular weights are stained by monoclonal antibody MPM-1 (Fig. 5). In general, the polypeptide antigens recognized by MPM-1 antibodies are found in

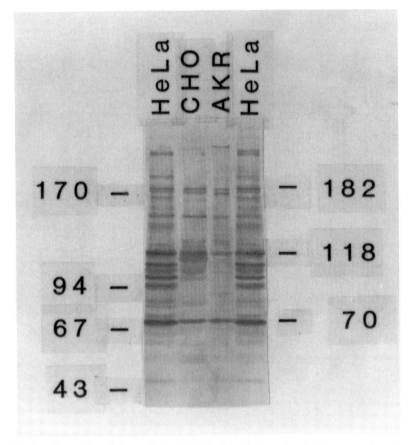

Fig. 5. Reactivity of the antibody MPM-1 with mitotic cells of different species. HeLa cells, CHO cells, and mouse (AKR) cells were synchronized in mitosis with Colcemid (0.1 μg/ml). The polypeptides were separated by electrophoresis in SDS–gradient polyacrylamide gels, electrophoretically transferred to nitrocellulose sheets, and stained with antibodies by an indirect immunoperoxidase procedure. Approximately 25 μg of protein per lane was analyzed (From Davis *et al.*, 1983.)

the molecular weight range of about 70,000 to over 200,000, although the molecular weight range does vary from one species to another. Some bands are more intensely stained than others. For example, human cells contain major bands at approximately 70,000, 118,000, and 182,000. The antibody-stained bands do not seem to comigrate with major protein bands stained in gels with Coomassie blue or on nitrocellulose transfers stained with Amido black or India ink. Both the number of antibody-stained bands and their relative intensity were unchanged by preparing the sample in the presence of a variety of protease inhibitors or mercap-toethanol, by immediate resuspension of the cells in a sample-loading buffer containing 30% SDS and boiling for 3 min, or by prolonged arrest of cells in mitosis with nitrous oxide or Colcemid. The relative intensity of staining of the major bands was not altered by decreasing the concentration of the primary immunoglobulin used for immunoblot staining, suggesting that the polypeptide bands may share the same epitope, rather than that various crossreacting epitopes with different affinities for the antibodies were present.

Cells in mitosis from every species tested have some antibody-stained bands. Cells from mammalian species, such as the Chinese hamster and the mouse, as shown in Fig. 5, have a family of polypeptide bands stained by the antibodies that is very similar to that family found in human cells. Cells from distantly related species, such as slime mold, nematode, or yeast, contain a family of polypeptide antigens that scarcely resembles that of human cells. In all cases, however, there are major bands and minor bands and nearly all of the polypeptide antigens are greater than 50,000 in molecular weight. We have seen no antibody reactivity with proteins that migrate in the region of the core histones.

B. Relationship between the Detection of Polypeptide Antigens and Cell Cycle Phases

The polypeptide antigens in cells synchronized in various phases of the cell cycle were also examined by the immunoblot method. As seen in Fig. 6, only cells in mitosis contained detectable amounts of MPM-1-reactive antigens. MPM-2 recognized a similar subset of antigens in mitotic cells. These results are in agreement with results obtained by indirect immuno-fluorescence of synchronized cells, whose fluorescence was measured with a single-cell photometer (Fig. 7). Metaphase cells had the highest levels of immunological reactivity which by anaphase/telophase was reduced by about 50%. Cells in S phase had the lowest level of fluorescence and seemed to contain about 10% of the immunologically reactive material of mitotic cells.

C. Time of Synthesis of Antigens during Cell Cycle

We wanted to know whether the antigens recognized by the MPM-1 and MPM-2 monoclonalantibodies were among those proteins synthesized in the last hour before mitosis. Therefore, we pulse-labeled cells synchronized in various phases throughout the cell cycle with a mixture of ^{3}H-amino acids, arrested these cells with Colcemid as they came into

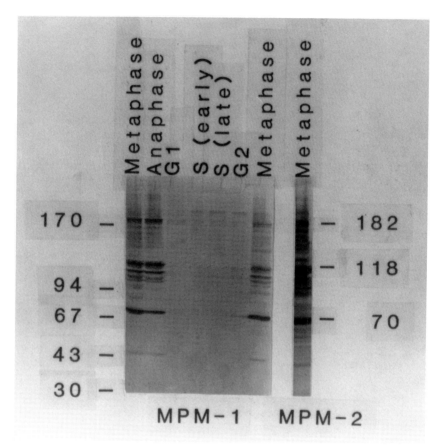

Fig. 6. Cell cycle specificity of polypeptides recognized by the antibodies. Polypeptides from whole HeLa cells synchronized in various cell cycle phases were separated in SDS–gradient polyacrylamide gels, electrophoretically transferred to nitrocellulose sheets, and stained with monoclonal antibodies MPM-1 and MPM-2 by an indirect immunoperoxidase procedure. Approximately 25 μg of protein per lane was loaded. M, Metaphase cells, A, anaphase cells 1 hr after reversal of N_2O block (Rao, 1968); G_1, cells in G_1 phase 3.5 hr after reversal of N_2O blockade; S_e, cells in early S phase 2 hr after reversal of second thymidine block (Rao and Engelberg, 1966); S_1, cells in late S phase 6.5 hr after reversal of second thymidine block; G_2, cells in G_2 phase 9 hr after reversal of second thymidine block. Sizes shown in kilodaltons are of marker proteins (on the left) and of the major antigens (on the right). (From Davis et al., 1983.)

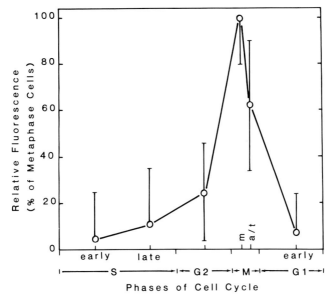

Fig. 7. Relative antigenic reactivity of synchronized HeLa cells. Cytocentrifuge preparations of HeLa cells synchronized as in Fig. 6 were stained by indirect immunofluorescence with antibody MPM-1. The fluorescence intensity of 25 cells from each population was measured with a Leitz MPV microscope photometer. The fluorescence from blank areas on the slide adjacent to the cells was subtracted from each measurement. The fluorescence intensity from the nonmetaphase populations was normalized to that from 25 metaphase cells included on the same slide. The mean fluorescence intensities were plotted, and the bars represent the standard deviation. Metaphase cells (m) fluoresced more intensely than anaphase–telephase cells (a/t), and mitotic cells fluoresced more intensely than interphase cells. (From Davis *et al.*, 1983.)

mitosis, prepared extracts from the cells, and precipitated immunoreactive polypeptides using MPM-2 antibodies and protein A (Pansorbin, Calbiochem). Surprisingly, these studies showed that the majority of MPM-2-reactive antigens were synthesized not in late G_2 or at the time of the G_2 to mitosis transition but primarily in the S phase, although a portion of these antigens were synthesized throughout the cell cycle (Fig. 8).

D. Phosphorylation of Antigens Reactive with MPM-1 and MPM-2

Since the antigens that react with MPM-1 and MPM-2 seem to be synthesized as polypeptides throughout the cell cycle but can be detected

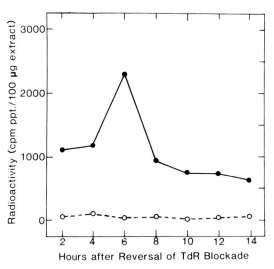

Fig. 8. Mitosis-specific antigens are synthesized primarily during S phase. HeLa cells synchronized by double thymidine blockade (Rao and Engelberg, 1966) were pulse-labeled for 2 hr during the cell cycle with ³H-labeled mixed amino acids, then washed and incubated in medium without radioactive amino acids and collected at mitosis using Colcemid. DNA synthesis continued for 8 hr after reversal of the block, and cells began to enter mitosis by 11 hr. Extracted proteins were precipitated using either MPM-2 IgG (—●—●—) or control mouse IgG (--○--○--) and Pansorbin.

either by immunofluorescence or by immunoblots only in mitosis, we felt that a posttranslational modification that gave rise to an antigenic epitope reactive with the MPM-1 and MPM-2 antibodies was possible, particularly because the phosphorylation of proteins is much higher in cells in mitosis, compared to cells in interphase. Phosphorylation of histones (Lake *et al.,* 1972; Bradbury *et al.,* 1973; Hanks *et al.,* 1983) and nonhistone chromosomal proteins (Sahasrabuddhe *et al.,* 1984), high mobility group (HMG) proteins (Bhorjee, 1981; Paulson and Taylor, 1982), nucleolar proteins (Shibayama *et al.,* 1982), nuclear-matrix (Henry and Hodge, 1983) and nuclear-lamin (Gerace and Blobel, 1980) proteins, ribosomal proteins (Kalthoff *et al.,* 1982; Maller *et al.,* 1977), and intermediate filament proteins (Robinson *et al.,* 1981; Evans and Fink, 1982; Bravo *et al.,* 1982) has been shown to be increased in cells in mitosis compared to cells in interphase. We therefore investigated the phosphorylation of polypeptide antigens reactive with MPM-2 in three types of experiments. First, cells synchronized in G_2 were pulse-labeled with ortho[³²P]phosphate. The phosphorylated proteins were separated on a one-dimensional SDS–

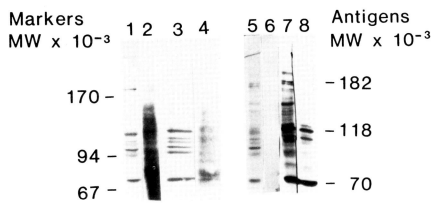

Fig. 9. Antigens recognized by MPM-1 and MPM-2 are phosphorylated polypeptides. Synchronized HeLa cells were labeled with ortho[^{32}P]phosphate during the G$_2$ phase, and cells were arrested in mitosis with Colcemid (0.05 μg/ml). An extract was prepared from one aliquot of the cells by sonication in a buffer consisting of 0.5% Nonidet P-40, PBS, and 1 mM phenylmethylsulfonyl fluoride. After immunoprecipitation or digestion with some extracts with alkaline phosphatase, the polypeptides were separated by electrophoresis, transferred to nitrocellulose, and stained by the indirect immunoperoxidase procedure with antibody MPM-1. Proteins of mitotic extracts (lane 1; lane 2, autoradiogram). Proteins immunoprecipitated from mitotic extracts by MPM-2 (lane 3, immunoblot; lane 4, autoradiogram). Immunoblots of proteins from mitotic extracts that had been incubated in the absence (lane 5) or the presence (lane 6) of alkaline phosphatase before gel electrophoresis. Immunoblots of proteins from mitotic extracts that were incubated in the absence (lane 7) or presence (lane 8) of alkaline phosphatase after transfer to nitrocellulose. Approximately 25 μg of protein per lane was loaded. Molecular weights are shown for marker proteins on the left and major antigens on the right. (From Davis *et al.*, 1983.)

polyacrylamide gel, and the ^{32}P-labeled bands identified by autoradiography were compared with the immunoreactive bands on immunoblots (Fig. 9). The immunoreactive polypeptides comigrated with the ^{32}P-labeled bands. Moreover, when proteins in soluble extracts prepared from the ^{32}P-labeled cells were precipitated with MPM-2 and Pansorbin, radioactivity was precipitated by the antibodies. When the precipitated phosphorylated proteins were separated on an SDS–polyacrylamide gel, the autoradiogram of the bands was very similar to the immunoblot. These studies showed that antigens reactive with MPM-1 and MPM-2 were phosphorylated polypeptides but that they were only a subset of the phosphorylated polypeptides detectable in cells in mitosis.

Two other types of experiments showed not only that polypeptides reactive with MPM-1 or MPM-2 could be phosphoproteins but that in fact the phosphorylation of the proteins was required for antibody recogni-

tion. In one type of experiment, extracts from cells synchronized in mitosis were incubated with alkaline or acid phosphatases from commercial
sources. Subsequent separation of the protein on SDS–polyacrylamide
gels, transfer to nitrocellulose, and staining by the immunoblot method
with MPM-1 or MPM-2 antibodies showed that the phosphatase treatment
was sufficient to remove all detectable MPM-1 (Fig. 9, lanes 5 and 6) or
MPM-2 reactivity (data not shown). In the other experiment, proteins
from cells synchronized in mitosis were separated on polyacrylamide gels
and transferred to nitrocellulose prior to treatment with phosphatases.
Again, phosphatases greatly decreased the MPM-1 and MPM-2 immuno-
reactivity (Fig. 9, lanes 7 and 8) but had no effect on recognition of
unrelated nonphosphorylated antigens by other antibodies. Subsequently,
incubation of fixed cells on slides with commercially purchased phosphatases has been shown to reduce greatly indirect immunofluorescence due
to MPM-1 or MPM-2 (data not shown).

E. *In Vitro* Phosphorylation and Dephosphorylation of Antigens

Because the polypeptides that react with MPM-1 and MPM-2 antibodies are present throughout the cell cycle but react with the antibodies only
when cells are in mitosis, it seemed there must be a mechanism for phosphorylating these proteins as cells enter mitosis and then rapidly dephosphorylating them as cells exit from mitosis. We asked whether the phosphorylation of antigens could occur *in vitro* in an extract from cells in
mitosis. The extract was incubated in the presence of $[\gamma\text{-}^{32}P]ATP$, and the
incorporation of radioactivity into material precipitable by antibody plus
Pansorbin increased over time. Kinetics of incorporation into antigens
were similar to the kinetics of incorporation of label into trichloroacetic
acid (TCA)-precipitable material. In fact, up to 7% of the ^{32}P incorporated
into TCA-precipitable material could be accounted for by incorporation
into MPM-2-reactive antigens (Fig. 10). In contrast, when extracts from
S-phase cells that are known to contain the polypeptides (in a nonphosphorylated, nonreactive state) are incubated in a similar *in vitro* phosphorylation system, no incorporation of label into antibody-reactive material
was observed, and decreased incorporation of label into TCA-precipitable
material was seen (data not shown). These studies suggest that there is a
kinase activity present specifically in mitotic cells that is able to phosphorylate the phosphopeptide antigens. Whether this kinase is related to
those described by Halleck *et al.* in Chapter 8, this volume, is unknown at
present.

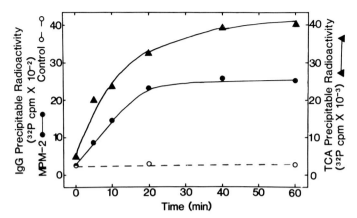

Fig. 10. *In vitro* phosphorylation of polypeptides reactive with MPM-2. [^{32}P]ATP (100 μCi/ml) was added to an isotonic extract from mitotic cells (4 mg protein/ml) and the solution was incubated at 37°C. At the indicated times, aliquots were removed and processed for total incorporation into TCA-precipitable material and for incorporation into MPM-2/Pansorbin-precipitable material.

Like the phosphorylation of nonhistone proteins extractable in 0.2 *M* sodium chloride that rapidly decreases as cells exit mitosis (Sahasrabuddhe *et al.*, 1984), detection of antigens by MPM-1 and MPM-2 also decreases rapidly as cells exit mitosis. Since the appearance of the antigens is related to their phosphorylation, we felt that the disappearance of the antigens might be related to their dephosphorylation. As an approach to the study of a putative intracellular phosphatase, we incubated the extracts of cells synchronized in mitosis with extracts from cells synchronized in the G_1 phase of the cell cycle. Surprisingly, even at a one-to-one ratio of mitotic- to G_1-phase protein concentrations in the mixture, there was nearly complete elimination of antibody-reactive material from the extract (Fig. 11). This depletion of immunoreactive material did not occur when an extract from cells synchronized in S phase was substituted for the extract of cells synchronized in the G_1-phase of the cell cycle (data not shown). During the incubation period, the amount of proteolytic activity in the cell extracts was not sufficient to change the Coomassie blue stained pattern of polypeptides separated in polyacrylamide gels. However, confirmation that the depletion of the reactivity of the antigens in extracts of cells synchronized in mitosis under the influence of extracts from cells synchronized in G_1 is due to a phosphatase activity will await the results of studies currently in progress, using antigens purified by MPM-2 immunoaffinity chromatography.

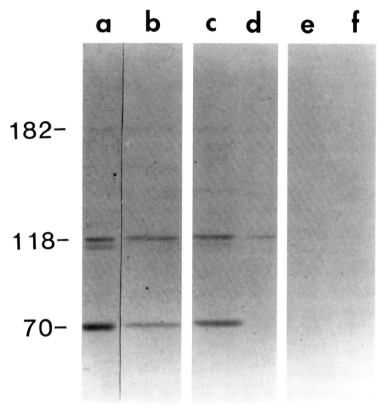

Fig. 11. Incubation of mitotic extract with G_1 extract destroys antigenic reactivity. Immunoblots of extracts from mitotic HeLa cells (a, b), and G_1 phase cells (e, f), and a 1 : 1 mixture of mitotic and G_1 phase cell extracts at equivalent protein concentrations (c, d). Extracts were incubated for 2 hr at 0°C (a, c, e) or 37°C (b, d, f) before separation by polyacrylamide gel electrophoresis and immunoblotting. The numbers on the left represent the apparent molecular weight ($\times$ 10^{-3}) of the major antigenic bands in HeLa cells.

IV. CYTOLOGICAL LOCALIZATION OF ANTIGENS REACTIVE WITH MPM-1 AND MPM-2

Our initial experiments to show the specificity of MPM-1 and MPM-2 antibodies for cells in mitosis utilized cytocentrifuge preparations that were subsequently dried, fixed, and stained with antibody. In these studies, material reactive with MPM-1 and MPM-2 was found throughout cells in mitosis, both on the chromosomes and in the cytoplasm (Fig. 1). Live cells incubated in suspension with antibodies showed no staining. Therefore, the antigens were intracellular, although their association with specific cellular components was not observed.

Subsequent studies by Vandre *et al.* (1984b) have demonstrated the preferential association of at least a subset of the antigens with discrete mitotic structures in PtK$_1$ cells. Unlike most cells cultured in the laboratory, PtK cells remain attached to plastic and are relatively flat even during the process of mitosis. PtK$_1$ cells were grown on coverslips and fixed from an aqueous environment in either methanol, glutaraldehyde, or a paraformaldehyde–lysine–periodate fixative. Cells fixed in methanol or glutaraldehyde (Fig. 12) show the presence of immunoreactive material throughout the cell as seen earlier with cytocentrifuge preparations of HeLa cells fixed with methanol.

Cells in mitosis showed increased fluorescence intensity at mitotic microtubule organizing centers including centrosomes, kinetochores, and midbodies, suggesting that the phosphorylation state of the protein components of the microtubule organizing centers may be involved in the reorganization of the microtubular network that occurs during the G$_2$ to mitosis transition. Fixation with the paraformaldehyde–lysine–periodate fixative (Fig. 13) depleted the more amorphous, immunologically reactive material present throughout the cytoplasm of the mitotic cell, and confirmed the predominant association of the immunoreactive material with microtubule organizing centers. Similar results could be obtained with methanol or glutaraldehyde fixation, if the cells were gently extracted with a detergent prior to fixation. The extraction procedure removed a substantial amount of immunoreactive material, leaving the detergent-insoluble immunoreactive material associated with microtubule organizing centers. Hecht *et al.* (1983, 1986) also reported the association of antigens recognized by MPM-1 with chromosomes and asters in the nematode embryo. Partial extraction of the more easily solubilized antigens during the squash preparation of embryos and during fixation in 70% ethanol may account for the localization observed.

The association of antigens with isolated components of the mitotic apparatus has also been investigated, using Taxol-stabilized mitotic spindles isolated from Chinese hamster ovary (CHO) cells (Vandre *et al.*, 1986) (Fig. 14). Antibody-reactive material was present at the spindle poles and along bundles of spindle fibers. An additional fluorescent region was located at the tips of some of the spindle fibers and possibly represents elements of the kinetochore. This was supported by the finding that the kinetochore of isolated CHO chromosomes was immunoreactive with the MPM-2 antibodies. Specific spindle pole staining with MPM-2 was observed whether or not centrioles were present at the spindle poles, even in cells having intranuclear spindles (Fig. 15). The mitosis-specific

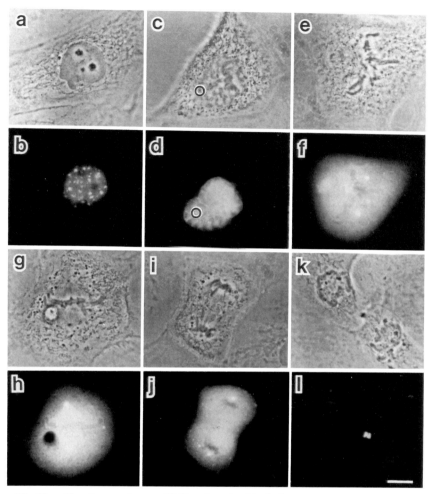

Fig. 12. Monolayer cultures of PtK$_1$ cells stained by indirect immunofluorescence with monoclonal antibody MPM-2. Phase-contrast and fluoresence photomicrograph pairs of cells at interphase (a and b), prophase (c and d), prometaphase (e and f), metaphase (g and h), anaphase (i and j), and telophase (k and l). An example of a phase dense spot corresponding to the kinetochore and its corresponding immunofluorescent pair are indicated by the circles in c and d; bar= 10 μm. (From Vandre *et al.,* 1984b.)

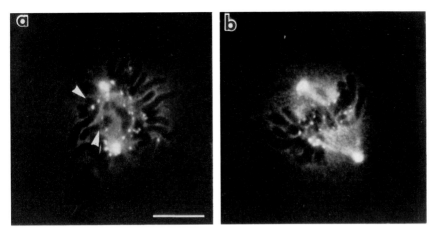

Fig. 13. PtK₁ cells fixed in periodate–lysine–paraformaldehyde and stained with monoclonal antibody MPM-2. Immunoreactive material is associated with kinetochores and centrosomes in both the prometaphase (a) and metaphase (b) cells. Two individual chromosomes with two associated immunoreactive spots are indicated by the arrowheads (a). Weak spindle fiber immunofluorescence remains (b), but cytoplasmic and chromosomal staining is absent; bar = 10 μm. (From Vandre *et al.*, 1984b.)

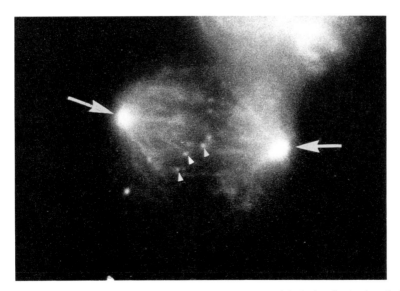

Fig. 14. Isolated spindles contain MPM-2 reactive material. Isolated mitotic spindle from CHO cell stained with monoclonal antibody MPM-2. Immunoreactive material is associated with each mitotic pole (arrows), along spindle fibers, and at small dots, possibly kinetochore remnants, at the end of spindle fibers (arrowheads). It should be noted that chromosomes are lost from the spindle during its isolation (From Vandre *et al.*, 1986).

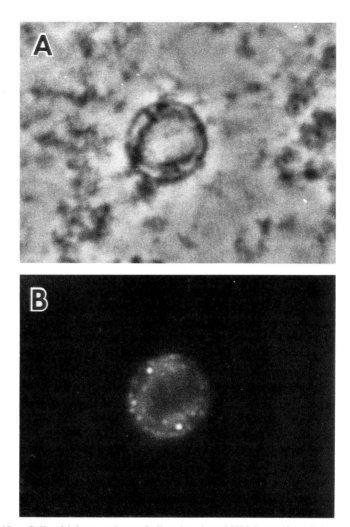

Fig. 15.   Cells with intranuclear spindles also show MPM-2-specific staining at mitosis. *Physarum* plasmodium smear with staining localized to the spindle poles within the metaphase nucleus. The nuclear envelope remains intact through the mitotic cycle in the plasmodia. Phase-contrast (A) and fluorescence (B) photomicrograph pair of cells stained by indirect immunofluorescence with monoclonal antibody MPM-2. (From Vandre *et al.*, 1986.)

TABLE II

Distribution of Phosphoproteins Reactive with MPM-2 during Mitosis as Determined by Indirect Immunofluorescence[a]

Species	Cell type	Prophase nucleus	Metaphase cytoplasm	Chromosome periphery	Mitotic MTOC[b]			
					c	k[c]	m	Other[d]
Human	356; foreskin fibroblast	+[e]	+	+	+	ND	+	−
	HeLa; cervical carcinoma	+	+	+	+	+	+	−
Mouse	N$_2$A; neuroblastoma	+	+	+	+	+	+	−
Chinese hamster	CHO; ovary	+	+	+	+	+	+	−
Indian Muntjac	Skin	+	+	+	+	ND	+	−
Rat kangaroo	PtK$_1$; kidney	+	+	+	+	+	+	−
Chicken	Intestinal tissue	ND	+	−	+	+	ND	−
	Primary retinal culture	+	+	+	+	+	+	−
Drosophila	Schneider's line 1; embryo	+	+	+	−	ND	+	−
	1182.4; embryo, lacks centriole	+	+	+	−	ND	+	−
Physarum	Myxamoebae	+	+	−	+	ND	ND	
	Plasmodium	−	−	−		ND		sp
Dictyostelium	Amebae	+	−	−		ND	+[f]	nab
Yeast	Saccharomyces cerevisiae	−	−	−		ND		spb
Onion	Root tip	+	+	+		ND		p

[a] From Vandre et al. (1986).

[b] Mitotic microtubule organizing centers (MTOCs) are abbreviated as follows: c, centrosome; k, kinetochore; m, midbody.

[c] The presence of kinetochore staining could not be observed in some intact cells, but was present on isolated chromosomes. In either case, a "+" was given to indicate staining.

[d] Other mitotic structures are abbreviated as follows: sp, spindle pole, nab, nuclear associated body; spd, spindle pole body, p, phragmoplast.

[e] +, Positive staining; −, staining not present; ND, not determined. (In the case of kinetochore staining the absence of staining in intact cells could not justify a "−" designation since staining was associated with kinetochores of isolated chromosomes from some cell types that did not show clear kinetochore staining in whole cells.)

[f] Indicates staining localized to the constriction between the two daughter nuclei.

staining of different cell types with MPM-2 antibody and the localization of the staining throughout mitotic cells is summarized in Table II.

In the course of a broad survey of many different types of cells, in which mitosis-specific staining with MPM-2 antibody was investigated, there were a few specific tissue and cell types in which MPM-2 staining was increased in nonmitotic cells (Table III). The interphase microtubule network of neuroblastoma cells and the basal bodies of flagellated cells in particular have been shown to be specific sites of MPM-2 antibody localization by indirect immunofluorescence (Fig. 16). Thus, specialized cytoskeletal structures responsible for nucleating microtubules, in addition to the mitotic microtubule organizing centers, are sites for the accumulation of MPM-2-reactive antigens. Specifically, microtubule-associated protein 1 (MAP-1), and its proteolytic degradation products present in twice-cycled porcine brain microtubules, were reactive with MPM-2 (Fig. 17). Recently, the microtubule-associated protein 2 (MAP-2) in mitotic CHO cells has also been shown to react with MPM-2 (R. Brady, personal communication).

Vandre *et al.* (1984a) also studied the subset of MPM-2-reactive polypeptides present in the mitotic apparatus isolated from CHO cells synchronized in various phases of mitosis. One of these phosphopolypeptides has been shown to be a component of the midbody.

The nuclear lamina proteins are specifically phosphorylated during mitosis (Gerace and Blobel, 1980), and there was a possibility that phosphorylated lamins would be recognized by the MPM antibodies. This has recently been shown not to be the case (Ottaviano and Gerace, 1985).

TABLE III

Distribution of Phosphoproteins Reactive with MPM-2 in Nonmitotic Cells[a]

Localization of stain	Cell types
Microtubule network	Mouse neuroblastoma cells
Nuclear patches	Human foreskin fibroblasts, HeLa, CHO, Indian muntjac, PtK$_1$, *Drosophila*
Microtubule organizing center	*Physarum* myxamoebae
Basal bodies	Paramecium, *Physarum* flagellates, sea urchin sperm

[a] From Vandre *et al.* (1986).

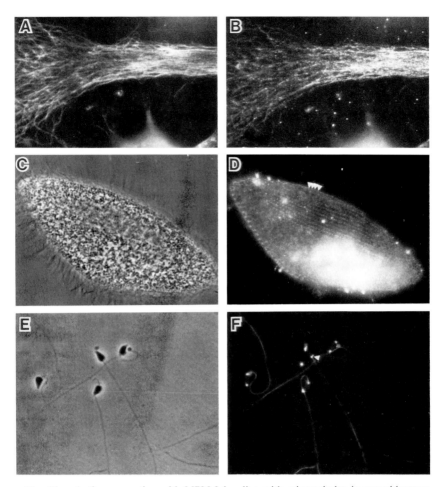

Fig. 16. Antigens reacting with MPM-2 localize with microtubules in neuroblastoma cells and are associated with basal bodies. Mouse neuroblastoma cells (A, B) were doubly stained with rabbit antityrosinylated tubulin (A) and monoclonal antibody MPM-2 (B). Immunofluorescence micrographs show a process extending from the cell body (located out of the field of view to the right). Microtubules are clearly seen extending the length of the process (A). The MPM-2 antibody is localized to the same microtubules in a punctate fashion (B).

Pairs of ciliated and flagellated cells were also stained with MPM-2 monoclonal antibody. A *Paramecium* cell (C, D) shows localization of MPM-2 to rows of basal bodies located on the top surface of the cell. In phase contrast (C), some residual cilia are apparent, and the basal bodies stained by MPM-2 antibodies can be seen in D (arrowheads). The lower right portion of the cell shows internal staining localized to the macronucleus. Sea urchin sperm (E, phase contrast; F, immunofluorescence) show intense MPM-2 staining of the acrosome and basal bodies, with faint staining along the flagellum. In some sperm two spots of MPM-2 staining could be observed at the base of the flagellum (arrowhead). (From Vandre *et al.*, 1986.)

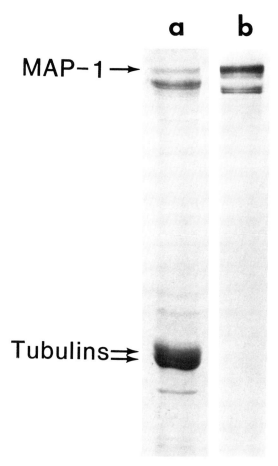

Fig. 17.   MPM-2 reactive polypeptides of neuroblastoma microtubules include MAP-1. SDS–polyacrylamide gel electrophoresis and immunoblot analysis of porcine brain microtubules. Lane a, Coomassie blue stained polyacrylamide gel. Lane b, corresponding nitrocellulose transfer and immunoblot. (From Vandre *et al.*, 1986.)

V. BIOLOGICAL ACTIVITY OF MPM-1 AND MPM-2
   ANTIBODIES

Both the presence of the phosphorylated polypeptide antigens in all cells in mitosis and the association of the antigens with structures known to be involved in the orderly segregation of chromosomes at mitosis suggest that the antigens are functionally important. Moreover, a study by

Yamashita *et al.* (1985) shows that the appearance of the antigens is associated with chromosome condensation in a temperature-sensitive mutant of baby hamster kidney cells that undergoes premature chromosome condensation at the restrictive temperature. Recently, Hecht *et al.* (1986) suggested that a lethal, temperature-sensitive embryonic mutation in nematodes is due to a temperature-sensitive kinase that phosphorylates the antigens recognized by MPM-1; at the nonpermissive temperature, the antigens are not phosphorylated and the cells are blocked in G_2. We studied the biological activity of the antigens recognized by MPM-1 and MPM-2 monoclonal antibodies more directly by microinjection.

A. Maturation of Xenopus Oocytes Not Inhibited by Injected Antibodies

Immature oocytes of *Xenopus laevis* can be induced to mature *in vitro*, either by incubation with progesterone or by the microinjection of extracts of mitotic or meiotic cells, as described by Sunkara *et al.* (1979). Oocyte maturation involves entry into meiosis and is characterized by the breakdown of the germinal vesicle, or nucleus, of the oocyte and chromosome condensation. Neither MPM-1 nor MPM-2 monoclonal antibodies, injected either prior to or subsequent to incubation of oocytes in progesterone, was able to inhibit or delay the progesterone-induced oocyte maturation that usually occurs within 8 hr after progesterone treatment (Masui, 1966) (Table IV).

Extracts from synchronized mitotic HeLa cells are able to induce oocyte maturation within 2.5 hr after injection into immature *Xenopus* oocytes (Sunkara *et al.*, 1979). The ability of the MPM-1 and MPM-2 antibodies to inhibit this process was tested both by preincubating the extract with antibodies and by removal of complexed antigens from the extracts by protein A-mediated precipitation. The protocols used for precipitation and removal of the antigens were sufficient to deplete all immunoreactive material detected on immunoblots of the extracts separated by electrophoresis in SDS–polyacrylamide gels. As with progesterone-induced oocyte maturation, the antibodies did not inhibit or delay oocyte maturation when preincubated mixtures of extracts and antibodies were injected (Table V). Moreover, the precipitation and removal of the antigens from the extracts also had no effect on the ability of the extracts to induce maturation even though the experiments were done at a concentration of maturation-promoting activity sufficient to cause maturation of only 60–90% of the oocytes injected and, therefore, more sensitive to decreases in activity.

TABLE IV

Progesterone-Induced Oocyte Maturation Is Not Inhibited by MPM-1 or MPM-2 Antibodies[a]

Immunoglobulin injected	Time of injection (hr) relative to progesterone treatment	Oocytes injected	Oocytes maturing (%)
None		10	100
Control mouse	−1	5	100
	+1	5	100
	+2	5	100
MPM-1	−1	5	100
	+1	5	100
	+2	5	100
MPM-2	−1	5	100
	+1	5	100
	+2	5	100

[a] Immunoglobulins (65 nl at 2 mg/ml in phosphate-buffered saline) were injected, and oocyte maturation was scored at 10 hr after the start of incubation in progesterone (10 μg/ml).

B. Effect of Antibodies on the G_2 to Mitosis Transition of Cultured Cells

Although oocyte maturation has been used to great advantage as a bioassay for mitotic factors, oocyte maturation may be somewhat different from the entry of cultured cells into mitosis, since the oocyte is a specialized cell containing sufficient macromolecules to accomplish several cell division cycles during cleavage in the absence of additional RNA or protein synthesis. We therefore investigated the effect of microinjection of MPM-1 or MPM-2 immunoglobulins into cultured HeLa cells. Random HeLa cells were injected manually using glass microneedles according to the method of Diacumakos *et al.* (1970), and 16–20 hr later, the mitotic index of cells injected with antibodies was compared with that of cells injected with control mouse immunoglobulins. Only cells in which antibody injection was confirmed by subsequent immunofluorescent staining with FITC anti-mouse immunoglobulin were scored. We found no decrease in the mitotic index of the cells injected with antibody; on the contrary, an increase in the mitotic index was observed (Table VI). Most of the cells scored as mitotic exhibited condensed chromosomes arranged on a metaphase plate.

TABLE V

Oocyte Maturation Induced by Mitotic Extracts Is Not
Inhibited by MPM-1 or MPM-2 Antibodies[a]

Immunoglobulin incubated with extracts	Pansorbin precipitation	Oocytes injected	Oocytes maturing (%)
None	−	14	86
	+	14	93
Buffer only	−	14	64
	+	14	57
Control mouse	−	15	87
	+	14	57
MPM-1	−	14	64
MPM-2	−	15	67
	+	15	87

[a] Aliquots (100 μg protein) of extracts of mitotic HeLa cells were incubated for 30 min at 37°C followed by 1–2 hr at 0°C with immunoglobulins (100 μg of protein) in a total volume of 150 μl. During the final 1 hr at 0°C, some mixtures were incubated with phosphate-buffered saline-washed pellets of 80 μl of Pansorbin. The mixtures were then pelleted by centrifugation for 2 min in a Beckman microfuge, and 65-nl aliquots of the supernatant were injected into $X.$ $laevis$ oocytes. Maturation was scored 2–4 hr later.

In other experiments, HeLa cells synchronized in S phase by double thymidine blockade were microinjected with antibodies by fusion of antibody-loaded red blood cell ghosts, according to the method of Schlegel and Rechsteiner (1978) as modified by Brown et $al.$ (1985). The mitotic index of the injected cells was scored 16.5 hr after injection in the presence and absence of Colcemid. Although the percentage of injected cells varied from experiment to experiment, only injected cells identified by using FITC anti-mouse immunoglobulin and immunofluorescence were scored for mitotic index. These studies showed that the majority of cells were able to enter mitosis after immunoglobulin injection since ≥86% of the injected cells in the Colcemid-treated cultures contained chromosomes but not nuclei when examined by phase contrast microscopy (see Table VI). The mitotic index of the antibody-injected cells that were incubated without Colcemid was also higher than that of the cells injected with control immunoglobulins, and many of the mitotic cells contained chromosomes on a metaphase plate.

TABLE VI

Entry of Cultured Cells into Mitosis Is Not Inhibited by Microinjection of MPM-1 or MPM-2 Immunoglobulins[a]

Microinjection method	Immunoglobulin injected	Colcemid treatment	Number of injected cells scored	Mitotic index of injected cells
Glass needles	Control mouse	−	47	4
	MPM-1	−	58	19
	MPM-2	−	26	15
Red blood cell-mediated	Control mouse	−	100	19
		+	100	86
	MPM-1	−	100	47
		+	100	94
	MPM-2	−	100	52
		+	100	90

[a] Immunoglobulins (20–30 mg/ml) were loaded into glass needles or red blood cell ghosts and injected into HeLa cells. Injected cells were identified by fixation and immunofluorescence staining using anti-mouse immunoglobulins. Some coverslips with injected cells were incubated continuously in the presence of Colcemid (0.05 μg/ml) starting at 3 hr after injection. The mitotic index of random cells injected using glass needles was scored at 20 hr after injection. The mitotic index of injected S-phase HeLa cells, synchronized by double thymidine blockade and injected by fusion with antibody-loaded red blood cell ghosts, was scored at 16.5 hr after injection. The mitotic index was scored by phase-contrast microscopy and was subsequently confirmed for some cells injected by glass needles by acetoorcein staining.

C. Inhibition of the Mitosis to G_1 Transition by Injected Antibodies

The experiments designed to investigate the effect of the microinjection of MPM-1 and MPM-2 antibodies on the entry of HeLa cells into mitosis suggested that the antibodies might be delaying the exit of cells from mitosis. This possibility was investigated directly by fusing red cells loaded with MPM-2 antibodies to mitotic HeLa cells and then determining the mitotic index of the injected cells at various times after reversal of the mitotic block. The mitotic index of the MPM-2 antibody-injected cells was 20 ± 5% at 16–19.5 hr after reversal compared to 1 ± 1% for cells injected with control antibodies (Fig. 18). Because each mitotic cell yields two G_1 cells, approximately 30% of the MPM-2 antibody-injected mitotic cells had not divided. Most of the mitotic cells (62 ± 18%) were in metaphase, as determined by the arrangement of the chromosomes on a metaphase plate. The kinetics of decrease of the mitotic index showed that

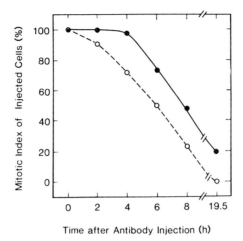

Time after Antibody Injection (h)

Fig. 18.   Microinjected monoclonal antibody MPM-2 delays the M/G_1 transition. MPM-2 and control mouse IgG (22 mg/ml) were separately loaded into human red blood cell ghosts and fused with N_2O-synchronized mitotic HeLa cells. At the indicated times after injection, cytocentrifuge slides from replicate cultures of injected cells were prepared and cells that had been injected with immunoglobulin were identified using FITC anti-mouse IgG. The mitotic index of cells injected with MPM-2 (●—●) or control mouse immunoglobulin (○ – – ○) was determined. A similar delay in exit from mitosis of the cells injected with MPM-2 IgG was observed in four experiments.

those cells injected with MPM-2 immunoglobulin that progressed into G_1 were delayed by 2–3 hr compared to cells injected with control mouse immunoglobulin (Table VII).

The ability of the MPM-1 and MPM-2 immunoglobulins to inhibit cleavage in frog embryos was also investigated. We injected one cell of *Rana pipiens* embryos at the two-cell stage with either control mouse IgG or MPM-2 IgG, and observed cleavage over the next 4–6 hr. Most of the embryos injected with control mouse IgG (93%; 28 of 30) showed continued cleavage in both hemispheres, without any apparent delay (Fig. 19). In contrast, only 7% (2 of 30) of the embryos injected with MPM-2 IgG exhibited continued cleavage in both hemispheres. Most of these showed continued cleavage only in the noninjected hemisphere, with no additional cleavage in the injected hemisphere. In some embryos no additional cleavage in either hemisphere was observed. These embryos may have been those in which the first cleavage was not quite complete at the time of injection. Sections of cells showing cleavage inhibition revealed that the cells retained chromosomes and spindles and were apparently arrested in metaphase.

TABLE VII

Mitotic Delay Induced by MPM-2 Injection

Experiment No.	Mean mitotic delay (hr)[a]	Mitotic index at 16 hr	
		Control Ig	MPM-2
1	2.8	ND[b]	ND
2	3	0	22.2
3	2	0	20
4	2	0	20

[a] Compared to cells injected with control mouse Ig.
[b] ND = not done.

D. Inhibition of Dephosphorylation of Antigens by Antibody Binding

Our previous experiments had shown that MPM-1 and MPM-2 monoclonal antibodies did not inhibit the entry of cells into mitosis but did inhibit the exit of cells from mitosis into interphase. Since MPM-1 and MPM-2 had previously been shown to bind to proteins only when those

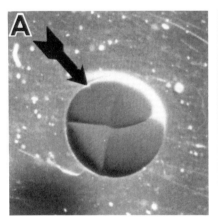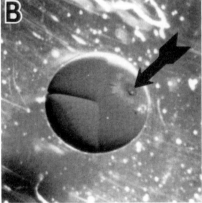

Fig. 19. MPM-2 inhibits cell division in cleaving *Rana* embryos. *R. pipiens* embryos at the two-cell stage were injected in one hemisphere with approximately 65 ng of either MPM-2 or control mouse immunoglobulin. Further cleavage in the cells of the injected embryos was observed over 2–4 hr after injection. (A), embryo injected with control mouse immunoglobulin showing continued cleavage in both hemispheres; (b), embryo injected with MPM-2 immunoglobulin, showing continued cleavage only in the uninjected hemisphere. Arrows, site of injection.

proteins were phosphorylated, and since the total amount of antigen detected in cells began to decrease as early as anaphase/telophase, we
thought that inhibition of the exit of cells from mitosis by MPM-1 and
MPM-2 antibodies might be a result of inhibition of the dephosphorylation
of the antigens that normally occurs as cells exit from mitosis into interphase. For this reason, the ability of the antibodies to inhibit dephosphorylation of antigens by alkaline phosphatase *in vitro* was investigated.
Extracts from mitotic cells that had been prelabeled during G_2 with
[^{32}P]orthophosphate were preincubated with either MPM-2 or control
mouse immunoglobulin prior to treatment with alkaline phosphatase. After this treatment, the antigens were precipitated using Pansorbin, and the
amount of precipitated radioactivity was determined. Precipitation of extracts not preincubated with immunoglobulins prior to alkaline phosphatase treatment resulted in an 88% reduction in ^{32}P-precipitable material
(Table VIII). Pretreatment with control mouse IgG was unable to protect
against phosphatase digestion. When extracts were pretreated with MPM-
2 IgG before alkaline phosphatase treatment, however, 85% of the label
initially present precipitated with antibodies. Thus, the MPM-2 IgG protected the antigens from dephosphorylation.

TABLE VIII

MPM-2 Binding Inhibits Dephosphorylation of Antigens[a]

IgG preincubated with extract	Alkaline phosphatase	^{32}P-cpm Precipitated $\pm$ SEM (% of control)
None	−	2,281 ± 324 (100)
	+	272 ± 98 (12)
MPM-2	−	2,445 ± 288 (107)
	+	1,936 ± 305 (85)
Control mouse	−	2,093 ± 238 (92)
	+	361 ± 139 (16)

[a] Aliquots (100 μg) of extract from synchronized mitotic cells that
had been prelabeled with ortho[^{32}P]phosphate during the G_2 phase were
incubated for 0.5 hr at 37°C and 1 hr at 5°C with 100 μg of either MPM-2
or control mouse IgG. Alkaline phosphatase (100 units/ml in 0.1 M Tris-
HCl, pH 8) was added to half of the aliquots to a final concentration
of 10 units/ml, while buffer only was added to the remaining tubes.
Digestion with alkaline phosphatase was for 0.5 hr at 37°C. Subsequently, antigens were precipitated by the addition of 100 μg MPM-2
and 200 μl Pansorbin, and the amount of pelleted radioactivity was
determined.

VI. DISCUSSION

We have shown that the monoclonal antibodies MPM-1 and MPM-2 react with a subset of cellular phosphoproteins greatly enriched in mitotic and meiotic cells from a wide variety of species. Moreover, these antibodies react specifically with structural components of microtubule organizing centers, including centrosomes, kinetochores, and midbodies. The antigens are generally restricted to cells in mitosis; however, cells of neuronal derivation and flagellated cells also showed specific localization of MPM-1 and MPM-2 antibodies to microtubule networks and basal bodies. On immunoblots, the MPM-1 and MPM-2 antibodies detected brain MAP-1 among a number of other phosphoproteins. In contrast, the MPM-1 and MPM-2 antibodies did not detect MAP-1 in extracts from mitotic cells. These results suggest that while the phosphoprotein components recognized by the MPM-1 and MPM-2 antibodies may be different for mitotic microtubule organizing centers, basal bodies, and other specialized cytoskeletal structures, the presence of a related phosphorylated domain on these proteins may be important for their proper function or interaction with microtubules.

The phosphorylated antigens in mitotic cells that are recognized by the MPM-1 and MPM-2 antibodies are synthesized as polypeptides throughout the cell cycle, but primarily in S phase. Since the antigenic epitope is not detected until the cells enter mitosis, the phosphorylation of the antigens may play a functional role in mitosis by mediating such processes as chromosome condensation, nuclear envelope breakdown, and reorganization of intermediate filaments.

Microinjection of the antibodies into cells inhibits the completion of the process of mitosis, but not its initiation. The mitosis-specific antigens first appeared to increase in the nucleus of cells in prophase and therefore may not be readily accessible to antibodies microinjected into the cytoplasm. Cells injected with antibodies are able to proceed to midmetaphase before the mitotic blockade becomes apparent. Since indirect immunofluorescence shows that the total amount or accessibility of antigen in mitotic cells begins to decrease at anaphase, it is tempting to speculate that some essential control element must be dephosphorylated for anaphase to proceed. Antibodies bound at or near the phosphorylated site could inhibit dephosphorylation. Because the antibodies are able to inhibit the alkaline phosphatase-induced dephosphorylation of the antigens *in vitro*, the inhibition of the completion of mitosis *in vivo* may be mediated by this mechanism. It is also possible, however, that bound antibodies may inhibit a phosphoprotein antigen that has an essential function. The multiplicity of phosphoprotein antigens recognized is yet another additional complication: Which antigen is localized in which structure in the mitotic cell is as yet unknown. However, Vandre *et al.* (1984b) showed that staining at kinetochores did not persist to late anaphase and that the immunoreactive material at the spindle poles decreased markedly in staining intensity after anaphase. More recently, immunoblots of spindles isolated throughout the process of mitosis have shown that many phosphorylated polypeptides are present in metaphase; but, by telophase, only a single major phosphopolypeptide can be detected, and is presumably present in the midbody (Vandre *et al.*, 1986). Therefore, the antigens localized in these structures could be targets for the action of the injected antibodies either as inhibitors of dephosphorylation or as inhibitors of an essential function of the phosphorylated polypeptide.

The MPM-1 and MPM-2 antibodies failed to recognize the maturation-promoting activity that is detected in extracts from mitotic cells by its ability to induce germinal vesicle breakdown and chromosome condensation. Miake-Lye *et al.* (1983) have proposed a phosphoprotein kinase cascade to link the activation of the maturation-promoting activity with increased phosphorylation of nuclear lamina proteins. Although our recent results with partially purified mitotic factors strongly suggest that mitotic factors from mammalian cells may not themselves be phosphorylated (Adlakha *et al.*, 1985), these studies also emphasized the importance of protein phosphorylation per se in oocyte maturation. Since the antibodies do not recognize the mitotic factors, the mitotic factors lack the phosphorylated epitope present in a family of phosphoproteins found in abundance specifically in mitotic cells. These results suggest that the activation of the mitotic factors and the phosphorylation of the mitosis-specific phosphoprotein antigens are different steps in the initiation of mitosis. It is possible that they could represent different steps in a phosphoprotein kinase cascade. If the maturation-promoting activity itself is a kinase, as suggested by other chapters in this volume, then candidate proteins phosphorylated by this kinase could be the mitosis-specific phosphoprotein antigens recognized by MPM-1 and MPM-2.

REFERENCES

Adlakha, R. C., Wright, D. A., Sahasrabuddhe, C. G., Davis, F. M., Prasad, N., Bigo, H., and Rao, P. N. (1985). Partial purification and characterization of mitotic factors from HeLa cells. *Exp. Cell Res.* **160,** 471–482.

Al-Bader, A. A., Orengo, A., and Rao, P. N. (1978). G2 phase-specific proteins of HeLa cells. *Proc. Natl. Acad. Sci. U.S.A.* **75,** 6064–6068.

Bhorjee, J. S. (1981). Differential phosphorylation of nuclear nonhistone high mobility group proteins HMG 14 and HMG 17 during the cell cycle. *Proc. Natl. Acad. Sci. U.S.A.* **78,** 6944–6948.

Bradbury, E. M., Inglis, R. J., Matthews, H. R., and Sarner, N. (1973). Histone H_1 phosphorylation in *Physarum polycephalum*. Correlation with chromosome condensation. *Eur. J. Biochem.* **33,** 131–139.

Bravo, R., and Celis, J. E. (1980) A search for differential polypeptide synthesis throughout the cell cycle of HeLa cells. *J. Cell Biol.* **84,** 795–802.

Bravo, R., Fey, S. J., Mose Larsen, P., and Celis, J. E. (1982). Modification of vimentin polypeptides during mitosis. *Cold Spring Harbor Symp. Quant. Biol.* **46,** 379–385.

Brown, D. B., Hanks, S. K., Murphy, E. C., Jr., and Rao, P. N. (1985). Early initiation of DNA synthesis in G1 phase HeLa cells following fusion with red cell ghosts loaded with S-phase cell extracts. *Exp. Cell Res.* **156,** 251–259.

Clevenger, C. V., and Epstein, A. L. (1984). Identification of a nuclear protein component of interchromatin granules using a monoclonal antibody and immunogold electron microscopy. *Exp. Cell Res.* **151,** 194–207.

Davis, F. M., and Rao, P. N. (1982). Antibodies specific for mitotic human chromosomes. *Exp. Cell Res.* **137,** 381–386.

Davis, F. M., Tsao, T. Y., Fowler, S. K., and Rao, P. N. (1983). Monoclonal antibodies to mitotic cells. *Proc. Natl. Acad. Sci. U.S.A.* **80,** 2926–2930.

Diacumakos, E. G., Holland, S., and Pecora, P. (1970). A microsurgical methodology for human cells in *vitro:* Evolution and applications. *Proc. Natl. Acad. Sci. U.S.A.* **65,** 911–918.

Drahovsky, D., and Kaul, S. (1984). Identification of proliferatively active cells by use of monoclonal antibodies against DNA-cytosine-5'-methyltransferase. *In* "International Cell Biology 1984" (S. Seno and Y. Okada, eds.), p. 392. Academic Press Japan, Inc., Tokyo.

Evans, R. M., and Fink, L. M. (1982). An alteration in the phosphorylation of vimentin-type intermediate filaments is associated with mitosis in cultured mammalian cells. *Cell (Cambridge, Mass.)* **29,** 43–52.

Franke, W. W., Grund, C., Kahn, C., Lehto, V.-P., and Virtanen, I. (1984). Transient change of organization of vimentin filaments during mitosis as demonstrated by a monoclonal antibody. *Exp. Cell Res.* **154,** 567–580.

Gerace, L., and Blobel, G. (1980). The nuclear envelope lamina is reversibly depolymerized during mitosis. *Cell (Cambridge, Mass.)* **19,** 277–287.

Hanks, S. K., Rodriguez, L. V., and Rao, P. N. (1983). Relationship between histone phosphorylation and premature chromosome condensation. *Exp. Cell Res.* **148,** 293–302.

Hecht, R. M., Berg-Zabelshansky, M., and Davis, F. M. (1983). Embryonic-arrest mutant of *C. elegans* fails to phosphorylate mitosis-specific phosphoproteins. *J. Cell Biol.* **97,** 254a.

Hecht, R. M., Berg-Zabelshansky, M., Rao, P. N., and Davis, F. M. (1986). Detection of a temperature-sensitive point in the cell cycle of an embryonic-arrest mutant of *Caenorhabditis elegans*. *J. Cell Sci.,* submitted.

Henry, S. M., and Hodge, L. D. (1983). Nuclear matrix: A cell-cycle-dependent site of increased intranuclear protein phosphorylation. *Eur. J. Biochem.* **133,** 23–29.

Jost, E., and Johnson, R. T. (1981). Nuclear lamina assembly, synthesis and disaggregation during the cell cycle in synchronized HeLa cells. *J. Cell Sci.* **47,** 25–53.

Kaltoff, H., Darmer, D., Towbin, H., Gordon, J., Amons, R., Moller, W., and Richter, D. (1982). Ribosomal protein S_6 from *Xenopus laevis* oocytes. *Eur. J. Biochem.* **122,** 439–443.

Lake, R. S., Goidl, J. A., and Salzman, N. P. (1972). F_1 histone modification at the metaphase in Chinese hamster cells. *Exp. Cell Res.* **73**, 113–121.

Maller, J. L., Wu, M., and Gerhart, J. C. (1977). Changes in protein phosphorylation accompanying maturation of *Xenopus laevis* oocytes. *Dev. Biol.* **58**, 295–312.

Masui, Y. (1966). Relative roles of pituitary follicle cells and progesterone in induction of oocyte maturation in *Rana pipiens. J. Exp. Zool.* **166**, 365–376.

Miake-Lye, R., Newport, J., and Kirschner, M. (1983). Maturation-promoting factor induces nuclear envelope breakdown in cycloheximide-arrested embryos of *Xenopus laevis. J. Cell Biol.* **97**, 81–91.

Ottaviano, Y., and Gerace, L. (1985). Phosphorylation of the nuclear lamins during interphase and mitosis. *J. Biol. Chem.* **260**, 624–632.

Paulson, J. R., and Taylor, S. S. (1982). Phosphorylation of histones 1 and 3 and nonhistone high mobility group 14 by an endogenous kinase in HeLa metaphase chromosomes. *J. Biol. Chem.* **257**, 6064–6072.

Petersen, D. F., Tobey, R. A., and Anderson, E. C. (1969). Synchronously dividing mammalian cells. *Fed. Proc. Fed. Am. Soc. Exp. Biol.* **28**, 1771–1779.

Pettijohn, D. E., Henzl, M., and Price, C. (1984). Nuclear proteins that become part of the mitotic apparatus: A role in nuclear assembly? *J. Cell Sci. Suppl.* **1**, 87–201.

Rao, P. N. (1968). Mitotic synchrony in mammalian cells treated with nitrous oxide at high pressure. *Science* **160**, 774–776.

Rao, P. N., and Engelberg, J. (1966). Effect of temperature on the mitotic cycle of normal and synchronized mammalian cells. *In* "Cell Synchrony: Studies in Biosynthetic Regulation" (I. L. Cameron and G. M. Padilla, eds.), pp. 332–352. Academic Press, New York.

Robinson, S. I., Nelkin, B., Kaufmann, S., and Vogelstein, B. (1981). Increased phosphorylation rate of intermediate filaments during mitotic arrest. *Exp. Cell Res.* **133**, 445–449.

Sahasrabuddhe, C. G., Adlakha, R. C., and Rao, P. N. (1984). Phosphorylation of nonhistone proteins associated with mitosis in HeLa cells. *Exp. Cell Res.* **153**, 439–450.

Schlegel, R. A., and Rechsteiner, M. (1978). Red cell-mediated microinjection of macromolecules into mammalian cells. *Methods Cell Biol.* **20**, 341–354.

Shibayama, T., Nakaya, K., Matsumoto, S., and Nakamura, Y. (1982). Cell cycle-dependent change in the phosphorylation of the nucleolar proteins of *Physarum polycephalum* in vivo. *FEBS Lett.* **139**, 214–216.

Sunkara, P. S., Wright, D. A., and Rao, P. N. (1979). Mitotic factors from mammalian cells induce germinal vesical breakdown and chromosome condensation in amphibian oocytes. *Proc. Natl. Acad. Sci. U.S.A.* **76**, 2799–2802.

Tobey, R. A., Petersen, D. F., and Puck, T. T. (1966). Life cycle analysis of mammalian cells. III. The inhibition of division in Chinese hamster cells by puromycin and actinomycin. *Biophys. J.* **6**, 567–581.

Vandre, D. D., Davis, F. M., Rao, P. N., and Borisy, G. G. (1984a). Widespread distribution of mitosis specific phosphoproteins. *J. Cell Biol.* **99**, 449a.

Vandre, D. D., Davis, F. M., Rao, P. N., and Borisy, G. G. (1984b). Phosphoproteins are components of mitotic microtubule organizing centers. *Proc. Natl. Acad. Sci. U.S.A.* **81**, 4439–4443.

Vandre, D. D., Davis, F. M., Rao, P. N., and Borisy, G. G. (1986). Distribution of cytoskeletal proteins sharing a conserved phosphorylated epitope. *Eur. J. Cell Biol.* **41**, 72–81.

Yamashita, K., Davis, F. M., Rao, P. N., Sekiguchi, M., and Nishimoto, T. (1985). Phosphorylation of non-histone proteins during premature chromosome condensation in a temperature-sensitive mutant, tsBN2. *Cell Struct. Funct.* **10**, 259–270.

10

Mitosis-Specific Protein Phosphorylation Associated with Premature Chromosome Condensation in a *ts* Cell Cycle Mutant

TAKEHARU NISHIMOTO,* KOZO AJIRO,†
FRANCES M. DAVIS,‡ KATSUMI YAMASHITA,*
RYOSUKE KAI,*,1 POTU N. RAO,‡
AND MUTSUO SEKIGUCHI*

* Department of Biology, Faculty of Science, Kyushu University 33
Fukuoka 812 Japan

† Laboratory of Cell Biology, Aichi Cancer Center Research Institute
Nagoya 464, Japan

‡ Department of Medical Oncology
The University of Texas System Cancer Center
M.D. Anderson Hospital and Tumor Institute
Houston, Texas 77030

I. INTRODUCTION

The cell cycle consists of four phases: G_1, S, G_2, and M. In the G_1 phase, there is a progressive decondensation of the chromosomes, which reaches its maximum at the beginning of the S phase, as demonstrated by

---

[1] Deceased (May 22, 1986).

MOLECULAR REGULATION OF NUCLEAR EVENTS
IN MITOSIS AND MEIOSIS

Johnson and Rao (1970) and Rao *et al.* (1976). On completion of the S phase, the chromosomes begin to recondense. The cell cycle can therefore be viewed as a cycle of chromosome condensation and decondensation.

During the G_2 phase, chromosome condensing factor is produced; it accumulates as a cell advances toward the M phase (Sunkara *et al.*, 1979). If completion of the S phase is inhibited by drugs such as bleomycin or neocarzinostatin (Rao and Rao, 1976; Ishida *et al.*, 1979), or by a temperature-sensitive mutation (Nishimoto *et al.*, 1980; Yasuda *et al.*, 1981), progression of the cell cycle halts at the S/G_2 boundary; this is known as the G_2 block. Therefore, at the S/G_2 boundary, there is a regulatory mechanism that recognizes the completion of the S phase and triggers a cascade of events leading to mitosis. Cells that are blocked in the G_2 phase do not initiate this cascade of events. Such a regulatory mechanism is important for cell proliferation because the activity of chromosomal DNA is severely suppressed by the condensation of chromosomes (Lewin, 1974). If the cascade of events leading toward mitotic chromosome condensation occurs during interphase, the cells will die as a result of inhibition of macromolecular synthesis.

The cell line, tsBN2, appears to have a temperature-sensitive (*ts*) defect in such a control mechanism for chromosome condensation (Nishimoto *et al.*, 1981). At the nonpermissive temperature, tsBN2 cells show premature chromosome condensation (PCC) similar to that reported by Johnson and Rao (1970) and Rao *et al.* (1976). Since the condensation of chromosomes can be induced by shifting from the permissive to the nonpermissive temperature, the tsBN2 cell line is a sensitive system for studying the cascade of events involved in chromosome condensation.

II. ISOLATION AND PRELIMINARY CHARACTERIZATION OF THE tsBN2 CELL LINE

The tsBN2 cell line was selected using a procedure designed to isolate *ts-dna*⁻ (DNA synthesis negative) mutants (Nishimoto and Basilico, 1978). However, subsequent studies revealed the defect to be somewhat more complex and interesting.

A. Selection of the tsBN2 Cell Line

BHK21/13 cells synchronized in $G_1(0)$ by serum starvation were exposed to 1 μg/ml *N*-methyl-*N*¹-nitro-*N*-nitrosoguanidine (MNNG) at 5 hr after serum addition and then incubated for 17 hr at 33.5°C. Following subculturing at a density of 4×10^5 cells/100-mm dish, cells were synchronized by serum starvation at 33.5°C, then shifted to 37.5°C immediately after release from the block with medium containing 20% calf serum. Just before initiation of cellular DNA synthesis, FUdR (25 µg/ml) was added to cultures. Cells able to replicate their DNA at 37.5°C, therefore, suffered lethal DNA damage. After 2 days the cells were washed free of FUdR and incubated in fresh medium containing 10% calf serum and 20 µg/ml of thymidine at 33.5°C. The tsBN2 mutant was isolated following this selection procedure (Nashimoto and Basilico, 1978).

At the nonpermissive temperature of 39.5–40.5°C, DNA replication ceases very rapidly in tsBN2 cells. Cells synchronized at the G_1/S boundary at the permissive temperature become halted in mid-S phase when released into complete medium at the nonpermissive temperature (Fig. 1). Initiation of DNA replication is blocked, but elongation of nascent chains is not inhibited (Eilen *et al.*, 1980).

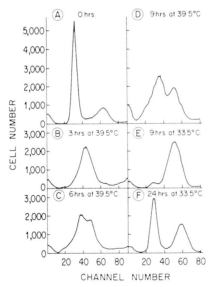

Fig. 1. DNA content of tsBN2 cells traversing the S phase at 39.5 and 33.5°C. Asynchronously growing cells at 2×10^5/100-mm dish were synchronized at the G_1/S boundary by sequential incubation in isoleucine-free (Ile⁻) medium and complete medium containing HU at 33.5°C (Nishimoto *et al.*, 1980). After washing and removal of HU, half of the cultures were shifted to 39.5°C. At the indicated times after release, cells were prepared for cytofluorographic analysis. Channel 30 corresponds to a G_1 DNA content. It should be noted that this cell population contained a large proportion of tetraploid cells (Nishimoto *et al.*, 1978).

B. Defect in G_1 Traverse of tsBN2 Cells

Progression of tsBN2 cells through the G_1 phase also seems to be defective at the nonpermissive temperature. Following synchronization in the $G_1(0)$ phase by isoleucine deprivation, cultures of tsBN2 cells were incubated at either 33.5 or 40.5°C, and labeled with [³H]uridine or a ³H-labeled amino acid mixture. At 40.5°C, incorporation of both [³H uridine and ³H-labeled amino acids decreased rapidly (Fig. 2), thereby suggesting that both RNA and protein synthesis were inhibited at the nonpermissive temperature in the G_1 phase (Nishimoto *et al.*, 1981). This inhibition was not as rapid in asynchronous cultures or in cells synchronized at the G_1/S boundary. Furthermore, the delay in entry into S phase following the reversal of the G_1 block was dependent on the length of time cells had been at 39.5°C (Nishimoto *et al.*, 1978) and was similar to that observed

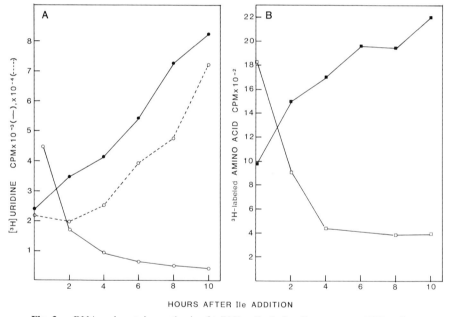

Fig. 2. RNA and protein synthesis of tsBN2 cells during G_1 traverse. tsBN2 cells were seeded at 2×10^4 cells/35-mm dish and synchronized at 33.5°C by Ile⁻ medium for 64 hr. Following isoleucine readdition, half of the dishes were shifted to 40.5°C. At the times indicated, a 30-min pulse of [5-³H]uridine (5 μCi/ml, 2.5 Ci/mmole) at 33.5 (-●-●-) and 40.5°C (-○-○-) or ³H-labeled amino acids (5 μCi/ml) at 33.5 (-■-■-) and 40.5°C (-□-□-) were given, after which acid-insoluble radioactivity was counted. As a control, cultures of wild-type BHK21 cells were seeded, synchronized similarly, and given a 30-min pulse of [5-³H]uridine at 40.5°C (-○-○-) (Nishimoto *et al.*, 1981).

when normal cells were treated during G_1 phase with an inhibitor of protein synthesis (Terasima and Yasukawa, 1966). Thus, at the nonpermissive temperature, the progression of tsBN2 cells through G_1 phase is halted, probably because of a defect in protein synthesis.

It appears that all of the *ts* characteristics of the tsBN2 cell line are linked to a single mutation because the spontaneous reversion frequency of these cells is 1.6×10^{-8}, a value consistent with that of mutants with a single mutated site (Luria and Delbruck, 1943).

III. INDUCTION OF PREMATURE CHROMOSOME CONDENSATION (PCC) IN tsBN2 CELLS AT THE NONPERMISSIVE TEMPERATURE

Fusion between interphase and mitotic cells results in the condensation of interphase chromatin into chromosomes (Johnson and Rao, 1970; Rao *et al.*, 1976). The prematurely condensed chromosomes (PCCs) of cells in G_1 phase are long and single stranded, while those of cells in S phase appear as pulverized chromatin masses. The G_2 PCCs are relatively longer than metaphase chromosomes and consist of two chromatids. A similar phenomenon was observed when tsBN2 cells were incubated at the nonpermissive temperature (Nishimoto *et al.*, 1978, 1981).

A. Morphology of PCCs

When tsBN2 cells in early S phase at the permissive temperature were shifted to 40.5°C, PCCs typical of S phase (pulverized appearance) were induced (Fig. 3A). Some cells with a morphology of late G_1 PCCs were also observed (Fig. 3B). Typical G_2 PCCs were observed when tsBN2 cells in late S phase were transferred to 40.5°C (Fig. 3C) (Nishimoto *et al.*, 1981). Although reversible synchronization of cells in G_2 phase is difficult, cells can be irreversibly arrested in G_2 phase by administering bleomycin or neocarzinostatin (Rao and Rao, 1976; Ishida *et al.*, 1979). When tsBN2 cells that had been arrested in G_2 phase at the permissive temperature by neocarzinostatin were shifted to 40.5°C, G_2 PCCs were observed (Ishida *et al.*, 1985). Since neocarzinostatin induces strand breaks in DNA (Ishida and Takahashi, 1978), many fragmented and double minute chromosomes were observed among the G_2 PCCs. Thus, the phenomenon of PCC, which occurs in interphase cells upon fusion with mitotic cells, can be induced in tsBN2 cells just by raising the temperature from 33.5°C to 39.5°C. Furthermore, tsBN2 cells showing PCC possess

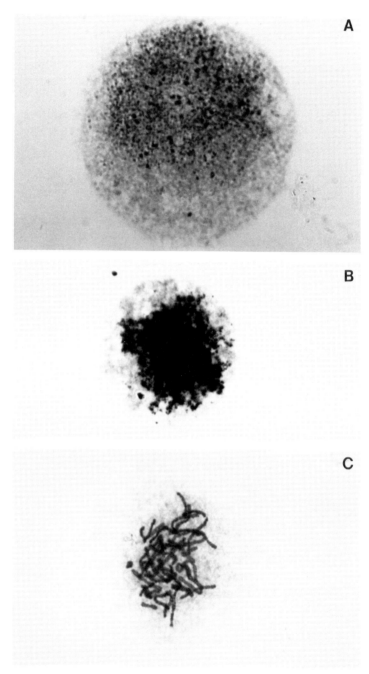

Fig. 3. Morphology of PCCs. tsBN2 cells were seeded and synchronized at the G_1/S boundary. Following washing and transfer to complete medium, the cultures were shifted to 40.5°C at 2 hr (A and B) or 6 hr (C). After incubation for 3 hr at 40.5°C, cells were collected, swollen, fixed, and stained with Giemsa. Slides were photographed on an Olympus photomicroscope at 250× using Fuji Mimicopy film (Nishimoto *et al.*, 1981).

the ability to induce condensation of the chromatin of interphase cells upon fusion (Hayashi *et al.*, 1982).

B. Dependence of PCC Induction on RNA and Protein Synthesis

The induction of PCC in tsBN2 cells shifted to the nonpermissive temperature is blocked by a low dose (0.5 μg/ml) of cycloheximide (Nishimoto *et al.*, 1981). This result is consistent with the idea that the chromosome condensing protein(s) that can trigger the cascade of events leading to mitosis may be newly synthesized at 39.5°C in tsBN2 cells showing PCC. Almost all of the proteins required for chromosome condensation are probably synthesized during the first 2 hr after the temperature shift, since cycloheximide had no effect on the appearance of PCCs after this period (Nishimoto *et al.*, 1981).

Induction of PCC was only partially blocked by a relatively high dose of actinomycin D (2 μg/ml), thereby suggesting that new RNA synthesis is not essential for the induction of PCC in tsBN2 cells at the nonpermissive temperature.

When tsBN2 cells blocked in G_1 are shifted to 39.5°C, PCCs are not induced. The ability to produce PCCs begins at the same time as cellular DNA synthesis with the frequency of cells showing PCC increasing progressively from the G_1/S boundary to the G_2 phase (Nishimoto *et al.*, 1981). These results suggest that the synthesis of mRNA required for the cascade of events leading to mitosis begins at the G_1/S boundary and that translation of the mRNA occurs during the G_2-M phase. Thus, chromosome condensation is assumed to be regulated in a posttranscriptional fashion.

Although tsBN2 cells begin to show PCC at the G_1/S boundary, DNA synthesis is not essential for the induction of PCC, because hydroxyurea and aphidocolin, strong inhibitors of DNA replication, do not inhibit PCC induction (Nishimoto *et al.*, 1981).

C. Macromolecular Synthesis During PCC Induction

Entry of cells into mitosis is accompanied by a cessation of RNA and protein synthesis (Lewin, 1974). A similar decrease in macromolecular synthesis occurs during the induction of PCC (Nishimoto *et al.*, 1980, 1981). When cultures of tsBN2 cells synchronized at the G_1/S boundary were shifted to the nonpermissive temperature, RNA synthesis was inhibited within 1 hr, and then DNA synthesis decreased. At 3 hr after the temperature shift, protein synthesis ceased. Thus, with initiation of chromosome condensation, DNA strands coiled progressively, resulting in the inhibition of macromolecular synthesis.

When tsBN2 cells synchronized in the $G_1(0)$ phase by isoleucine deprivation were incubated at 40.5°C, RNA and protein synthesis were inhibited (Fig. 2) but PCC did not occur. However, a slight phosphorylation of histones H1 and H3 was observed, thereby suggesting that early events related to chromosome condensation do occur in tsBN2 cells in G_1 phase at the nonpermissive temperature (Ajiro *et al.*, 1983). Even a slight condensation of chromosomes might be sufficient to inhibit both RNA and protein synthesis of cells in the early G_1 phase, because chromatin is still in a relatively more condensed state in this phase, compared to the S phase (Rao *et al.*, 1976). Thus, we consider that the inhibition of both RNA and protein synthesis in tsBN2 cells in the $G_1(0)$ phase (at the nonpermissive temperature) may be caused by slight changes in chromosome condensation.

IV. INDUCTION OF MITOSIS-SPECIFIC PHOSPHORYLATION IN tsBN2 CELLS BY TEMPERATURE SHIFT

With a shift from the permissive to the nonpermissive temperatures, both histone and nonhistone proteins of tsBN2 were phosphorylated, as is the case during the entry of cells into mitosis (Ajiro *et al.*, 1983; Yamashita *et al.*, 1985). This phosphorylation appeared simultaneously with the induction of chromosome condensation and disappeared simultaneously 4 hr later. Cycloheximide completely inhibited histone phosphorylation as well as PCC formation. These events were partially inhibited by actinomycin D (Ajiro *et al.*, 1983).

A. Phosphorylation of Histones H1 and H3 in tsBN2 Cells Showing PCC

In mitotic cells, histones H1 and H3 are specifically phosphorylated (Bradbury *et al.*, 1974; Gurley *et al.*, 1974, 1978; Ajiro *et al.*, 1981a). While the extent of phosphorylation of histone H1 changes dynamically during the cell cycle and is at a maximum in the mitotic (M) phase, histone H3 is phosphorylated only in the M phase. In addition, the sites phosphorylated on histone H1 molecules differ between S and M phases (Ajiro *et al.*, 1981b).

To date, a casual relationship between histone phosphorylation and mitotic chromosome condensation has not been established. However, since chromosome condensation is induced in tsBN2 cells by a shift to the nonpermissive temperature, this offers a unique system for investigating the relationship between histone phosphorylation and chromosome condensation.

1. H1 and H3 Phosphorylation

Cultures of tsBN2 cells were synchronized at the G_1/S boundary by sequential incubation in isoleucine-free (Ile$^-$) medium and medium containing hydroxyurea (HU) (Nishimoto *et al.*, 1980). Cells were labeled with ortho[^{32}P]phosphate, and acid-soluble nuclear proteins extracted from the labeled cells were analyzed by electrophoresis into polyacrylamide gels containing acid–urea and Triton X-100 (Ajiro *et al.*, 1983).

Electrophoretic patterns of the total proteins were the same for extracts from cultures at the permissive and nonpermissive temperatures (Fig. 4 and 5). Histone H3 has three distinct variants (H3.1, H3.2, and H3.3) that can be resolved using this gel system (Fig. 5). Since H3.1 has a rather low electrophoretic mobility, only H3.2 and H3.3 were analyzed in the experiments shown in Figs. 4 and 5 (Ajiro *et al.*, 1983). At 33.5°C, incorporation of ^{32}P into histone H1 was very low and that into histone H3 was not detectable at the G_1/S boundary (in the presence of HU). Following removal of HU, histone H1 was phosphorylated during S phase, and the level increased from G_2 through M phase. However, H3 was phosphorylated only in M phase, as reported previously (Gurley *et al.*, 1974).

At 40.5°C, histone H1 was heavily phosphorylated at all time points. Histone H3 was only phosphorylated at the time of appearance of PCC. This observation was confirmed by using only a short pulse of ortho[^{32}P]-phosphate (Fig. 5). At 2 hr after the temperature shift, the frequency of tsBN2 cells showing PCC increased significantly (Nishimoto *et al.*, 1981). At the same time, the level of histone H1 phosphorylation was increased and the phosphorylation of histone H3 began. The level of phosphorylation of histones H1 and H3 continued to increase until 3 hr after the temperature shift and then decreased rapidly, although the frequency of PCC increased further (see below, Fig. 11). Why such a reduction of histone phosphorylation occurs has yet to be determined.

Since histone H1 is phosphorylated in both S and M phases, but the pattern of H1 phosphopeptides differs between S and M phases (Ajiro *et al.*, 1981b), the phosphorylation pattern of histone H1 in tsBN2 cells showing PCC was compared with that of mitotic cells. Histone H1 was extracted from cells labeled with ^{32}P during M phase at 33.5°C and during

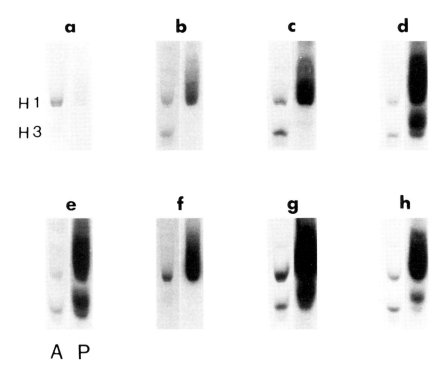

Fig. 4. Acid–urea–Triton X-100 gel electrophoresis showing histone H1 and H3 phosphorylation of tsBN2 cells at permissive (a–d) and nonpermissive (e–h) temperatures. Cultures of tsBN2 cells were synchronized at the G_1/S boundary, as in Fig. 1. Cells in (a) and (e) were labeled with ortho[^{32}P]phosphate, in the presence of HU, as follows: (a) for 4 hr at 33.5°C and (e) for 3 hr at 40.5°C. Following release from HU, cells were labeled with ortho[^{32}P]phosphate at 33.5°C as follows (b) 0–4 hr (S); (c) 4–8 hr (G_2); (d) 8–12 hr (M); and, at 40.5°C as follows: (f) 0–3 hr; (g) 3–6 hr; (h) 6–9 hr. A, Amido black staining; P, ^{32}P-labeled autoradiography. Mitotic indexes at 33.5°C were 0.2% (b), 4% (c), and 27.7% (d); PCC indexes at 40.5°C were 1.6% (f), 37.1% (g), and 35.4% (h). The PCC indexes of (a) and (e) were 0.1 and 44.1%, respectively (Ajiro *et al.,* 1983).

PCC induction at 40.5°C. The ^{32}P-labeled H1 was digested with trypsin, and the peptides were resolved by electrophoresis and thin layer chromatography (TLC). The pattern of H1 phosphopeptides in mitotic cells and in tsBN2 cells showing PCC showed no significant differences in either the number or the position of the ^{32}P-labeled spots (Fig. 6). Thus, the mitosis-specific phosphorylation of histone H1 appears to be induced in the tsBN2 cells by the temperature shift.

Phosphorylation of histones was also observed when tsBN2 cells cells in G_1 phase were transferred to 40.5°C, as mentioned earlier. These cells

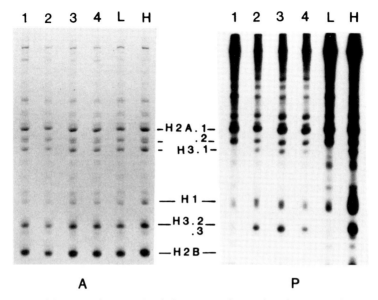

Fig. 5. Acid–urea–Triton X-100 gel electrophoresis showing phosphorylation of acid-soluble nuclear proteins of tsBN2 cells at the nonpermissive and permissive temperature. Cultures of tsBN2 cells synchronized at the G_1/S boundary were labeled every hour after the temperature shift to 40.5°C in the presence of HU as follows: (1) 0–1 hr, (2) 1–2 hr, (3) 2–3 hr, (4) 3–4 hr. As a control, one set of cultures was labeled for 4 hr at 33.5°C (L) and another at 40.5°C (H). A, Amido black-stained proteins; P, ^{32}P-labeled autoradiography (Nishimoto *et al.*, 1985).

show a slight but significant increase in the phosphorylation of both histones H1 and H3, thereby suggesting that the chromosomes may be slightly condensed, although the nuclear membrane remains intact (Ajiro *et al.*, 1983). This observation is consistent with the rapid decrease in total RNA and protein synthesis following a temperature shift to 40.5°C (Nishimoto *et al.*, 1981; Fig. 2).

2. Effect of Various Drugs on Phosphorylation of Histones H1 and H3

Cycloheximide at 10 μg/ml inhibited the phosphorylation of histones H1 and H3 in tsBN2 cells at 40.5°C, while actinomycin D at 2 μg/ml was only partially inhibitory (Fig. 7). Inhibitors of DNA synthesis did not affect the phosphorylation of histones H1 and H3 in tsBN2 cells at the nonpermissive temperature, suggesting that their phosphorylation at 40.5°C is not the result of a rapid transit of cells through the cell cycle at

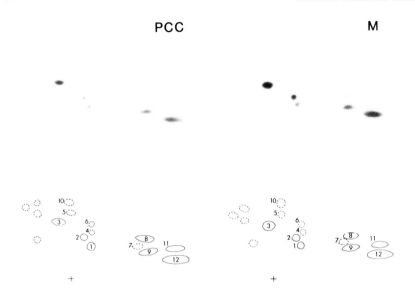

Fig. 6. Autoradiographs of ^{32}P-labeled tryptic phosphopeptides of H1 from PCCs and mitotic cells. Cultures of tsBN2 cells were synchronized at the G_1/S boundary. PCC sample: the cultures were shifted to 40.5°C and labeled with ^{32}P in the presence of HU for 3 hr. M sample: after release from HU treatment, cultures were incubated at 33.5°C and labeled with ^{32}P from the seventh to the twelfth hour (Mitotic phase). Following the labeling, H1 was extracted and purified by chromatography on Amberlite CG-50 resin. Fifty micrograms of the H1 fraction from the PCC and M sample were digested with 2% trypsin and then resolved by electrophoresis and chromatography on TLC plates (Ajiro *et al.*, 1983).

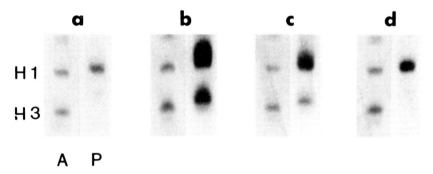

Fig. 7. Acid–urea-Triton X-100 gel electrophoresis showing the effect of metabolic inhibitors on H1 and H3 phosphorylation. A series of cultures synchronized at the G_1/S boundary were labeled with ortho[^{32}P]phosphate in the presence of HU at 33.5°C (a) or at 40.5°C (b–d) for 3 hr. (b) HU only, (c) HU plus 2 μg/ml or actinomycin D, (d) HU plus 10 μg/ml of cycloheximide. A, Amido black staining; P, ^{32}P-labeled autoradiography (Ajiro *et al.*, 1983).

the high temperature. The presence of HU also did not affect phosphorylation of histones H1 and H3 at either 33.5 or 40.5°C (Ajiro *et al.*, 1983). Thus, the effects of these drugs on the phosphorylation of histones were the same as those shown for PCC induction in tsBN2 cells at 40.5°C (Nishimoto *et al.*, 1981), suggesting that there is a close correlation between the phosphorylation of histones and the induction of chromosome condensation.

It is of interest that histone H1 is phosphorylated at 40.5°C to the same extent as at 33.5°C in the presence of cycloheximide and that histone H2A is also phosphorylated at the same level, either in the presence or absence of cycloheximide. It is possible that these phosphorylation reactions may be carried out by a protein kinase that is different from the kinase required for mitotic-specific phosphorylations. Thus, more than two different protein kinases may be involved in the phosphorylation of histone H1, consistent with the finding that W-7, a specific inhibitor of calmodulin (Hidaka *et al.*, 1981), inhibits mitosis-specific phosphorylation but not the phosphorylation of histone H2A nor the residual phosphorylation of histone H1 (Nishimoto *et al.*, 1985).

However, there remains another possibility: condensation of chromosomes might make histones H1 and H3 accessible to a protein kinase that phosphorylates histones H2A and H1 in S phase. Hence, histones H1 and H3 are phosphorylated in a mitosis-specific fashion. If this is the case, the mitotic phosphorylation of histones H1 and H3 may be a secondary process due to chromosome condensation.

B. Phosphorylation of Nonhistone Proteins in tsBN2 Cells Showing PCC

A specific set of nonhistone proteins is phosphorylated at mitosis, demonstrated using the monoclonal antibody, MPM-2, which reacts specifically with mitotic cells and recognizes phosphorylated proteins present only in the mitotic phase (Davis *et al.*, 1983). Upon treatment with phosphatase, the mitotic antigens lose their reactivity with the antibody.

The presence of MPM-2-reacting antigens in tsBN2 cells showing PCC was investigated by reacting the cells with the antibody. A culture of tsBN2 cells was synchronized at the G_1/S boundary, held there, and incubated at 33.5 or 40.5°C. Following incubation for 3 hr, cells were detached using a Teflon scraper, deposited on glass slides by cytocentrifugation, fixed in methanol, and then stained for indirect immunofluorescence using the MPM-2 antibody. DNA in nuclei or in chromosomes of S-phase PCCs

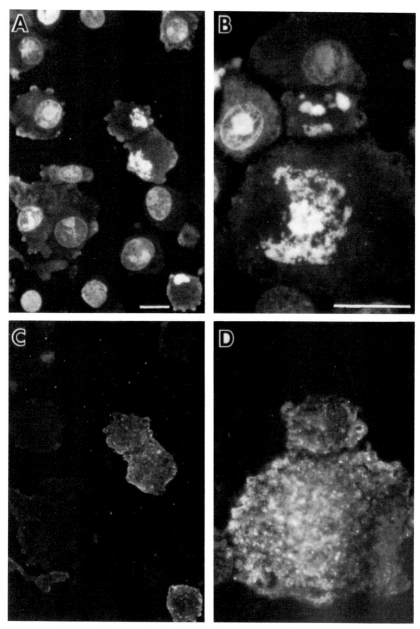

Fig. 8. MPM-2 reactivity of tsBN2 cells showing PCC at the nonpermissive temperature. Cultures of tsBN2 cells synchronized at the G₁/S boundary were incubated at 40.5°C for 3 hr in the presence of HU, deposited on slides using a cytocentrifuge, and fixed in methanol. Cells were stained using MPM-2 antibodies and FITC-conjugated rabbit anti-mouse IgG (Miles, Laboratories, Elkhart, IN). DNA in nuclei, chromosomes, and PCCs was

was stained with propidium iodide. The antibody MPM-2 stained cells containing PCCs but did not stain cells with nuclei (Fig. 8).

Although MPM-2 is a monoclonal antibody, it appears to recognize a set of phosphorylated proteins in mitotic cells (Davis *et al.*, 1983). Therefore, proteins reactive with MPM-2 in tsBN2 cells showing PCC were compared with those of mitotic cells. Following synchronization at the G_1/S boundary, cultures of the tsBN2 cell line and the wild-type BHK21/13 cell line were incubated either at 33.5 or 40.5°C. As a control, mitotic cells of the tsBN2 cell line were collected with Colcemid. Whole cell extracts from these cells were electrophoresed on gradient SDS-polyacrylamide gels and analyzed by the Western blotting method (Burnette, 1981) using MPM-2. The composition of MPM-2-reactive antigens in tsBN2 cells showing PCC was the same as that in mitotic cells (Fig. 9). The extract from tsBN2 cells incubated at 33.5°C or from BHK21 cells incubated at 40.5°C reacted only slightly with antibody MPM-2. As has been shown for normal mitotic cells, the MPM-2-reactive antigens in tsBN2 cells showing PCC were also phosphorylated (Yamashita *et al.*, 1985). These results demonstrate that nonhistone proteins in tsBN2 cells showing PCC are phosphorylated in the same manner as in mitotic cells.

MPM-2-reactive antigens appeared at the same time as PCC following the temperature shift (Fig. 9). The frequency of cells having MPM-2-reactive antigens was maximal at 3 hr after the temperature shift and then decreased rapidly, although the frequency of cells with PCC increased further (Yamashita *et al.*, 1985). All MPM-2-reactive antigens appeared synchronously at 2 hr after the temperature shift, when chromosome condensation is initiated. The levels of MPM-2-reactive antigens reached a maximum at 3 hr after the temperature shift and then decreased. At 5 hr after the temperature shift, the MPM-2-reactive antigens were undetectable. These changes in MPM-2-reactive antigens were the same as those seen with phosphorylation of histones H1 and H3 (Fig. 5). Probably, phosphorylation of both histones and nonhistone proteins is involved in chromosome condensation. This idea is consistent with the finding that the calmodulin inhibitor W-7 inhibits both PCC induction in tsBN2 cells at 40.5°C and the mitosis-specific phosphorylation of histones and nonhistones (Nishimoto *et al.*, 1985; Yamashita *et al.*, 1985).

These studies indicate a close correlation between histone and nonhis-

---

stained by adding 0.3 μg/ml of propidium iodide to the last phosphate-buffered saline wash before mounting. A and B, propidium iodide red fluorescence; C and D, antibody FITC green fluorescence of the same cells as in A and B. The bars represent 20 μm (Yamashita *et al.*, 1985).

M   0 0.5 1 1.5 2 2.5 3   4   5   M

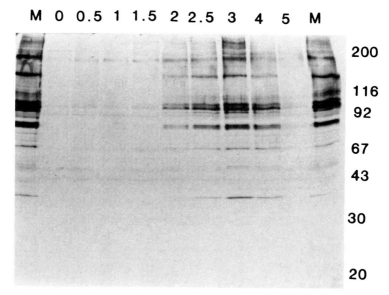

200

116
92

67

43

30

20

Fig. 9.   Induction of antigens detected by immunoblotting. Cultures of tsBN2 cells synchronized at the beginning of S phase were incubated at 40.5°C. At 0, 0.5, 1, 1.5, 2, 2.5, 3, 4, and 5 hr following the temperature shift, cells were collected and antigens were analyzed by 7.5–15% linear gradient SDS–polyacrylamide gel electrophoresis followed by immunoblotting with MPM-2. The molecular weights of proteins are indicated as $M_r \times 10^{-3}$. (M) = tsBN2 mitotic cells at 33.5°C on the left, and wild-type BHK21 mitotic cells on the right. PCC indexes of the samples for lanes (0) through (5) were 0, 0.3, 1.4, 5, 24, 40, 47, 60, and 55%, respectively (Yamashita *et al.*, 1985).

tone phosphorylation and chromosome condensation. We suggest, therefore, that the phosphorylation of proteins may trigger chromosome condensation and nuclear membrane breakdown.

V. NEWLY SYNTHESIZED PROTEIN(S) IN tsBN2 CELLS SHOWING PCC

As mentioned earlier, the induction of chromosome condensation in tsBN2 cells is completely inhibited by a low dose of cycloheximide, suggesting that new protein synthesis is required for chromosome condensation. Since tsBN2 cells showing PCC also possess the ability to induce PCC in interphase cells to which they are fused (Hayashi *et al.*, 1982), the chromosome condensing factor(s) that normally accumulates during the G_2 phase and reaches a maximum at metaphase (Sunkara *et al.*, 1979) can be assumed to be produced in tsBN2 cells at the nonpermissive temperature (Hayashi *et al.*, 1982). Therefore, we attempted to identify a protein(s) that may be responsible for chromosome condensation in tsBN2 cells showing PCC.

A. An Acidic 35K Phosphoprotein

tsBN2 cells were synchronized at the beginning of S phase, held there, and labeled with [³H]methionine, either at 33.5 or 40.5°C, for 4 hr. As a control, cultures of the BHK21/13 cell line synchronized at the beginning of S phase were labeled with [³H]methionine at 40.5°C. Total cellular proteins extracted from these cells were then analyzed by two-dimensional gel electrophoresis (Yamashita *et al.*, 1984). As shown in Fig. 10, an acidic 35K protein was found to be produced specifically at 40.5°C. Although two other proteins of 38 and 54K were also produced in tsBN2 cells at 40.5°C, they were considered to be a type of heat-shock protein because they were also observed in the wild-type BHK21 cells incubated at 40.5°C. This 35K protein appears only in the G_2–M phase at the permissive temperature and is observed in the chromosome fraction of both mitotic cells and tsBN2 cells showing PCC. At the physiological ionic concentration of 150 mM NaCl, this protein is bound to chromatin (Yamashita *et al.*, 1984).

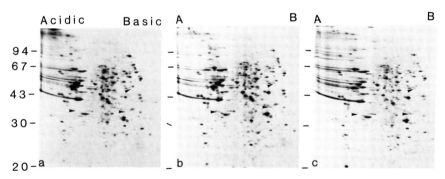

Fig. 10. Two-dimensional gel electrophortic patterns of [³⁵S]methionine-labeled total cellular proteins in tsBN2 and BHK21 cells. Cultures of tsBN2 or BHK21 cells were synchronized at the G_1/S boundary and labeled for 4 hr with [³⁵S]methionine at 33.5°C, (a) tsBN2; or 40.5°C, (b) tsBN2 and (c) BHK21. The PCC index of tsBN2 at 33.5°C was less than 2% and at 40.5°C was more than 50%, while that of BHK21 at 40.5°C was less than 1%. The molecular weights of protein standards are indicated as $M_r \times 10^{-3}$. Arrowheads indicate proteins specifically labeled in cells exposed to the nonpermissive temperature. 5×10^5 cpm of each sample was applied to the first-dimension gels (Yamashita *et al.*, 1984).

Like histone and nonhistone proteins in mitotic cells, the 35K protein was also phosphorylated because this acidic species moved to a more basic position on two-dimensional gels if it was pretreated with bacterial alkaline phosphatase. The 35K protein is, therefore, one of the mitosis-specific phosphorylated nonhistone proteins, and this protein may correspond to the 35K protein that is recognized by MPM-2 in tsBN2 cells showing PCC (Fig. 9) (Yamashita *et al.*, 1985). As to the function of this 35K protein, the following two possibilities have to be considered. This protein might bind stoichiometrically to a chromatin component, such as histone H1, and such binding may lead to a conformational change of the chromatin toward chromosome condensation. Phosphorylation may be necessary to make this protein accessible to chromatin. Another possibility is that this 35K protein may have protein kinase activity, since these enzymes are usually autophosphorylated. This idea is consistent with the finding that cyclohcximide inhibits both the mitosis-specific phosphorylation of histones H1 and H3 and the induction of PCC. Also, in *Schizosaccharomyces pombe,* the product of the *cdc 2* gene, which is a 35K protein with protein kinase activity, initiates mitosis (Nurse and Thuriaux, 1980; Nurse and Bisset, 1981; Hindley and Phear, 1984). In light of these observations it is reasonable to assume that in animal cells, some protein kinases may be involved in the initiation of mitosis.

B. Role of Calmodulin in Chromosome Condensation

These seems to be a correlation between chromosome condensation factor(s) and the intracellular concentration of calmodulin, which is at its lowest in G_1 phase, increases in two steps during the cell cycle, first at the G_1/S boundary and then in late G_2 phase (Sasaki and Hidaka, 1982). In addition, Wasserman and Smith (1981) also found that germinal vesicle breakdown (GVBD) and chromosome condensation in immature *Xenopus* oocytes could be induced by the injection of calmodulin.

When PCC were induced in G_1/S-phase tsBN2 cells by temperature shift, the cellular content of calmodulin increased before the initiation of chromosome condensation (Fig. 11). At 3 hr after the temperature shift, the content of calmodulin was maximal and then decreased rapidly, as was the case with the level of phosphorylation of histone and nonhistone proteins (Nishimoto *et al.*, 1985). These data suggest that calmodulin may have a role in chromosome condensation and are consistent with the finding that W-7 inhibits both mitosis-specific phosphorylation of proteins and formation of PCCs.

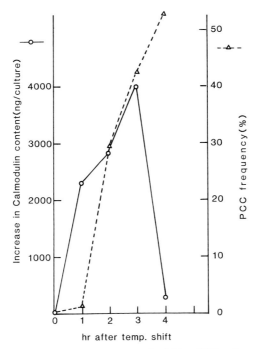

Fig. 11. Induction of camodulin synthesis and PCC in tsBN2 cells at the nonpermissive temperature. tsBN2 cells were seeded at 1×10^5 cells/50-mm dish and 6×10^5 cells/100-mm dish and synchronized at the G_1/S boundary. After HU treatment, cultures were washed with TD (Tris-buffered saline, without $CaCl_2$ and $MgCl_2$), fed serum-free DME medium containing 2.5 m*M* HU and then incubated at 40.5°C. Every hour, following the temperature shift, cells were collected from six 100-mm dishes to examine the increase in calmodulin (CaM) content (ng/culture) (-○-○-) and from one 50-mm dish to examine the PCC frequency (-△-△-). Increase in calmodulin content was estimated by subtracting CaM at time 0 from CaM at time t. CaM at time 0 was 8901 ng/culture. The specific content of calmodulin (CaM/total proteins) was 0 hr = 3.97 ng/μg; 1 hr = 4.76 ng/μg; 3 hr = 5.32 ng/μg; 4 hr = 3.61 ng/μg (Nishimoto *et al.*, 1985).

VI. DISCUSSION

In *S. pombe,* the *cdc 2* gene product initiates the mitotic phase and has homology with protein kinases (Nurse and Thuriaux, 1980; Hindley and Phear, 1984). It has been proposed that *cdc 2* gene activity is regulated by the *wee 1* gene product (Nurse and Thuriaux, 1980). The effect of the *wee 1* mutant is to advance cells through G_2 into mitosis earlier than normal and to diminish the size of *S. pombe,* possibly due to increased chromatin condensation caused by the continuous expression of the *cdc 2* gene. This behavior is analogous to the PCC behavior described in this chapter. Therefore, *wee 1* and *cdc 2* gene functions in *S. pombe* may be related to the tsBN2 mutant function in hamster cells. The interrelationships of the genes regulating mitosis in *S. pombe* (Fantes, 1981) are shown in Fig. 12. In animal cells, on the other hand, chromosome condensation in interphase, that is, PCC, is lethal to cells, and hence cells with mutations like *wee 1* cannot survive (Nishimoto *et al.,* 1981). These mutants can be isolated only as *ts* mutants in animal cells. No other mutant like the tsBN2 cell line has been isolated.

By shifting tsBN2 cells from the permissive to the nonpermissive temperature, phenomena that are similar to normal mitotic events are induced, such as chromosome condensation and mitosis-specific phosphorylation of histone and nonhistone proteins. These mitosis-related phenomena never appear in normal cells if progression of cells through S and G_2 phases is inhibited. Thus, there appears to be a tight coupling between the completion of DNA synthesis and preparation for mitosis during the normal cell cycle.

The mitosis-specific phosphorylation of histone and nonhistone proteins was induced at the same time as the appearance of chromosome condensation (Nishimoto *et al.,* 1985; Yamashita *et al.,* 1985). Both chromosome condensation and mitosis-specific phosphorylation of proteins were inhibited with a low dose of cycloheximide (Nishimoto *et al.,* 1981; Ajiro *et al.,* 1983). Therefore, it is reasonable to assume that with a temperature shift some protein kinase(s) or an activator of protein kinase may be newly synthesized in tsBN2 cells. By analogy with *S. pombe,* the "BN2" gene may activate gene *X,* corresponding to the *cdc 2* gene of *S. pombe,* whose product may possess protein kinase activity (Fig. 12). Such a protein kinase may initiate chromosome condensation by phosphorylating histone and nonhistone proteins, although it remains to be determined how the phosphorylation of histone and nonhistone proteins induces chromosome condensation. It is still possible that such phosphorylation are a secondary effect of the chromosome condensing factor(s).

It seems likely that the product of the tsBN2 gene is a *ts* regulator for the cascade of events leading to mitosis. Recently, using DNA-mediated gene transfer (Scango and Ruddle, 1981), we cloned a human DNA fragment of about 40 kb from *ts*[+] secondary transformants (Kai *et al.,* 1986). This fragment converts the *ts* phenotype of tsBN2 cells to *ts*[+] with an efficiency 100 times greater than transfection with total human DNA, indicating that the fragment contains a biologically active gene complementing the tsBN2 mutation and proving that tsBN2 cells have a *ts* muta-

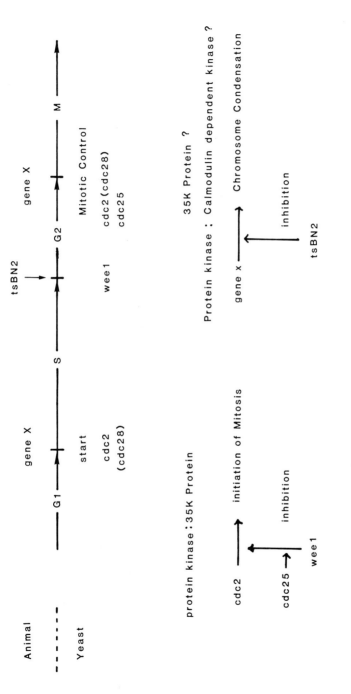

Fig. 12. Model for the regulatory mechanism of chromosome condensation comparing animal cells with yeast.

tion in a single gene. By identifying the product of the cloned human gene, the regulatory mechanism required for the cascade of events leading to mitosis will be elucidated.

ACKNOWLEDGMENTS

We thank P. Nurse (Imperial Cancer Research Fund Laboratories, London) and M. Ohara (Kyushu University) for comments on the manuscript.

REFERENCES

Ajiro, K., Borun, T. W., and Cohen, L. H. (1981a). The phosphorylation states of different histone 1 subtypes and their relationship to chromatin functions during the HeLa S-3 cell cycle. *Biochemistry* **20,** 1445–1454.
Ajiro, K., Borun, T. W., Shulman, S. D., McFadden, G. M., and Cohen, L. H. (1981b). Comparison of the structures of human histones 1A and 1B and their intramolecular phosphorylation sites during the HeLa S-3 cell cycle. *Biochemistry* **20,** 1454–1464.
Ajiro, K., Nishimoto, T., and Takahashi, T. (1983). Histone H1 and H3 phosphorylation during premature chromosome condensation in a temperature-sensitive mutant (tsBN2) of baby hamster kidney cells. *J. Biol. Chem.* **258,** 4534–4538.
Bradbury, E. M., Inglis, R. J., and Matthews, H. R. (1974). Control of cell division by very lysine rich histone (f1) phosphorylation. *Nature (London)* **247,** 257–261.
Burnette, W. N. (1981). "Western blotting": Electrophoretic transfer of proteins from sodium dodecyl sulfate-polyacrylamide gels to unmodified nitrocellulose and radiographic detection with antibody and radioiodinated protein A. *Anal. Biochem.* **112,** 195–203.
Davis, F. M., Tsao, T. Y., Fowler, S. K., and Rao, P. N. (1983). Monoclonal antibodies to mitotic cells. *Proc. Natl. Acad. Sci. U.S.A.* **80,** 2926–2930.
Eilen, E., Hand, R., and Basilico, C. (1980). Decreased initiation of DNA synthesis in a temperature-sensitive mutant of hamster cells. *J. Cell. Physiol.* **105,** 259–266.
Fantes, P. A. (1981). Isolation of cell size mutants of a fission yeast by a new selection method: Characterization of mutants and implications for division control mechanisms. *J. Bacteriol.* **146,** 746–754.
Gurley, L. R., Walters, R. A., and Tobey, R. A. (1974). Cell-cycle specific changes in histone phosphorylation associated with cell proliferation and chromosome condensation. *J. Cell Biol.* **60,** 356–364.
Gurley, L. R., D'Anna, J. A., Barham, S. S., Deaven, L. L., and Tobey, R. A. (1978). Histone phosphylation and chromatin structure during mitosis in Chinese hamster cells. *Eur. J. Biochem.* **84,** 1–15.
Hayashi, A., Yamamoto, S., Nishimoto, T., and Takahashi, T. (1982). Chromosome condensing factor(s) induced in tsBN2 cells at a nonpermissive temperature: Evidence for transferable material by cell fusion. *Cell Struct. Funct.* **7,** 291–294.
Hidaka, H., Sasaki, Y., Tanaka, T., Endo, T., Ohno, S., Fuji, Y., and Nagata, T. (1981). N-(16-aminohexyl)-5-chloro-1-naphthalenesulfonamide, a calmodulin antagonist, inhibits cell proliferation. *Proc. Natl. Acad. Sci. U.S.A.* **78,** 4354–4357.

Hindley, J., and Phear, G. A. (1984). Sequence of the cell division gene CDC 2 from *Schizosaccharomyces pombe;* patterns of splicing and homology to protein kinases. *Gene* **31,** 129–134.

Ishida, R., and Takahashi, T. (1978). Role of mercaptoethanol in *in vitro* DNA degradation by neocarzinostatin. *Cancer Res.* **38,** 2617–2620.

Ishida, R., Nishimoto, T., and Takahashi, T. (1979). DNA strand scission by neocarzinostatin and its relation to the inhibition of cell-cycle traverse and DNA synthesis. *Cell Struct. Funct.* **4,** 235–250.

Ishida, R., Takahashi, T., and Nishimoto, T. (1985). Chromosomes of G2-arrested cells are easily analyzed by use of the "tsBN2" mutation. *Cell Struct. Funct.* **10,** 417–420.

Johnson, R. T., and Rao, P. N. (1970). Mammalian cell fusion. II. Induction of premature chromosome condensation in interphase nuclei. *Nature (London)* **226,** 717–722.

Kai, R., Ohtsubo, M., Sekeguchi, M., and Nishimoto, T. (1986). Molecular cloning of a human gene that regulates chromosome condensation and is essential for cell proliferation. *Mol. Cell Biol.* **6,** 2027–2032.

Klee, C. B., Crouch, T. H., and Richman, P. G. (1980). Calmodulin. *Annu. Rev. Biochem.* **49,** 489–515.

Lewin, B. (1974). Control of transcription. *In* "Gene Expression-2" (B. Lewin, ed.), Vol. 2, pp. 320–335. Wiley, New York.

Luria, S., and Delbruck, M. (1943). Mutations of bacteria from virus sensitivity to virus resistance. *Genetics* **28,** 491–511.

Nishimoto, T., and Basilico, C. (1978). Analysis of a method for selecting temperature-sensitive mutants of BHK cells. *Somatic Cell Genet.* **4,** 323–340.

Nishimoto, T., Eilen, E., and Basilico, C. (1978). Premature chromosome condensation in a ts DNA⁻ mutant of BHK cells. *Cell (Cambridge, Mass.)* **15,** 475–483.

Nishimoto, T., Takahashi, T., and Basilico, C. (1980). A temperature-sensitive mutation affecting S-phase progression can lead to accumulation of cells with a G2 DNA content. *Somatic Cell Genet.* **6,** 465–476.

Nishimoto, T., Ishida, R., Ajiro, K., Yamamoto, S., and Takahashi, T. (1981). The synthesis of protein(s) for chromosome condensation may be regulated by a post-transcriptional mechanism. *J. Cell. Physiol.* **109,** 299–308.

Nishimoto, T., Ajiro, K., Hirata, M., Yamashita, K., and Sekiguchi, M. (1985). The induction of chromosomal condensation in tsBN2 a temperature-sensitive mutant of BHK21, inhibited by the calmodulin antagonist, W-7, *Exp. Cell Res.* **156,** 351–358.

Nurse, P., and Bisset, Y. (1981). Gene required in G1 for commitment to cell cycle and in G2 for control of mitosis in fission yeast. *Nature (London)* **292,** 558–560.

Nurse, P., and Thuriaux, P. (1980). Regulatory genes controlling mitosis in the fission yeast *Schizosaccharomyces pombe. Genetics* **96,** 627–637.

Rao, A. P., and Rao, P. N. (1976). The cause of G2-arrest in Chinese hamster ovary cells treated with anticancer drugs. *J. Natl. Cancer Inst. (U.S.)* **57,** 1139–1143.

Rao, P. N., Wilson, B., and Puck, T. T. (1976). Premature chromosome condensation and cell cycle analysis. *J. Cell. Physiol.* **91,** 131–142.

Sasaki, Y., and Hidaka, H. (1982). Calmodulin and cell proliferation. *Biochem. Biophys. Res. Commun.* **104,** 451–456.

Scango, G., and Ruddle, F. H. (1981). Mechanisms and applications of DNA-mediated gene transfer in mammalian cells—A review. *Gene* **14,** 1–10.

Sunkara, P. S., Wright, D. A., and Rao, P. N. (1979). Mitotic factors from mammalian cells induce germinal vesicle breakdown and chromosome condensation in amphibian oocytes. *Proc. Natl. Acad. Sci. U.S.A.* **76,** 2799–2802.

Terasima, T., and Yasukawa, M. (1966). Synthesis of G1 protein preceding DNA synthesis in cultured mammalian cells. *Exp. Cell Res.* **44,** 669–672.

Wasserman, W. J., and Smith, L. D. (1981). Calmodulin triggers the resumption of meiosis in amphibian oocytes. *J. Cell Biol.* **89,** 389–394.

Yamashita, K., Nishimoto, T., and Sekiguchi, M. (1984). Analysis of protein associated with chromosome condensation in baby hamster kidney cells. *J. Biol. Chem.* **259,** 4667–4671.

Yamashita, K., Davis, F. M., Rao, P. N., Sekiguchi, M., and Nishimoto, T. (1985). Phosphorylation of nonhistone proteins during premature chromosome condensation in a temperature-sensitive mutant, tsBN2. *Cell Struct. Funct.* **10,** 259–270.

Yasuda, H., Matsumoto, Y., Mita, S., Marunouchi, T., and Yamada, M. (1981). A mouse temperature-sensitive mutant defective in H1 histone phosphorylation is defective in deoxyribonucleic acid synthesis and chromosome condensation. *Biochemistry* **20,** 4414–4419.

11

Chromatin Structure and Histone Modifications through Mitosis in Plasmodia of *Physarum polycephalum*

HIDEYO YASUDA,[1] REINHOLD D. MUELLER,[2] AND E. MORTON BRADBURY

Department of Biological Chemistry
School of Medicine
University of California
Davis, California 95616

I. INTRODUCTION

During the cell cycle, chromosomes undergo a condensation process required for the control of the very large amounts of eukaryotic DNA during the orderly separation of sister chromatids at mitosis. In understanding this process we are concerned with both the structural changes which chromatin undergoes during condensation and the biochemical control of these changes. Major structural problems are posed by the enormous lengths of DNA which are 10^3- to 10^4-fold longer than metaphase chromosomes. This large compaction ratio led to an early view that many orders of linear coilings were required to package DNA into the

---

[1] Present address: Faculty of Pharmaceutical Sciences, Kanazawa University, Takara - Machi, Kanazawa 920, Japan

[2] Present address: Department of Biology, Marquette University, Milwaukee, Wisconsin 53233

319

length of the metaphase chromosome. Over the past 10 years, however, there have been advances in our understanding of chromosome structure and new ideas have emerged which have to be considered in the process of chromsome condensation.

II. CHROMATIN STRUCTURE AND ORGANIZATION

A. Chromosome Organization

1. Chromatin Domains

It is generally accepted that a chromatid contains a single DNA molecule (see Gall, 1981). Evidence has emerged to suggest that this DNA molecule is subjected to long range interactions that constrain the DNA into domains or loops. Suggestive evidence for these domains comes from biophysical and biochemical studies of chromosome structure. Electron micrographs of interphase nuclei and of metaphase chromosomes show that these states are composed largely of 25- to 30-nm thick fibrils which on the periphery of the chromosomes appear to be coiled into 50 to 60-nm loops (Benyajati and Worcel, 1976; Paulson and Laemmli, 1977; Marsden and Laemmli, 1979). A striking demonstration of these loops was evidenced by electron micrographs of histone-depleted metaphase chromosomes which showed a "halo" of DNA loops around a matrix of nonhistone protein, the scaffold proteins (Laemmli *et al.,* 1978). The paths of some of the loops could be followed and they were observed to emerge from and return to the same point on the scaffold or nuclear matrix. A similar organization was observed for interphase mouse nuclei (Hancock and Hughes, 1982). Loop sizes were estimated to be in the range of 50 to 100 kbp of DNA. Biochemical evidence suggesting chromatin domains in interphase chromatin comes from micrococcal nuclease and restriction nuclease digestion of chromatin in rat liver nuclei (Igo-Kemenes and Zachau, 1977; Igo-Kemenes *et al.,* 1977). Restriction and micrococcal nuclease digestion results accord with a domain model whereby a soluble chromatin component derives from the loops and an insoluble fraction from the chromatin attached to the scaffold or nuclear matrix proteins. In this model, because the DNA ends of the chromatin loops are fixed by their attachments to scaffold or matrix protein, DNA supercoiling has to be considered a parameter in the control of chromosome structure. Sedimentation behaviors of chromatins from lysed cells from *Drosophila* (Benyajati and Worcel, 1976), mouse (Ide *et al.,* 1975), HeLa (Cook and Brazell,

1976; Levin *et al.*, 1978), and yeast (Pinon and Salts, 1977) showed a biphasic response after exposure to intercalating agents, which suggested supercoiled DNA domains. In relating chromatin domains to functional units a relationship has been found between replicon size and supercoiled loop domains in the eukaryotic genome (Buongiorno-Nardelli *et al.*, 1982).

Two or three groups of proteins are thought to be involved in the long-range organization of chromosomes in interphase nuclei and metaphase chromosomes. The first group, lamins, are structural protein components of the nuclear membrane (Gerace *et al.*, 1978; Gerace and Blobel, 1981) which are reversibly depolymerized following phosphoryation at mitosis (Gerace and Blobel, 1980). The proteins involved in the intranuclear organization of interphase nuclei, the nuclear matrix (Barrack and Coffey, 1982; Pienta and Coffey, 1984), are probably related to the proteins involved in the long-range organization of the metaphase chromosomes, the scaffold proteins (Laemmli *et al.*, 1978; Lewis *et al.*, 1984). There are two views of the functional roles of the nuclear matrix. First, the proteins involved in DNA functions such as replication and transcription are located on or in the nuclear matrix and chromatin loops are drawn through the nuclear matrix during processing (Pienta and Coffey, 1984). The second view is that whereas there may be changes in the patterns of loops during differentiation or the cell cycle (Laemmli, 1985), the processes of DNA replication and transcription proceed around the loops. This would be directly analogous to the transcriptional processing of the giant Balbiani rings of *Chironomus tentans* (Daneholt, 1982; Andersson *et al.*, 1984). Although both models are based on chromatin loops or domains, different functional views may result from the different biochemical procedures used in these studies. In studies that support the view that DNA functions are located on or in the nuclear matrix, chromatin was depleted of histones by 2 M NaCl, raising the possibility of salt-induced slippage of DNA relative to the matrix proteins. In the second approach, histones are depleted by a low-salt detergent procedure (Laemmli, 1985), and specific DNA sequences have been found in association with the scaffold proteins after restriction enzyme digestion. In the 5-kbp *Drosophila* histone gene cluster, a single attachment site was found on an AT-rich 657 bp spacer between the histone H1 and H3 genes. Because the 5-kbp cluster is tandemly repeated, each repeat could form a loop. Sequences of the scaffold attachment DNA regions showed the presence of a topoisomerase II binding site, consistent with the identification of one of the nuclear matrix or scaffold proteins as topoisomerase II (Earnshaw *et al.*, 1985; Earnshaw and Heck, 1985) located at the bases of the radial loops of histone-depleted metaphase chromosomes. Attachment sites have been found upstream from the transcribed sequences of several different genes. Relevant to the problem of chromosome condensation was the finding that the 5' attachment sites of the histone and heat shock protein 70 genes were maintained in both metaphase chromosomes and in the nuclei of *Drosophila* embryos, showing that some chromosome regions maintain the same domains through the cell cycle (Laemmli, 1985). The indication that one gene or a group of linked genes (e.g. histones) are contained within one chromatin domain is consistent with the findings of Judd and Young (1974) that one band in a *Drosophila* chromosome contains no more than a few complementation groups. Even the giant puffs of *Chironomus tentans* forming Balbiani rings I and II may contain only one transcriptional unit (Daneholt, 1982), and a model describing the chromatin changes occurring during transcription around the loop of the Balbiani rings has been proposed (Andersson *et al.*, 1984). As discussed by Gall (1981), a general model for the long-range organization of eukaryotic chromosomes might include a relatively constant number of chromatin domains, comparable to the 5000 bands in the polytene chromosomes of *Drosophila* (see Ashburner and Novitski, 1976). Each domain would be equivalent to a functional unit and contain one or a few transcriptional units. The size of the domains or loops would vary over a very wide range to accommodate the widely different DNA contents of eukaryotic genomes.

2. Chromatin Domains and Chromosome Condensation

Based on the general chromosome model outlined above, the process of chromosome condensation would involve both structural changes within a condensing chromatin loop and changes in the arrangements and packing of condensed loops through mitosis. During the condensation process, topological changes in the DNA constrained in loops are most probably controlled by topoisomerases. Yeast cells with mutants in the topoisomerase II gene are unable to segregate tangled daughter chromosomes and die in mitosis (Goto and Wang, 1984; Di Nardo *et al.*, 1984; Uemura and Yanagida, 1984). Thus, topoisomerase II identified as a scaffold protein (Earnshaw *et al.*, 1985; Earnshaw and Heck, 1985) is involved in mitosis and possibly in other chromatin functions where DNA topology is involved, e.g. transcription.

3. Chromatin Structure in Chromsome Loops

It is now well-established that chromatin consists of a repeating subunit structure, the nucleosome (Hewish and Burgoyne, 1973; Olins and Olins, 1974; Kornberg, 1974). Nucleosomes contain variable lengths of DNA depending on tissue and organism although for most somatic tissues the DNA repeat is 195 ± 5 bp. The range of DNA contents is from 165 bp for lower eukaryotes to 241 bp for sea urchin sperm (see Compton *et al.,* 1976). Micrococcal nuclease digestion of chromatins with different DNA contents gave well-defined subnucleosomal particles: the chromatosome with 168 bp DNA, the histone octamer $[(H2A,H2B)_2(H3_2H4_2)]$, and one H1 molecule (Simpson, 1978a), the nucleosome core particle with 146 ± 2 bp of DNA and the histone octamer. The chromatosome and the core particle are the constant structural units of chromosomes, whereas the linker DNA joining chromatosomes is variable. The variability is thought to involve the more variable H2A, H2B, and particularly H1 histones (Stein and Bina, 1984).

B. Nucleosome Structure

1. Core Particle Structure

The regularity of the structure of the core particle has allowed its structure to be determined in solution by neutron scatter techniques (Bradbury *et al.,* 1976; Hjelm *et al.,* 1977; Suau *et al.,* 1977; Braddock *et al.,* 1981) and in crystals by X-ray (Richmond *et al.,* 1984) and neutron diffraction (Bentley *et al.,* 1984). All these studies give essentially the same structure for the core particle of a flat disk, 11.0 nm × 5.5 to 6.0 nm with 1.7 ± 0.2 turns of DNA of pitch about 3.0 nm coiled on the outside of the histone octamer 7.0 nm × 5.5 to 6.0 nm (Fig. 1A). The arrangement of histones in the core particle is based very largely on histone–DNA chemical crosslinking data (Mirzabekov *et al.,* 1978; Shick *et al.,* 1980). Very recently a new model for the core particle has been proposed based on the crystal structure determination of the histone octamer (Burlingame *et al.,* 1985) and is a globular structure 11.0 nm × 11.0 nm in which the DNA is coiled on the outside of a histone octamer of dimensions 11 nm × 6.5 to 7.0 nm. This model does not accord with the detailed neutron scatter results nor with the X-ray crystal determination of the core particle itself. It does not accord with structural parameters obtained for extended chromatin (Suau *et al.,* 1979). It is possible that the crystal structure of the octamer is of another structural arrangement of histones showing evidence of histone polymorphism, which would be of considerable interest.

2. Chromatosome Structure

The proposed structure of the chromatosome is based on the structure of the core particle, nuclease digestion kinetics of chromatin and reconsti-

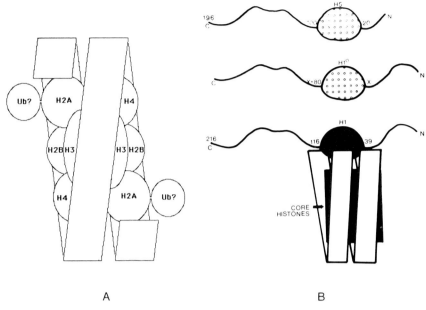

Fig. 1. (A) Outline structure for the nucleosome core particle (11.0 nm × 5.5 to 6.0 nm) with proposed location of ubiquitin covalently attached to H2A. (B) Proposed mode of binding of very lysine-rich histones to the nucleosome.

tuted chromatins, and the structure of very lysine-rich histones. In the nuclease digestion of chromatin there is a pause at 168 bp of DNA in the digestion kinetics before digestion proceeds to 146 bp of DNA and histone H1 is released. If the 146 bp of the core particle is contained in 1.7 ± 0.2 turns of DNA of pitch about 3.0 nm, then the 168 bp of DNA of the chromatosome would require two turns of DNA. These two turns are sealed off by histone H1 located on the side of the chromatosome (Simpson, 1978a; Thoma *et al.*, 1979).

3. Structure of Histone H1

The structure and properties of histone H1 are central to our understanding of chromatin structure and chromosome condensation. Mammalian histone H1 has very unusual structural properties and consists of three well-defined domains: a flexible N-terminal domain of about 40 residues, a globular central domain of about 80 residues, and a flexible C-terminal domain of about 100 residues (Bradbury *et al.*, 1975a,b,c; Chapman *et al.*, 1976, 1978; Hartman *et al.*, 1977). The structural properties accord with the sequence in which nearly all of the apolar and aromatic residues are located in the central globular domain; the N- and C-terminal domains are very basic and contain a very high proportion of helix-de-stabilizing residues such as glycine and proline, e.g., the C-terminal domain contains more than 90% of lysine, alanine, and proline together with serines and threonines (Rall and Cole, 1971; von Holt *et al.*, 1979). Neutron scatter analysis of the globular domain peptide of histone H5, an H1-like histone discussed in Section V, showed that it had a diameter of 2.9 nm (Aviles *et al.*, 1978). The structural properties of H1 together with the model for the chromatosome led to the proposal that the globular domain of H1 sealed off two turns of DNA (Crane-Robinson *et al.*, 1980; Bradbury *et al.*, 1981). Strong evidence in support of this model came from the demonstration that only peptides containing the globular domain of H1 were able to protect 168 bp DNA against nuclease digestion (Allan *et al.*, 1980).

The proposed model for the chromatosome is given in Fig. 1B. Also included in this model are the outline structures for the H1° and H5 classes of very lysine-rich histones (Cary *et al.*, 1981). All classes of very lysine-rich histones have a globular domain of about 80 residues although their sequences can be quite different. The constant physical size of the globular domains is clearly required for their binding in a cage of DNA formed between DNA at the entry and exit to the chromatosome and the central coil of DNA. The modes of binding of N- and C-terminal domains are not understood and considerable conceptual difficulties are provided by their lengths and flexible nature. However, the unusual structural properties of the very lysine-rich histones and the location of their central globular regions at the entry and exit DNA points of the nucleosome are fully consistent with their major involvement in the control of higher order chromatin structures.

C. Orders of Chromatin Structure

1. Extended Chromatin: The 11-nm Fibril

Under conditions of low ionic strengths, isolated chromatin is in the form of an 11-nm diameter fibril. In electron micrographs this can be seen as a somewhat irregular linear array of nucleosomes with the flat faces lying down on the substrate (Olins and Olins, 1974; McKnight and Miller, 1976). Neutron scatter studies of chromatin solutions gave a direct measurement of the mass/unit length of the extended chromatin fibril which is equivalent to about 1 nucleosome per 11 nm, i.e., a DNA packing ratio of

6 to 7 : 1 (Suau *et al.*, 1979). This technique also gave the transverse radii of gyration of both the DNA (3.4 nm) and the histone (2.1 nm). With the flat disk structure for the nucleosome in Fig. 1B, these values for the mass per unit length and the transverse radii of gyration can be accommodated only by a model for the 11-nm fibril with the disks roughly edge-to-edge (Fig. 2A). It should be pointed out that the transverse radius of gyration of the histone moiety of 2.1 nm is in accord with the histone core of the nucleosome of 7.0 nm × 5.5 to 6.0 nm but not with the recent model in which the histone core is 11.0 × 6.5 to 7.0 nm (Burlingame *et al.*, 1985). Thus, contrary to the suggestions of these authors this structure is not found in chromatin.

2. 30-nm Chromatin Fibril

In electron micrographs of nuclei and metaphase chromosomes, the bulk of the chromatin is in the form of a 25- to 30-nm fibril. Scanning electron micrographs of metaphase chromosomes show thick 50- to 60-nm fibrils on the periphery of the chromosome which, from serial sectioning, are made up of interwound 30-nm fibrils (Marsden and Laemmli, 1979). The 30-nm fibril is a folded or coiled form of the 11-nm fibril. The transition from the 11-nm fibril to the 25 to 30-nm fibril with increase of ionic strength has been followed by neutron scatter techniques (Suau *et al.*, 1979) and by electron microscopy (Thoma *et al.*, 1979). Neutron scatter

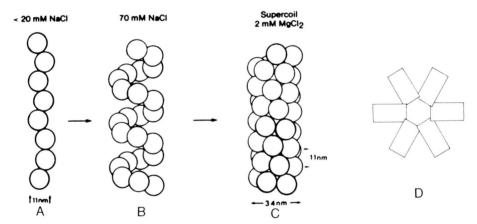

Fig. 2. Fibril (11 nm) with an edge-to-edge arrangement of nucleosome disks (A) undergoing a salt-induced transition (B) to the 34-nm diameter supercoil of six to seven nucleosomes per turn and pitch 11 nm (C). The nucleosome disks in this supercoil are arranged approximately radially (D).

techniques have the advantage that they can give direct measurements of mass per unit length, transverse radii of gyration, and the diameter of the hydrated form of this higher order chromatin structure. The mass per unit length was equivalent to a DNA packing ratio of 40 to 50:1 for this structure, and its diameter, from the distance distribution function, was 34 nm. This is larger than the 25- to 30-nm fibril measured in electron micrographs, but the hydrated form would be expected to give a larger dimension than the dehydrated form. The 11-nm semimeridional arc observed in fiber X-ray diffraction studies has been attributed to the pitch of a supercoil of nucleosomes (Carpenter *et al.*, 1976) or a solenoid (Finch and Klug, 1976). This semimeridional arc contains off-meridian maxima consistent with a helical arrangement of nucleosomes of pitch 11 nm (Carpenter *et al.*, 1976). The diameter of 34 nm and pitch of 11 nm corresponds to six to seven nucleosomes per turn which accords with the direct measurement of mass per unit length measurement of 34-nm fibril. Of the three possible orthogonal arrangements of the flat nucleosome disks, the radial or approximately radial arrangement is the only one which accords with both neutron (Suau *et al.*, 1979) and electron dichroism (McGhee *et al.*, 1980, 1983; Mitra *et al.*, 1984) measurements. A model for the 34-nm fibril is given in Fig. 2B.

Neutron scatter analysis gives a measurement of the diameter of the supercoil of nucleosomes of 34 nm; the nucleosome disks of 11 nm in diameter are arranged radially, which leaves a 12-nm diameter hole down the axis of the supercoil or solenoid. Such a hole, if present, in a regular supercoil or solenoid should result in a discontinuity in the pair distance distribution functions obtained from the neutron scatter curves. A discontinuity has not been observed in these functions from either chicken erythrocyte or calf thymus chromatins (Suau *et al.*, 1979; J. P. Baldwin, personal communication, 1985). This has two possible explanations: the regularity of the supercoil of nucleosomes extends only over short distances, and/or the 12-nm hole along the supercoil axis is not empty but contains nucleosomal components such as histone H1 and linker DNA. Based on electron microscope studies of the salt-induced 11- to 34-nm fibril transition of oligonucleosomes, it has been suggested that H1 is located on the inside of the solenoid (Thoma *et al.*, 1979). To accommodate variable linker lengths of DNA, Butler (1984) has proposed that the DNA linker forms a reverse helix sense loop, i.e., compared to the helix sense of core particle DNA, on the inside of the solenoid. Location of H1 and linker DNA on the inside of the solenoid would give an explanation for the absence of a discontinuity in the pair distance distribution function as would the considerable irregularity observed in the 34-nm supercoil of nucleosomes in electron micrographs.

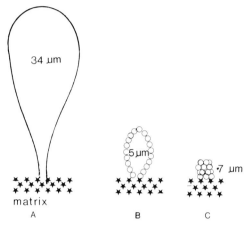

Fig. 3.   Contraction of a 34-μm loop of DNA (A) into a 5-μm loop of nucleosomes (B) which is further coiled into a 0.7-μm loop of the 34-nm supercoild (C).

3. Condensation in Relation to Chromatin Domains

In the chromosome domain model, loops of chromatin are arranged in transverse sections around the chromatids of metaphase chromosomes. Thus the process of chromosome condensation can be envisaged as the contraction of transverse loops and their packaging into the thickness of a metaphase chromosome. A loop of 100,000 bp of DNA which is 34 μm long would be reduced to about 5 μm of the 11-nm fibril and to about 0.7 μm of the 34-nm fibril using the mass per unit lengths obtained for these fibrils from neutron scatter studies (Suau *et al.*, 1979). Only one further stage of condensation may be required to account for the thickness of metaphase chromosomes, and fibrils 50–60 nm thick made up of intertwined 30-nm fibrils have been observed in metaphase chromosomes (Marsden and Laemmli, 1979). An outline of this process is given in Fig. 3.

III. CONTROL OF CHROMOSOME CONDENSATION

Histone Modification

In looking for mechanisms that control the process of chromosome condensation, attention must be directed first to reversible changes that affect chromosomal proteins, in particular the histones that are the major structural proteins in chromatin. Histones together with ubiquitin, also a chromosomal protein, have the most rigidly conserved sequences in nature. The reasons for this are not understood, but the obvious implication is that almost each and every residue, particularly in ubiquitin, H3, H4, and the central apolar domains of H1, H2A and H2B, are essential to their functions in chromosomes. Although histone sequences are conserved, they are subjected to reversible chemical modifications which change the character of the modified residue and in effect "relax" this sequence conservation for functional requirements. The reversible chemical modifications are of two types: acetylation of lysines in H2A, H2B, H3, and H4, which convert a basic lysine to an uncharged ε-N-acetyllysine, and phosphorylations of serines and threonines in histones H1 and H3, which convert uncharged residues to negatively charged serine and threonine phosphates. In addition to acetylation and phosphorylation, there is a very unusual modification of H2A and H2B which involves the covalent attachment of the 89-residue globular protein ubiquitin through the amino group of a lysine to give a bifurcated ubiquitinated H2A (uH2A) and H2B (uH2B). This was first found for H2A by Goldknopf *et al.* (1975) and then for H2B (Wu *et al.*, 1981; West and Bonner, 1980). Details of these modifications and their locations in the histone sequences have been reviewed (Allfrey, 1980; Matthews and Bradbury, 1982; Matthews and Waterborg, 1985).

1. Histone Modifications in Relation to Histone Structures

Similar to the three domain structure shown for the very lysine-rich histones in Fig. 1B, the core histones also have well-defined domains. Histones H2A and H2B are each made up of three domains: a very basic flexible N-terminal domain of about 30 residues, a central structured domain of 80–90 residues, and a basic C-terminal tail (Moss *et al.*, 1976a). Histones H3 and H4 each have two domains: a basic N-terminal domain and an apolar structural central and C-terminal domain (Moss *et al.*, 1976b; Bohm *et al.*, 1977; Bradbury *et al.*, 1978). It is notable that all of the reversible chemical modifications of histones are located in their basic flexible N- and C-terminal domains: acetylation of the core histones in their N-terminal domains (see Allfrey, 1980; Matthews and Waterborg, 1985); ubiquitination of H2A in its C-terminal tail (Goldknopf *et al.*, 1977) and probably also in this location for H2B (Wu *et al.*, 1981; West and Bonner, 1980); phosphorylation of H3 in the N-terminal domain and hyperphosphorylations of very lysine-rich histones in their N- and C-terminal domains (see Gurley *et al.*, 1981; Langan *et al.*, 1981). Both acetylations and phosphorylations are sufficient to change the character of the chemically modified N- and C-terminal domains. For example, it has been shown that the binding of the N-terminal peptide (residues 1–23) of H4 to DNA can be suppressed when it is acetylated at the four sites which are modified *in vivo* (Cary *et al.*, 1982).

2. Effect of Histone Modifications on Histone Interactions and Chromatin Structure

Reversible modifications of the basic flexible N-terminal domains of histones can be viewed as mechanisms for the reversible modulation of chromatin structure. In relation to nucleosome and chromatin structure, it has been shown that the N-terminal domains of H2A and H2B are not bound within the 146-bp core particle and that the N-terminal domains of H3 and H4 are released between 0.3 and 0.6 M NaCl, before the salt-induced unfolding of core particles (Cary *et al.*, 1978). In the 7-Å resolution structure of the core particle, the major histone–DNA interactions involve helical segments most probably located in the central and C-terminal regions (Richmond *et al.*, 1984). Thus the conserved structural unit of eukaryotic chromosomes is the core particle containing a constant 146 bp of DNA and generated by interactions involving the rigidly conserved sequences of H3 and H4 and of the central structured regions of H2A and H2B. Hyperacetylation has been shown not to affect the gross structural parameters of the core particle (Vidali *et al.*, 1978; Imai *et al.*, 1986). Although the structure of the fully acetylated core particles have yet to be studied, the lack of an effect of hyperacetylation on core particle structure led to the suggestion that acetylation may be required for the destabilization of the 34-nm supercoil prior to DNA processing (Vidali *et al.*, 1978; Simpson, 1978b). Acetylation of the core histones could alter the path of the DNA at the entry and exit to the nucleosome and weaken the binding site for the globular domain of the very lysine-rich histones.

Ubiquitinated H2A has been found to be an integral component of nucleosomes (Goldknopf *et al.*, 1977; Martinson *et al.*, 1979). Nucleosome core particles can incorporate two uH2As without affecting the kinetics of DNase I digestion compared to native core particles (Kleinschmidt and Martinson, 1981). Thus, using biochemical probes little effect of ubiquitination has been observed on core particle structure. The probable location of the ubiquitin moiety is on each face of the disk-shaped core particle (Kleinschmidt and Martinson, 1981) adjacent to H2A as indicated in Fig. 1A. In the radial disk model for the 34-nm supercoil, this would place the ubiquitins between adjacent faces of consecutive core particles. The effects of ubiquitination of uH2A and uH2B on the stability and structure of the 34-nm supercoil are major unanswered questions. There seems little doubt that this ubiquitin modification would perturb the higher order structure and could provide a mechanism for "labeling" specific regions of chromatin.

In contrast to acetylations and ubiquitinations which modify only a small subset of the core histones, in keeping with their proposed roles in replication and transcription, the mitotically related phosphorylation of the very lysine-rich histones and H3 involve all of these molecules in *Physarum* (Mueller *et al.*, 1985a; Yasuda *et al.*, 1985b) and in other cell types (Gurley *et al.*, 1981; Langan *et al.*, 1981). It would be expected from physical chemistry that phosphorylation would weaken the binding of the N- and C-terminal regions to DNA, and this has been observed in phosphorylated H1–DNA interactions (Rattle *et al.*, 1977; Adler *et al.*, 1971a,b; Fasy *et al.*, 1979; D'Anna *et al.*, 1979; Langan, 1982; Lennox *et al.*, 1982). The effects of H1 and H3 phosphorylations on the structure of extended chromatin, the 34-nm supercoil, and higher order structures are not known.

IV. CELL CYCLE STUDIES OF HISTONE MODIFICATIONS

A. *Physarum polycephalum*

We have used the lower eukaryote *Physarum polycephalum* extensively as a model system for studies of the biological functions of histone modifications. The life cycle of *Physarum polycephalum* is shown in Fig. 4A, adapted from diagrams of Alexopoulous and Mims (1979) and Rusch (1980). As a model system it has two major advantages: (1) it has a small number of well-defined stages in its life cycle and is suitable for studies of chromatin changes during differentiation; and (2) in growth stage the plasmodium is multinuclear and all the nuclei come under the same cell cycle controls. A plasmodium can be grown easily to 14 nm diameter and contains 10^9 nuclei that all divide within 2–3 min of each other in a 9 to 11-hr cell cycle. The 10^9 nuclei provide about 1 mg of DNA, 1 mg of histones, and 1 mg of nonhistone proteins, which are biochemically useful amounts for studies of histone modifications. There is no G_1 phase in the *Physarum* nuclear division cycle. Events through mitosis can be recognized in the phase contrast microscope with a time resolution of minutes (Fig. 4B).

Physarum Histones

Physarum histones have been characterized and shown to be analogous to mammalian histones, particularly histones H3 and H4 (Côté *et al.*, 1982; Mende *et al.*, 1983). H2A and H2B are larger than the analogous

A

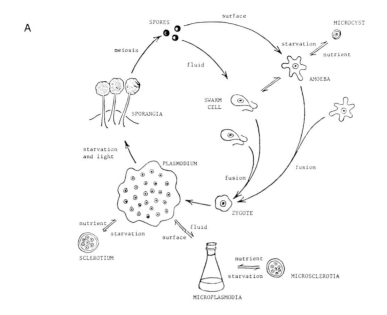

B

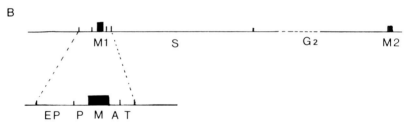

Fig. 4. (A) Life cycle of *Physarum polycephalum* (from Cooney, 1984, with permission). (B) Cell cycle of the plasmodial stage of *Physarum polycephalum*. EP, early prophase; P, prophase; M, metaphase; A, anaphase; T, telophase.

mammalian histones. In keeping with the divergence observed in the N-terminal domains of H2A and H2B (see von Holt *et al.*, 1979), both *Physarum* H2A and H2B differ from analogous mammalian histones in the sequences of their N-terminal ends. *Physarum* H1, however, exhibits the most divergence when compared with mammalian H1 (Chambers *et al.*, 1983; Mende *et al.*, 1983). It has a molecular weight of 30,700 compared to about 21,000 for mammalian H1s (Mende *et al.*, 1983). Chymotrypsin cuts *Physarum* H1 at a single phenylalanine similar to calf thymus H1. For calf H1 this cut produces an N-terminal peptide (residues 1–106) and a C-terminal peptide (residues 107–212). For *Physarum* H1, whereas the N-terminal peptide comigrates with the N-terminal calf H1 peptide (residues 1–106), the C-terminal peptide comigrates close to the intact calf H1. Thus, in relation to the structure of the very lysine-rich histones of Fig. 1B, it appears that the increase in size of *Physarum* H1 compared to calf H1 is accommodated by a doubling of the length of the C-terminal domain presumably through sequence duplication. Treatment of phosphorylated *Physarum* H1 with alkaline phosphatase to remove heterogeneity produced by phosphorylation shows that there is only one species of H1 molecules (Mueller *et al.*, 1985a).

B. Histone Acetylation

In studies of the acetylation of histone H4 through the cell cycle of *Physarum* it has been shown that the di-, tri-, and tetraacetylated states of H4 reached their minimum values at metaphase with a corresponding increase in the non- and monoacetylated states (Chahal *et al.*, 1980). More detailed cell cycle studies of acetylation of all core histones demonstrate at least three patterns of behavior (Fig. 5) (Waterborg and Matthews, 1983, 1984). Histones H2A and H2B are acetylated in the lower states of one or two acetates per molecule in S phase only; acetates on histones H3 and H4 also turn over in the lower states in S phase but, in addition, in the higher states of acetylation in G_2 phase. The S-phase acetylations of all four core histones are associated with both replication and transcription whereas the G_2-phase turnover of acetate in the higher states of H3 and H4 is associated only with transcriptional activity. Through the process of chromosome condensation, transcriptional activity decreases and little or no transcription is observed at metaphase. In parallel with these changes, steady-state acetylation (Chahal *et al.*, 1980) and acetate turnover (Waterborg and Matthews, 1983) decrease to a very low level but do not appear to completely turn off. This is consistent with the findings that certain genes are in a potentially active state of chromatin in metaphase chromosomes (Stadler *et al.*, 1978; Gazit *et al.*, 1982; Wilhelm *et al.*, 1982), possibly because they are required immediately after metaphase or because they may be required to protect cells against injury.

C. Histone Ubiquitination

uH2A and uH2B have been shown to be absent in metaphase chromosomes (Matsui *et al.*, 1979; Wu *et al.*, 1981), and it has been proposed that deubiquitination is a general factor in chromosome condensation (Matsui

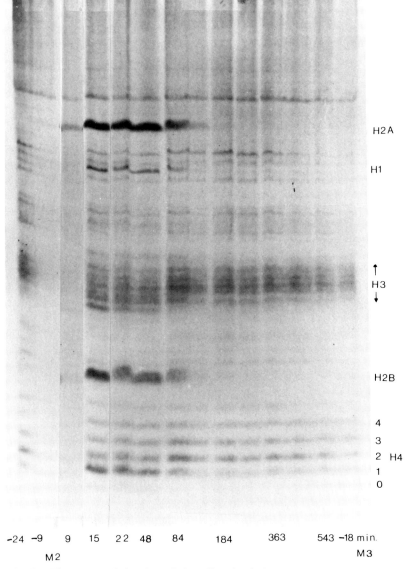

Fig. 5.   Histone acetylation through the cell cycle of *Physarum polycephalum*. The four states of H4 acetylation are indicated. In this *Physarum* cell cycle prophase begins at $M_2 -$ 24 min; telophase at $M_2 + 7$ min; chromatin decondenses by $M_2 + 15$ min. Histones from 0.2 macroplasmodium were electrophoresed in parallel lanes in a 0.5-mm thick acetic acid– urea–Triton X-100 (AUT) gel. Migration was from top to bottom. Fluorography was for 31 days. (From Waterborg and Matthews, 1983.)

et al., 1979). Ths absence of uH2A at metaphase has been questioned because inhibitors of isopeptidase were not used during the isolation procedure (Finley *et al.*, 1984). *Physarum* is a particularly suitable model system in which to study this question because at different times in the cell cycle the plasmodium is arrested by dropping it into liquid N_2. It is scraped from the frozen state into 6 *M* guanidine chloride and homogenized, a procedure that would stop the activity of any enzyme.

Temperature-sensitive mutant cells *ts85,* derived from the FM3A mouse mammary carcinoma cell line, arrest mainly in G_2 at the nonpermissive temperature, and there is no uH2A nor phosphorylated H1 (Mita *et al.*, 1980). At the permissive temperature, H2A is ubiquitinated followed by H1 phosphorylation showing a progression of these events in G_2 phase leading to mitosis (Yasuda *et al.*, 1981; Matsumoto *et al.*, 1980). It has been suggested that ubiquitination may be involved in some aspects of transcriptional control or in potentially active regions of chromatin (Goldknopf *et al.*, 1978, 1980; Levinger and Varshavsky, 1982). A rapid turnover has been observed for ubiquitin in uH2A (Wu *et al.*, 1981; Seale, 1981) showing that it is regulated by an ATP-dependent ubiquitin conjugation enzyme (Matsumoto *et al.*, 1983; Ciechanover *et al.*, 1980) and a uH2A isopeptidase (Andersen *et al.*, 1981; Matsui *et al.*, 1982).

Ubiquitinated histones uH2A.1, uH2A.2, and uH2B were identified in the basic nuclear proteins of the plasmodial stage of *Physarum polycephalum* by peptide mapping, by cross-reaction with antiubiquitin antibody and by uH2A and uH2B isopeptidase cleavage (Mueller *et al.*, 1985b). In microplasmodia, uH2A amounts to 7% of H2A and uH2B to 6% of H2B. Detailed studies of mitosis in the macroplasmodium show that in early prophase, which lasts 15 min, both the uH2As and uH2B are strongly present, whereas minutes later in metaphase they disappear. When the nuclei enter anaphase, which lasts 3 min, the uH2As and uH2B reappear. This sequence of events, presented in Fig. 6, shows that deubiquitination of the uH2As and uH2B is a very late, possibly final event in the process of chromosome condensation to metaphase chromosomes and that ubiquitination is a very early event in the process of decondensation. At the nonpermissive temperature *ts85* cells arrest in early G_2 phase and there are no ubiquitinated H2A molecules. At the permissive temperature these histones are ubiquitinated in early G_2 even though they will be deubiquitinated prior to metaphase. It appears that following S phase it is essential for the cell to label a chromatin subcomponent with ubiquitin which is then packaged very late into the metaphase chromosome following deubiquitination. Based on these results and the earlier reported association of ubiquitinated histones with potentially active or "poised" chro-

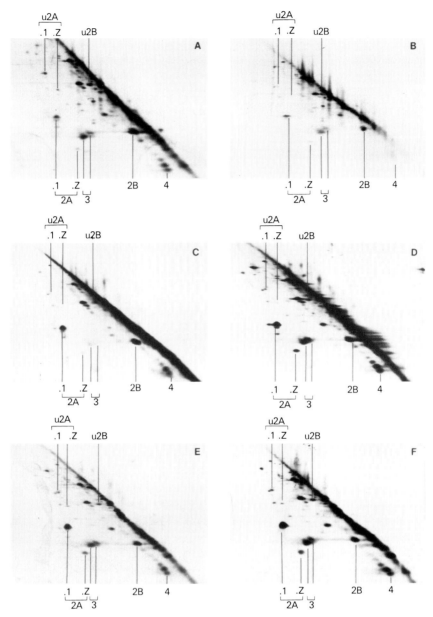

Fig. 6. Cell cycle changes in the ubiquitination of histones H2A.1, H2A.Z, and H2B shown on AUT-AUC two-dimensional gels (acid–urea–Triton X-100 in the first dimension, acid–urea–cetyltrimethylammonium bromide in the second dimension). (A) Mid-G$_2$ phase (M + 6 hr); (B) late prophase (M − 5 min); (C) metaphase; (D) anaphase (M + 3 min); (E) midreconstruction (M + 45 min); (F) late reconstruction (M + 1 hr). (From Mueller *et al.*, 1985b.)

matin, it was proposed that uH2A and uH2B label a specific subcomponent of chromatin which contains an important subset of genes, e.g., stress genes. The cell needs these genes to be available for as long as possible and consequently they are packaged into the metaphase chromosomes for the shortest possible times.

D. Phosphorylation of the Very Lysine-Rich Histones and H3

1. Cell Cycle Changes in H1, H1°, and H3 Phosphorylations

As shown in Fig. 1B, the very lysine-rich histones are located in the nucleosome through their conserved central globular domain which completes the nucleosome. Their flexible basic N- and C-terminal regions are required for the salt-induced 11- to 34-nm fibril transition (Allan *et al.*, 1980) and they are probably involved also in even higher order chromatin structures. The C-terminal domain of mammalian H1 molecules is about 100 residues which, as an extended polypeptide chain, could stretch 35.0 nm; for *Physarum* H1 the C-terminal domain is about 200 residues and could extend 70 nm. Their N-terminal domains are about 40 residues and could extent 14 nm. There is therefore considerable potential for the N- and C-terminal domains to be involved in long-range interactions in addition to a possible involvement with linker DNA between chromatosomes. These long-range interactions could be between adjacent nucleosomes, between nucleosomes on adjacent gyres of the supercoil of nucleosomes, or could involve the topology of the whole loop of a chromatin domain.

The N- and C-terminal domains, particularly the C-terminal domains, are subjected to extensive reversible mitotic phosphorylations. In early studies we have shown that *Physarum* H1 phosphorylation increases through G_2 phase to a hyperphosphorylated state just prior to metaphase (Bradbury *et al.*, 1973, 1974a). This increase in H1 phosphorylation paralleled the chromosome condensation process and led to the proposal that G_2-phase/metaphase H1 phosphorylation initiated and controlled chromosome condensation, though other later events were thought to be involved in the final packaging of metaphase chromosomes (Bradbury *et al.*, 1974a). Similar behaviors of H1 phosphorylations have also been observed for synchronized Chinese hamster ovary (CHO) cells (Gurley *et al.*, 1974, 1978a,b), rat hepatoma cells (Langan *et al.*, 1980), and HeLa cells (Ajiro *et al.*, 1981). Histone H3 also undergoes a late metaphase phosphorylation which has been correlated with the late stages of chromosome condensation (Gurley *et al.*, 1978a,b). Recently the identifications of temperature-sensitive G_2-phase cell mutants *ts85* and *FT210* have added considerable support for an involvement of H1 and H3 phosphorylations in chromosome condensation and mitosis (Matsumoto *et al.*, 1980, 1983; Yasuda *et al.*, 1981). Another type of temperature-sensitive mutant tsBN2 exhibits premature chromosome condensation (PCC) at the non-permissive temperature, and at any time in the cell cycle this temperature-induced PCC is paralleled exactly by H1 and H3 phosphorylations at the mitosis-related sites (Ajiro *et al.*, 1983).

Thus far in all cell types studied, H1 undergoes an increase in phosphorylation through G_2 phase to a hyperphosphorylated state at metaphase and is then dephosphorylated following metaphase. Contrary to these studies a subsequent study of H1 phosphorylation in *Physarum* reported that although H1 was subjected to G_2-phase/metaphase hyperphosphorylation, there was no dephosphorylation of H1 following mitosis (Fischer and Laemmli, 1980). There are two reasons why this conclusion concerning H1 dephosphorylation was erroneous, and the issue is now resolved as discussed in detail below: The first reason is the low resolution of the protein gels; the second was the assumption in these earlier studies that *Physarum* H1 was phosphorylated at a similar number of sites, 5–6, as found for mammalian H1. *Physarum* H1 has now been shown to be 50% larger than mammalian H1 (Chambers *et al.*, 1983; Mende *et al.*, 1983) with a C-terminal domain of about 200 residues (Chambers *et al.*, 1983) which raised the strong possibility that *Physarum* H1 may be phosphorylated to higher states than mammalian H1s.

2. Cell Cycle Phosphorylations of Physarum Histones in Macroplasmodia

Figure 7 shows the cell cycle changes for all *Physarum* histones (Yasuda *et al.*, 1987). Histone H1 is subjected to phosphorylation throughout the cell cycle which increase through G_2 phase to a hyperphosphorylated state at metaphase; H1° (see below, Section V) is phosphorylated through G_2 phase to metaphase; H2B is subjected to lower levels of phosphorylation through S phase; and H3 is phosphorylated late in G_2 phase just prior to metaphase. Except for H2B these behaviors have been reported for a range of mammalian cell types. In mammalian cells H2A phosphorylations are observed instead of the H2B phosphorylation in *Physarum* (Gurley *et al.*, 1978a,b). Either these low level H2A and H2B phosphorylations are promiscuous or H2B phosphorylations in *Physarum* substitute for H2A phosphorylations in mammalian cells.

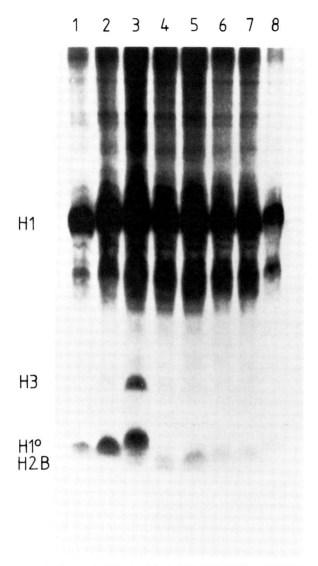

Fig. 7. Cell cycle changes in ^{32}P pulse-labeled H1, H1°, H2B, and H3 histones separated on AUT gel: (1) G$_2$ phase (M + 6.25 hr); (2) late G$_2$ (M + 7 hr); (3) metaphase; (4) S phase (M + 1.75 hr); (5) late S phase (M + 2.75 hr); (6) early G$_2$ phase (M + 3.75 hr); (7) G$_2$ phase (M + 4.75 hr); (8) G$_2$ phase (M + 5.75 hr).

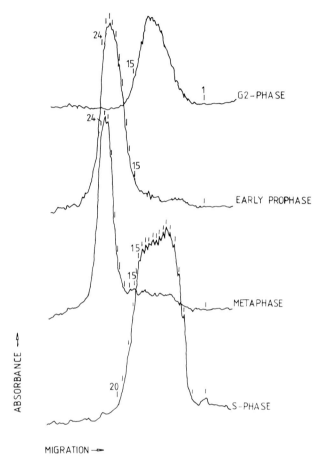

Fig. 8. Densitometer scans of *Physarum* H1 phosphorylated at S phase, G₂ phase, early prophase, and metaphase. (From Mueller *et al.*, 1985a.)

3. *H1 Phosphorylation*

H1 shows the most dramatic cell cycle dependent phosphorylations of the histones (Mueller *et al.*, 1985a; Jerzmanowski and Maleszewski, 1985). Figure 8 gives scans of the negatives of acid–urea gels of perchloric acid extracted H1 from S phase, G₂ phase, early prophase, and metaphase (Mueller *et al.*, 1985a). These scans show the high levels and complexities of H1 phosphorylations: Each inflection and peak in the scans correspond to distinct bands of the different states of H1 phosphorylations. In S phase a complex pattern of bands corresponding to 1–20 states of phosphorylation is observed with one group centered about band 7 overlapping with another group centered about band 13. As the cell cycle progresses into G_2 phase, the bands in the lower region are further phosphorylated so that the envelope of bands in G_2 phase is centered on band 13. A probable interpretation of these behaviors is that in the S-phase scan the lower group of bands is from newly synthesized H1 undergoing S-phase phosphorylations, whereas the upper group of bands is from the dephosphorylated states of "old" H1 previously phosphorylated at metaphase. As the cell cycle progresses, the "new" H1 molecules are increasingly phosphorylated until they overlap with the "old" phosphorylated H1 molecules. Further into G_2 phase all H1 molecules are increasingly phosphorylated until in early prophase the bands have moved to the higher states of 15–24 phosphorylations centered about band 22; at metaphase the level of phosphorylations increases further to give a group of bands between 20 and 24–25 centered about band 23. In this metaphase scan there is a small component of S-phase phosphorylation. This probably results from a late

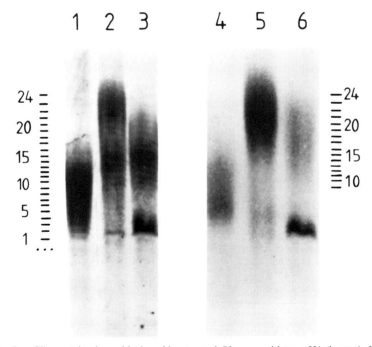

Fig. 9. Silver-stained perchloric acid-extracted *Physarum* histone H1 (lanes 1–3) and their respective autoradiograms (lanes 4–6). Macroplasmodia were pulse labeled for 30 min to study phosphate turnover rate at G_2 phase (1 and 4), late metaphase (2 and 5), and S phase (3 and 6). (From Mueller *et al.*, 1985a.)

timing of the metaphase harvest although, because the plasmodial nuclear division cycle has no G_1 phase (Fig. 4B), it is possible that some preparation for S phase is already underway in metaphase. An example of a probable late metaphase harvest is shown in the polyacrylamide gels of Fig. 9 where H1 from G_2 phase (M + 6.5 hr), metaphase (M), and S phase (M + 1 hr) with the corresponding autoradiographs from 30-min pulses of ^{32}P pulses applied immediately prior to harvest. The G_2-phase lane shows bands 1–16 coincident with the ^{32}P label. The metaphase lane shows stained bands 9–16 and 20–24, but most of the ^{32}P label is located in the upper group of bands. Thus bands 9–16 contain nonradioactive ^{31}P and come from the dephosphorylation of H1 which was phosphorylated before the 30-min pulse. This strongly implies a specific mitotic pattern of H1 phosphorylations. The S-phase lane shows stain in bands 1–3 and 9–17, whereas the ^{32}P autoradiograph shows that only the lower bands 1–3 are strongly labeled and there is a very weak labeling of the upper bands which does not coincide with the stained bands. There are three patterns of early S-phase phosphorylations: (1) rapid ^{32}P turnover in the lowest bands 1–3 which come from the phosphorylation of newly synthesized H1 molecules; (2) an intermediate group of bands 9–16 containing largely nonradioactive ^{31}P which result from the dephosphorylation of the "old" H1 molecules to intermediate states of phosphorylation; and (3) an upper group of bands 14–22 coming from the dephosphorylation of metaphase H1 to intermediate levels of phosphorylation and containing residual ^{32}P label. The major conclusion of the S-phase phosphorylation behavior is that metaphase phosphorylated H1 is dephosphorylated to intermediate states of phosphorylation while newly synthesized H1 is phosphorylated to higher levels. Progressing through S phase, increasingly phosphorylated new H1 molecules overlap with the dephosphorylated old H1 to give the late S-phase group of bands 1–16. H1 phosphorylation increases with progression through G_2 phase to the prophase and metaphase distributions of phosphorylated H1 shown in Fig. 8. It is to be noted that the maximum state of H1 phosphorylation is at metaphase and not 20 min prior to metaphase as we reported earlier (Bradbury *et al.*, 1973, 1974a).

The behaviors described above provide an explanation for the results of Fischer and Laemmli (1980) which appeared to show that metaphase H1 was not dephosphorylated, and S-phase H1 became increasingly phosphorylated until it merged with the previously metaphase phosphorylated H1. These experiments were carried out using lower resolution gels which probably showed the newly phosphorylated S-phase H1 molecules merging with the H1 dephosphorylated to the S-phase intermediate states, and the very high states of phosphorylation of metaphase H1 were not observed.

V. HISTONE H1°

In mammalian cells the H1 class of very lysine-rich histones is the most abundant. Some specialized cells contain other classes of very lysine-rich histones (see von Holt *et al.*, 1979). In avian erythrocytes H1 is largely replaced by H5; this replacment has been associated with the transcriptional inactivation of erythrocytes during erythropoiesis (Neelin *et al.*, 1964; Hnilica, 1964; Bradbury *et al.*, 1972). A minor very lysine-rich histone called H1° has been isolated from mammalian cells (Panyim and Chalkley, 1969a,b) which has a 70% sequence homology in its globular domain with the analogous H5 domain (Smith *et al.*, 1980). It has been proposed that the three globular domains have the same binding site (Cary *et al.*, 1981) (Fig. 1B). H1° has been associated with several functions. Amounts of H1° have been correlated inversely with mitotic activity (Panyim and Chalkley, 1969b). Regenerating liver has a much reduced level of H1° which returns to its normal level following regeneration (Benjamin, 1971; Balhorn *et al.*, 1972). In cell cultures H1° levels increase on contact inhibition or following serum deprivation leading to the suggestion that H1° blocks DNA replication (Pehrson and Cole, 1980). Another proposal is that H1° suppresses gene expression in a manner similar to H5 (Keppel *et al.*, 1977).

We have identified H1° in *Physarum polycephalum* and followed changes in H1° through the cell cycle and through the transition from the mitotically active plasmodial stage to the inactive, resting sclerotial stage (see Fig. 4A) (Yasuda *et al.*, 1986, 1987). *Physarum* H1° was identified by the cross reaction of an antiserum specific against the globular domain of bovine H1° and then characterized. Unlike *Physarum* H1 which is 50% larger than mammalian H1, *Physarum* H1° comigrated with bovine H1°, showing that they have similar sizes. A cell cycle study similar to that described above for H1 has been carried out for H1° (Yasuda *et al.*, 1987). H1° is subjected to an increase in phosphorylation through G_2 phase to reach a maximum at metaphase of five to six phosphates per molecule, similar to mammalian H1 and H1° (D'Anna *et al.*, 1980). The H1°/H1 ratio in *Physarum* plasmodia is 0.67, which is very high compared to the amounts of H1° in mitotically inactive mammalian cells, from which studies it was proposed that H1° was involved in mitotic inactivation. The high level of H1° in the mitotically active plasmodium argues against a unique role of H1° in the inactivation of cell division.

When a plasmodium is starved, it cycles into a sclerotium. H1° levels have been compared in both states, and the H1°/H1 ratio of 0.67 for the plasmodium was found to increase to 1.33 for the sclerotium. Further, sclerotial H1° was in a highly phosphorylated state. These results suggest that H1° is associated with transcriptionally competent but quiescent chromatin. In the sclerotial stage, H1° increases because the proportion of competent, nonexpressed chromatin increases. Hyperphosphorylation of H1° in sclerotia is probably required to maintain an inactive state of chromatin which can be reversed by dephosphorylation to allow transcriptionally competent chromatin to become available for expression.

VI. HISTONE H1 KINASES

The protein kinase involved in the G_2-phase/metaphase hyperphosphorylation of histone H1 was first described in CHO cells by Lake and Salzman (1972). This kinase has now been detected in many proliferating cells (Schlepper and Knippers, 1975; Langan, 1978). Called the growth-associated H1 kinase, kinase GR, it has been found to be cAMP-independent and bound to chromatin. The specific activity of kinase GR increases through G_2 phase to metaphase, and it has been shown to be specific for histone H1, phosphorylating the same H1 sites *in vitro* as *in vivo* (Lake and Salzman, 1972; Langan, 1978). In *Physarum* we have observed a 15-fold increase in nuclear kinase activity from its S-phase level to a maximum some 2 hr before metaphase (Bradbury *et al.*, 1974a). This increase in activity was shown to be due largely to kinase activation or transport of the kinase into the nucleus and not to its synthesis (Mitchelson *et al.*, 1978). Evidence has been found with Novikoff hepatoma cells for kinase activation (Zeilig and Langan, 1980). The maximum of *Physarum* kinase activity coincides with the maximum rate of H1 phosphorylation. It has been shown (Bradbury *et al.*, 1974a) that there is a very strong correlation between the cell cycle changes in H1 kinase activity and the published data on the effects of heat shock and plasmodial fusion on mitosis (Rusch *et al.*, 1966; Brewer and Rusch, 1968; Chin *et al.*, 1972).

Chromatography of *Physarum* nuclear extracts on DEAE-cellulose reproducibly gave an unbound, run-through component, kinase R, and two major bound components, kinase A and kinase B (Chambers *et al.*, 1983). This was in accord with Hardie *et al.* (1976) who reported the cell cycle dependences of kinases A and B. In this original report kinase R was not characterized.

Physarum kinase A, which eluted from DEAE-cellulose at 0.05 M NaCl, was not affected by cAMP in the assay medium. It was inhibited by the heat-stable inhibitor of mammalian cAMP-dependent kinase and gave an H1 phosphopeptide very similar to a phosphopeptide obtained from H1 phosphorylated by cAMP-dependent kinase. From these observations

(Chambers *et al.*, 1983) it is probable that kinase A is the catalytic subunit of the *Physarum* cAMP-dependent kinase.

Physarum kinase B eluted from DEAE-cellulose at 0.14 *M* NaCl and, similar to kinase A, it was not affected by cAMP but was inhibited by the heat-stable inhibitor of mammalian cAMP-dependent kinase. Unlike kinase A which phosphorylates H1, H2B, and protamine, kinase B phosphorylates only H1 (Hardie *et al.*, 1976) at a site or sites in the N-terminal peptide (1–106). The relationship between kinase B and mammalian kinases is not known at present.

Physarum kinase R does not bind to DEAE-cellulose and is not inhibited by heat-stable inhibitor of cAMP-dependent kinase. Kinase R phosphorylates calf thymus H1 in both the N- and C-terminal domains with a pattern of phosphorylation sites very similar to the pattern generated by Ehrlich ascites kinase GR (Chambers *et al.*, 1983). *Physarum* kinases A and B do not phosphorylate the H1 C-terminal domain, suggesting that mitotic H1 C-terminal phosphorylation is due largely to kinase R. The sites phosphorylated by kinase R include those phosphorylated by kinase GR which has been shown to phosphorylate H1 at mitosis (Langan, 1978; Lake, 1973). The phosphorylation of the same sites in calf thymus H1 by *Physarum* kinase R and mammalian GR demonstrates a high degree of evolutionary conservation of both kinases and the sites of mitotic H1 phosphorylation.

In the cell cycle behaviors of these kinases, Kinase A has maximum activity in late G_2 phase (Hardie *et al.*, 1976) and appears to be closely related to the catalytic subunit of mammalian cAMP-dependent kinase. Because the site (serine 37) phosphorylated by the cAMP-dependent kinase is not phosphorylated through mitosis (T. A. Langan, personal communication, 1985), it is very unlikely that kinase A is involved in the mitosis-related phosphorylation of H1. Kinase B also shows its maximum activity in late G_2 phase and is specific for H1 *in vitro* (Hardie *et al.*, 1976). Whether it phosphorylates H1 *in vivo* and is involved in the mitosis-related phosphorylation of H1 remains to be shown.

Advancement of *Physarum* Mitosis by Kinase GR Activity

As a direct test of the proposal that the H1 kinase activity triggers and controls chromosome condensation, the effect of kinase GR activity from Ehrlich ascites cells on the timing of mitosis in the plasmodium of *Physarum* polycephalum was investigated (Bradbury *et al.*, 1974b). Heterologous GR kinase was used because of its availability and as discussed previously the control of H1 phosphorylation and chromosome condensation must be a general feature of eukaryotes. Plasmodia were treated with solutions containing GR kinase activity on control solutions. In all cases when GR kinase activity was added during the normal increase of histone kinase activity, it was very effective in advancing mitosis (Bradbury *et al.*, 1974b; Inglis *et al.*, 1976). When added before or after the cell cycle increase in kinase activity, there was no effect. Also when the kinase activity was inactivated by freezing and thawing, there was no advancement of mitosis. If other factors are involved, they must also be inactivated by freezing. The data are fully consistent with changes in kinase activity triggering and controlling chromosome condensation.

VII. DISCUSSION

Physarum polycephalum is a particularly useful lower eukaryote to use as a model system for studies of changes in chromatin through the cell cycle and in the limited differentiation stages of this organism. Its genome is only 12 times that of yeast, but it contains well-characterized histones and nucleosomes. Patterns of core histone modifications, acetylations, and ubiquitinations are amplified in the *Physarum* cell cycle presumably because different functional states of chromatin comprise a larger proportion of its genome. A very large amplification is observed for H1 phosphorylation, and this is attributed to the larger size of the flexible C-terminal domain which is twice that of mammalian H1 (Chambers *et al.*, 1983), whereas the N- and C-terminal domains are the same size. For comparison, *Physarum* H1°, which is similar in size to mammalian H1°, undergoes similar levels of phosphorylation. Summaries of the cell cycle modifications of *Physarum* histone are given in Figs. 10 and 11. The core histone modifications, acetylations, and ubiquitinations affect only a small proportion of the core histones in keeping with their association with the DNA functions of replication and transcription. Immediately following mitosis, all four core histones are acetylated largely in the lower states of one and two acetates per molecule (Fig. 10A). Through S phase all of the core histones will undergo the acetylations associated with chromatin replication and, for those histones associated with S-phase expressed genes, the additional acetylations required by transcriptional activity. Histones H2A and H2B are acetylated only in S phase; in G_2-phase acetate turns over in small proportions of H3 and H4 in the upper states of acetylation of three and four acetates per molecule. These acetylations are associated with transcriptional activity in G_2 phase. Through the process of chromosome condensation, transcriptional activity decreases and

A

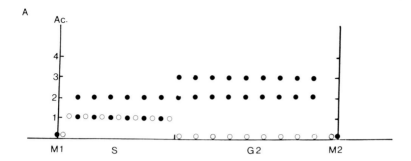

B

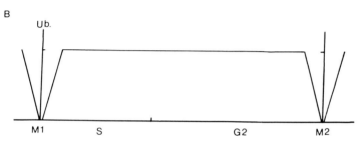

Fig. 10. (A) Outline of cell cycle changes in histone acetylation. ○, histones H2A and H2B; ●, histones H3 and H4. (B) Outline of cell cycle changes of histone H2A and H2B ubiquitination.

this is paralleled by a decrease in H3 and H4 acetylation. It is to be noted that treatment of Chinese hamster metaphase chromosomes with low levels of DNase I followed by incubation with DNA polymerase I and radioactive nucleotides labels DNA in DNase I-hypersensitive regions of chromatin, showing that sensitivity is carried over into metaphase chromosomes (Kuo and Plunkett, 1985).

Figure 10B shows the very late prophase deubiquitination of uH2A and uH2B and the very early ubiquitination of H2A and H2B in anaphase. It has been proposed that the deubiquitination of uH2A is a general factor in chromosome condensation (Matsui *et al.*, 1979). Such a model requires that uH2A be located at regular intervals in chromatin. The observation that nucleosomes containing heat shock genes are enriched 50% in uH2A (Levinger and Varshavsky, 1982) argues that ubiquitin labels specific regions of chromatin of, as yet, unknown functions.

The observations that acetylation of the core histones reaches a minimum level and uH2A and uH2B are completely deubiquitinated at metaphase strongly suggest that these modifications perturb chromatin struc-

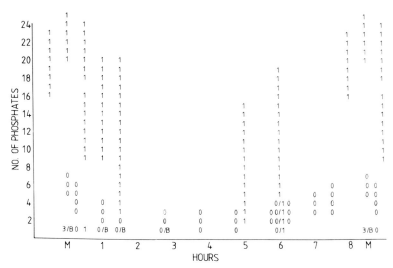

Fig. 11. Outline of cell cycle changes in phosphorylation of histone H1 (1); H1° (0); H2B (B); H3 (3).

ture for functional reasons, and that these perturbations have to be removed to allow the correct packaging of chromatin into the metaphase chromosomes. Changes in histones which parallel the process of chromosome condensation are the phosphorylations of the very lysine-rich histones H1 and H1° and the core histone H3 as shown in Fig. 11. As demonstrated in Figs. 7–9, these phosphorylations affect all of the H1, H1°, and H3 molecules and are clearly general factors affecting all of the chromatin in eukaryotic cells in the progression of the cell cycle through G_2 phase to metaphase. It has been proposed (Bradbury *et al.*, 1973, 1974a) that the phosphorylations of very lysine-rich histones initiate and control chromosome condensation, though other late events may be required for the final packaging of metaphase chromosomes, e.g., the phosphorylation of H3 is a late event affecting all nucleosomes, and there is a strong possibility that other mitotic factors, e.g., calmodulin (Nishimoto *et al.*, 1985), are involved. However, there are three observations that suggest very strongly that the phosphorylations of very lysine-rich histones are probably the primary event in chromosome condensation. First, the observation that mitosis in *Physarum* macroplasmodia can be advanced by up to 1 hr when extracts containing H1 kinase activity are added 3 hr before mitosis (Bradbury *et al.*, 1974b; Inglis *et al.*, 1976). The second observation is that in the temperature-sensitive mutant cell line tsBN2, PCC can be induced at any stage of the cell cycle by increasing the

temperature to the nonpermissive temperature (Ajiro *et al.*, 1983). H1 and
H3 isolated from these prematurely condensed chromosomes are found to
be hyperphosphorylated at the mitotic-related sites of phosphorylation.
The third observation comes from another temperature-sensitive cell mutant FT210 derived from the mouse mammary carcinoma cell line FM3 by
Yamada's laboratory (H. Yasuda and M. Yamada, personal communication, 1985). At the nonpermissive temperature the cells stop in early G_2;
H2A and H2B are ubiquitinated, but H1 is not phosphorylated. On reduc-
ing the temperature to the permissive temperature there is a pause and
then H1 is phosphorylated and cells progress to metaphase. Evidence so
far suggests that the temperature-sensitive lesion in FT210 cells involves
histone H1 kinase. Although a direct cause and effect remains to be
demonstrated, the evidence for an involvement of histones H1 and H3
phosphorylations in the initiation and control of chromosome condensation is considerable. Other factors are clearly involved in this process.
Recent results suggest that calmodulin-dependent protein kinase may be
involved in the mitotic phosphorylations of histones H1 and H3 (Nishimoto *et al.*, 1985). An involvement of calmodulin is supported by the
observations that Ca^{2+} stimulates H1 and H3 phosphorylations (Whitlock
et al., 1980, 1983).

Figure 12 shows nucleosomes with phosphorylated *Physarum* H1 and
also phosphorylated mammalian H1 molecules or H1° molecules. These
are drawn to scale and show the mode of binding of H1 to the nucleosome
through its globular domain sealing off two turns of DNA. The basic
flexible domains are examples of poorly understood protein behaviors;
their locations and modes of interaction in chromatin are not known.
Their high base charge and flexible nature strongly suggest a major in-

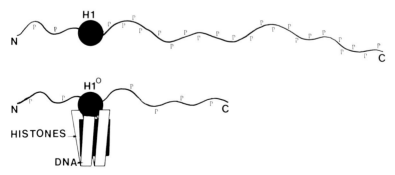

Fig. 12.   Model showing the binding of metaphase phosphorylated *Physarum* H1 and
H1°. These outline models are drawn to scale.

volvement of electrostatic interactions with DNA. As discussed earlier, stabilization of the 34-nm supercoil appears to involve H1 and the N-terminal domains of the core histones, and it has been proposed that core histone acetylation may destabilize the 34-nm supercoil (Vidali *et al.*, 1978; Simpson, 1978b). The stable form of the 34-nm supercoil is thought to be the inactive form of chromatin, and it comprises the major component in interphase nuclei and metaphase chromosomes. The chromatin domain is an attractive model for both structural and functional reasons. Each chromatin domain could be a functional unit and, drawing an analogy with the puffs of polytene chromosomes or the loops of Balbiani rings, the physical process of gene activation would involve the unfolding of the chromatin domain into a loop to make genes available for recognition and control. As can be seen in Fig. 3, a 30-μm loop of DNA can be contracted into a 0.6-μm loop of 34 nm supercoil. In its most extended form the loop will extend out only 0.3 μm from the axis of the chromatid, which is comparable to the thickness of a chromatid of a metaphase chromosome. Thus, in the transverse packing of loops into a metaphase chromosome no more than one order of packing above the 34-nm supercoil is required to give the transverse dimensions of a chromatid. This next order packing could be another higher order coil of the 34-nm supercoil, e.g., the 50 to 60-nm fibril observed by Marsden and Laemmli (1979) and probably the closer packing of these loops brought about by changes in the matrix or scaffold proteins (Laemmli *et al.*, 1978; Barrack and Coffey, 1982; Pienta and Coffey, 1984; Lewis *et al.*, 1985) and lamins (Gerace and Blobel, 1980; Gerace *et al.*, 1978). It is known that lamins undergo mitotic depolymerization probably through phosphorylation (Gerace and Blobel, 1980). Whether the kinases that phosphorylate the very lysine-rich histones or H3 also phosphorylate lamins is not known, but it raises the attractive possibility that chromosome condensation and mitosis are controlled by a cascade of phosphorylations involving histones and other mitotic proteins.

What are the possible mechanisms by which very lysine-rich histones and H3 phosphorylations induce chromosome condensation? The first considerations are electrostatic effects. In chromatin the high negative charge density of the DNA is largely, but not completely, balanced by the basic charges of the histone. The introduction of negative phosphates into the basic flexible flanking regions, particularly the C-terminal domain, must modulate and weaken interactions of these regions with DNA in chromatin and other possible interactions with acidic proteins. Whether under physiological ionic conditions this is sufficient to release the C-terminal domain is unknown. As can be seen from its amino acid composition (Table I), the C-terminal domain contains 88% of just five residues;

TABLE I

Amino Acid Composition of *Physarum* **Plasmodial Histone H1**[a]

Amino acid	Composition (mol%)		
	H1	N-T[b]	C-T[b]
Asx	3.98	6.85	1.90
Thr	6.70	6.86	6.38
Ser	8.71	7.19	11.57
Glx	6.78	11.84	2.57
Pro	10.23	7.07	18.20
Gly	4.62	6.47	2.66
Ala	18.07	11.77	27.74
Val	3.74	4.51	2.16
Met	0.80	1.00	0
Ile	2.78	3.40	0.32
Leu	3.70	4.97	0.97
Tyr	1.25	1.35	0
Phe	1.63	2.34	0
His	3.78	2.56	0
Lys	17.56	15.39	23.96
Arg	5.19	6.42	1.30
Cys[c]	0.20	nd[d]	nd[d]

[a] From Mende *et al.* (1983).
[b] N-T is the amino-terminal chymotryptic fragment of H1. C-T is the carboxy-terminal chymotryptic fragment of H1.
[c] Cys was determined as cysteic acid after performic acid oxidation.

alanine, lysine, proline, serine, and threonine (Mende *et al.*, 1983). This corresponds to 36 serines and threonines, sufficient to incorporate the 24–25 mitotic phosphates located largely in the C-terminal domain (Mueller *et al.*, 1985a; Jerzmanowski and Maleszewski, 1985). Mitotic phosphorylations introduce 48–50 negative charges into the C-terminal domain. This is almost equivalent to the positive charges of the 48 lysines and 2–3 arginines and must weaken considerably the interaction of the C-terminal domain of *Physarum* H1 with DNA in mitotic chromosomes.

There are several possible effects of H1 phosphorylation on chromatin structure through the release or weakening of a constraint applied by the C-terminal domain of H1 to the linker DNA between adjacent nucleosomes and/or between nucleosomes on adjacent gyres of the 34-nm supercoil. This release of a constraint may (1) result in a rearrangement of the nucleosomes in the 34-nm supercoil necessary for their mitotic packing; (2) change the DNA supercoiling in a chromatin loop resulting in a higher order coiling of the 34-nm supercoil, e.g., the 50 to 60-nm fibril; (3) release or weaken the binding of the H1 C-terminal domain to allow subsequent interactions of divalent cations, e.g., Ca^{2+} with DNA or between the serine and threonine phosphates of the phosphorylated H1; (4) release the C-terminal domain to allow other mitotic factors to bind to chromatin, though such factors remain to be identified.

The other well-characterized mitotic modification of histones is the phosphorylation of all of the histone H3 molecules at metaphase. Because of the rigid conservation of the H3 sequence, this phosphorylation would be expected to be an important late mitotic event in the packaging of nucleosomes in the metaphase chromosomes.

ACKNOWLEDGMENTS

These studies have been supported by grants from the American Cancer Society (NP 473), the National Science Foundation (PCM 8319215), and the National Institutes of Health (GM 26901-05).

REFERENCES

Adler, A. J., Langan, T. A., and Fasman, G. D. (1971a). Complexes of DNA with lysine rich (F1) histone phosphorylated at two separate sites: Circular dichroism studies. *Arch. Biochem. Biophys.* **153,** 769–774.

Adler, A. J., Schaffhausen, B., Langan, T. A., and Fasman, G. D. (1971b). Altered conformational effects of phosphorylated lysine-rich histone (f-1) in f-1-deoxyribonucleic acid complexes. Circular diochroism and immunological studies. *Biochemistry* **10,** 909–913.

Ajiro, K., Borun, T. W., and Cohen, L. H. (1981). Phosphorylation states of different histone 1 subtypes and their relationship to chromatin functions during the HeLA S-3 cell cycle. *Biochemistry* **20,** 1445–1454.

Ajiro, K., Nishimoto, T., and Takahashi, T. (1983). Histone H1 and H3 phosphorylation during premature chromosome condensation in a temperature sensitive mutant tsBN2 of baby hamster kidney cells. *J. Biol. Chem.* **258,** 4534–4538.

Alexopoulous, C. J., and Mims, C. W. (1979). "Introductory Mycology," 3rd ed. Wiley, New York.

Allan, J., Hartman, P. G., Crane-Robinson, C., and Aviles, F. J. (1980). The structure of histone H1 and its location in chromatin. *Nature (London)* **288,** 675–679.

Allfrey, V. G. (1980). Molecular aspects of the regulation of eukaryotic transcription; nucleosome proteins and their post-synthetic modifications in the control of DNA conformation and template function. *In* "Cell Biology" (L. Goldstein and D. M. Prescott, eds.), Vol. 3, pp. 347–437. Academic Press, New York.

Andersen, M. W., Ballal, N. R., Goldknopf, I. L., and Busch, H. (1981). A24 lysine activity in nucleoli of thioacetamide-treated rat liver releases histone H2A and ubiquitin from conjugated protein A24. *Biochemistry* **20**, 1100–1104.

Andersson, K., Björkroth, B., and Daneholt, B. (1984). Packing of a specific gene into higher order structures following repression of RNA synthesis. *J. Cell Biol.* **98**, 1296–1303.

Ashburner, M., and Novitski, E. (1976). "The Genetics and Biology of *Drosophilia*," Vol. 1A. Academic Press, New York.

Aviles, F. J., Chapman, G. E., Kneale, G. G., Crane-Robinson, C., and Bradbury, E. M. (1978). The conformation of histone H5. Isolation and characterization of the blogular segment. *Eur. J. Biochem.* **88**, 363–371.

Balhorn, R., Chalkley, R., and Granner, D. (1972). Lysine-rich histone phosphorylation. A positive correlation with cell replication. *Biochemistry,* **11**, 1094–1098.

Barrack, E. R., and Coffey, D. S. (1982). Biological properties of the nuclear matrix. *Recent Prog. Horm. Res.* **38**, 133–195.

Benjamin, W. B. (1971). Selective *in vitro* methylation of rat chromatin associated histone after partial hepatectomy. *Nature (London)* **234**, 18–20.

Bentley, G. A., Finch, J. T., Lewit-Bentley, A., and Roth, M. (1984). The crystal structure of the nucleosome core particle by contrast variation. *Basic Life Sci.* **27**, 105–117.

Benyajati, C., and Worcel, A. (1976). Isolation, characterization and structure of the folded interphase genome of *Drosophila melanogaster. Cell (Cambridge, Mass.)* **9**, 393–407.

Bohm, L., Hayashi, H., Cary, P. D., Moss, T., Crane-Robinson, C., and Bradbury, E. M. (1977). Sites of histone-histone interactions in the H3-H4 complex. *Eur. J. Biochem.* **77**, 487–493.

Bradbury, E. M., Crane-Robinson, C., and Johns, E. W. (1972). Specific conformations and interactions in chicken erythrocyte histone F2C. *Nature (London), New Biol.* **238**, 262–264.

Bradbury, E. M., Inglis, R. J., Matthews, H. R., and Sarner, N. (1973). Phosphorylation of the very lysine-rich histone in *Physarum polycephalum;* correlation with chromosome condensation. *Eur. J. Biochem.* **33**, 131–139.

Bradbury, E. M., Inglis, R. J., and Matthews, H. R. (1974a). Control of cell division by very lysine-rich histone phosphorylation. *Nature (London)* **247**, 257–261.

Bradbury, E. M., Inglis, R. J., Matthews, H. R., and Langan, T. A. (1974b). Molecular basis of control of mitotic cell division in eukaryotes. *Nature (London)* **249**, 553–556.

Bradbury, E. M., Cary, P. D., Chapman, G. E., Crane-Robinson, C., Danby, S. E., Rattle, H. W. E., Bouslik, M., Palan, J., and Aviles, F. X. (1975a). The conformation of histone H1. *Eur. J. Biochem.* **52**, 605–613.

Bradbury, E. M., Danby, S. E., Rattle, H. W. E., and Giancotti, V. (1975b). Studies on the role and mode of operation of the very lysine rich histone H1 in eukaryotic chromatin and H1: DNA complexes. *Eur. J. Biochem.* **57**, 97–105.

Bradbury, E. M., Chapman, G. E., Danby, S. E., Hartman, P. G., and Riches, P. L. (1975c). Studies on the role and mode of operation of the very lysine rich histone H1 in eukaryotic chromatin. *Eur. J. Biochem.* **57**, 521–528.

Bradbury, E. M., Baldwin, J. P., Carpenter, B. G., Hjelm, R. P., Hancock, R., and Ibel, K. (1976). Neutron scattering studies of chromatin. *In* "Neutron Scattering for the Analysis of Biological Structures," Vol. 4, pp. 97–117. Natl. Tech. Inf. Serv., U.S. Department of Commerce, Washington, D.C.

Bradbury, E. M., Moss, T., Hayashi, H., Hjelm, R. P., Suau, P., Stephens, R. M., Baldwin, J. P., and Crane-Robinson, C. (1978). Nucleosomes, histone interactions and the role of histones H3 and H4. *Cold Spring Harbor Symp. Quant. Biol.* **42**, 277–286.

Bradbury, E. M., Matthews, H. R., and MacLean, N. (1981). "DNA, Chromatin and Chromosomes." Blackwell, Oxford.

Braddock, G. W., Baldwin, J. P., and Bradbury, E. M. (1981). Neutron scattering studies of the structure of chromatin core particles in solution. *Biopolymers* **20**, 327–343.

Brewer, E. N., and Rusch, H. P. (1968). Effect of elevated temperature shocks on mitosis and on the initiation of DNA replication in *Physarum polycephalum*. *Exp. Cell Res.* **49**, 79–86.

Buongiorno-Nardelli, M., Micheli, G., Carri, M. T., and Marilley, M. (1982). A relationship between replicon size and supercoiled loop domains in the eukaryotic genome. *Nature (London)* **298**, 100–102.

Burlingame, R. W., Love, W. E., Wang, B.-C., Hamlin, R., Xuong, N.-H., and Moudriana-kis, E. N. (1985). Crystallographic structure of the octameric histone core of the nucleosome at a resolution of 3.3 Å. *Science* **228**, 546–553.

Butler, P. J. G. (1984). A defined structure of the 30 nm chromatin fibre which accommodates different nucleosomal repeat lengths. *EMBO J.* **3**, 2599–2604.

Carpenter, B. G., Baldwin, J. P., Bradbury, E. M., and Ibel, K. (1976). Organization of subunits in chromatin. *Nucleic Acids Res.* **3**, 1739–1746.

Cary, P. D., Moss, T., and Bradbury, E. M. (1978). High resolution proton magnetic resonance studies of chromatin core particles. *Eur. J. Biochem.* **89**, 475–482.

Cary, P. D., Hines, M. L., Bradbury, E. M., Smith, B. J., and Johns, E. W. (1981). Conformational studies of histone H1° in comparison with histones H1 and H5. *Eur. J. Biochem.* **120**, 371–377.

Cary, P. D., Crane-Robinson, C., Bradbury, E. M., and Dixon, G. H. (1982). Effect of acetylation on the binding of the N-terminal peptide of histone H4 to DNA. *Eur. J. Biochem.* **127**, 137–143.

Chahal, S. S., Matthews, H. R., and Bradbury, E. M. (1980). Acetylation of histone H4 and its role in chromatin structure and function. *Nature (London)* **287**, 76–79.

Chambers, T. C., Langan, T. A., Matthews, H. R., and Bradbury, E. M. (1983). H1 histone kinase from nuclei of *Physarum polycephalum*. *Biochemistry* **22**, 30–37.

Chapman, G. E., Hartman, P. G., and Bradbury, E. M. (1976). Isolation of the globular and non-globular regions of the histone H1 molecule. *Eur. J. Biochem.* **61**, 69–75.

Chapman, G. E., Hartman, P. G., Cary, P. D., and Lee, D. R. (1978). A nuclear magnetic resonance study of the globular structure of the H1 histone. *Eur. J. Biochem.* **86**, 35–44.

Chin, B., Friedrich, P. D., and Bernstein, I. A. (1972). Stimulation of mitosis following fusion of plasmodia in the myxomycetes of *Physarum polycephalum*. *J. Gen. Microbiol.* **71**, 93–101.

Ciechanover, A., Heller, H., Elias, S., Haas, A. L., and Hensko, A. (1980). ATP-dependent conjugation of reticulocyte proteins with the polypeptide required for protein degradation. *Proc. Natl. Acad. Sci. U.S.A.* **77**, 1365–1368.

Compton, J. L., Bellard, M., and Chambon, P. (1976). Biochemical evidence of variability in the DNA repeat lengths in the chromatin of higher eukaryotes. *Proc. Natl. Acad. Sci. U.S.A.* **73**, 4382–4386.

Cook, P. R., and Brazell, I. A. (1976). Conformational constraints in nuclear DNA. *J. Cell Sci.* **22**, 287–302.

Cooney, C. A. (1984). 5-Methyldeoxycytidine in eukaryotic DNA. Ph.D. Thesis, University of California, Davis.

Côté, S., Nadeau, P., Neelin, J. M., and Pallotta, D. (1982). Isolation and characterization of histones and other acid-soluble chromosomal proteins from *Physarum polycephalum*. *Can. J. Biochem.* **60**, 263–271.

Crane-Robinson, C., Bohm, L., Puigdomenech, P., Cary, P. D., Hartman, P. G., and Bradbury, E. M. (1980). Structural domains in histones. *In* "FEBS DNA—Recombination Interactions and Repair" (G. Zadrazil and J. Sponar, eds.), pp. 293–300. Pergamon, Oxford.

Daneholt, B. (1982). Structural and functional analysis of Balbiani ring genes in the salivary glands of *Chironomus tentans*. *In* "Insect Ultrastructure" (R. C. King and H. Akai, eds.), pp. 382–401. Plenum, New York.

D'Anna, J. A., Strniste, G. F., and Gurley, L. R. (1979). Circular dichroic and sedimentation studies of phosphorylated H1 from Chinese hamster cells. *Biochemistry* **18**, 943–951.

D'Anna, J. A., Gurley, L. R., Becker, R. R., Barham, S. S., Tobey, R. A., and Walters, R. A. (1980). Amino acid analysis and cell cycle dependent phosphorylation of an H1-like butyrate enhanced protein (BEP, H1°, IP$_{25}$) from Chinese hamster cells. *Biochemistry* **19**, 4331–4341.

Di Nardo, S., Voelkel, K., and Sternglanz, R. (1984). DNA topoisomerase II mutant of *Saccharomyces cerevisiae:* Topoisomerase II is required for segregation of daughter molecules at the termination of DNA replication. *Proc. Natl. Acad. Sci. U.S.A.* **81**, 2616–2620.

Earnshaw, W. C., and Heck, M. M. (1985). Localization of topoisomerase II in mitotic chromosomes. *J. Cell Biol.* **100**, 1716–1725.

Earnshaw, W. C., Halligan, B., Cooke, C. A., Heck, M. M., and Liu, F. (1985). Topoisomerase II is a structural component of mitotic chromosome scaffolds. *J. Cell Biol.* **100**, 1706–1715.

Fasy, T. M., Inoue, A., Johnson, E. M., and Allfrey, A. G. (1979). Phosphorylation of H1 and H5 histones by cyclic AMP-dependent protein kinase reduces DNA binding. *Biochim. Biophys. Acta* **564**, 332–334.

Finch, J. T., and Klug, A. (1976). Solenoidal model for superstructure in chromatin. *Proc. Natl. Acad. Sci. U.S.A.* **73**, 1897–1901.

Finley, D., Ciechanover, A., and Varshavsky, A. (1984). Thermolability of ubiquitin-activating enzyme from the mammalian cell cycle mutant ts85. *Cell (Cambridge, Mass.)* **37**, 43–55.

Fischer, S. G., and Laemmli, U. K. (1980). Cell cycle changes in *Physarum polycephalum* histone H1 phosphate: Relationship to DNA binding and chromosome condensation. *Biochemistry* **19**, 2240–2246.

Gall, J. G. (1981). Chromosome structure and the C-value paradox. *J. Cell Biol.* **91**, 35–145.

Gazit, B., Cedar, H., Lever, I., and Voss, R. (1982). Active genes are sensitive to deoxyribonuclease I during metaphase. *Science* **217**, 648–650.

Gerace, L., and Blobel, G. (1980). The nuclear envelope lamina is reversibly depolymerized during mitosis. *Cell (Cambridge, Mass.)* **19**, 277–287.

Gerace, L., and Blobel, G. (1981). Nuclear lamina and the structural organization of the nuclear envelope. *Cold Spring Harbor Symp. Quant. Biol.* **46**, 967–978.

Gerace, L., Blum, A., and Blobel, G. (1978). Immuno-cytochemical localization of the major polypeptides of the nuclear core complex-lamina fraction. *J. Cell Biol.* **79**, 546–555.

Goldknopf, I. L., Taylor, C. W., Baun, R. M., Yeoman, L. C., Olson, M. O. J., Prestayko, A., and Busch, H. (1975). Isolation and characterization of protein A24, a histone-like nonhistone protein. *J. Biol. Chem.* **250**, 7182–7187.

Goldknopf, I. L., French, M. F., Musso, R., and Busch, H. (1977). Presence of protein A24 in rat liver nucleosomes. *Proc. Natl. Acad. Sci. U.S.A.* **74**, 5492–5495.

Goldknopf, I. L., French, M. F., Daskal, Y., and Busch, H. (1978). A reciprocal relationship between contents of free ubiquitin and protein A24, its conjugate with histone 2A, in chromatin fractions obtained by the DNase II, Mg^{++} procedure. *Biochem. Biophys. Res. Commun.* **84,** 786–793.

Goldknopf, I. L., Wilson, G., Ballal, N. R., and Busch, H. (1980). Chromatin conjugate protein A24 is cleaved and ubiquitin is lost during chicken erythropoiesis. *J. Biol. Chem.* **255,** 10555–10558.

Goto, T., and Wang, J. C. (1984). Yeast DNA topoisomerase II is encoded by a single-copy, essential gene. *Cell (Cambridge, Mass.)* **36,** 1073–80.

Gurley, L. R., Walters, R. A., and Tobey, R. A. (1974). Cell cycle-specific changes in histone phosphorylation associated with cell proliferation and chromosome condensation. *J. Cell Biol.* **60,** 356–364.

Gurley, L. R., D'Anna, J. A., Barham, S. S., Deaven, L. L., and Tobey, R. A. (1978a). Histone phosphorylation and chromatin structure during mitosis in Chinese hamster cells. *Eur. J. Biochem.* **84,** 1–15.

Gurley, L. R., Tobey, R. A., Walters, P. A., Hildebrand, C. E., Hohman, P. G., D'Anna, J. A., Barham, S. S., and Deaven, L. L. (1978b). Histone phosphorylation and chromatin structure in synchronized mammalian cells. *In* "Cell Cycle Regulation" (J. R. Jeter Jr., I. L. Cameron, G. M. Padilla, and A. M. Zimmerman, eds.), pp. 37–60. Academic Press, New York.

Gurley, L. R., D'Anna, J. A., Halleck, M. S., Barham, S. S., Walters, R. A., Jeff, J. J., and Tobey, R. A. (1981). Relationships between histone phosphorylation and cell proliferation. *Cold Spring Harbor Conf. Cell Proliferation* **8,** 1073–1093.

Hancock, R., and Hughes, M. E. (1982). Organization of DNA in the interphase nucleus. *Biol. Cell.* **44,** 201–212.

Hardie, D. G., Matthews, H. R., and Bradbury, E. M. (1976). Cell cycle dependence of two nuclear histone kinase enzyme activities. *Eur. J. Biochem.* **66,** 37–42.

Hartman, P. G., Chapman, G. E., Moss, T., and Bradbury, E. M. (1977). The three structural regions of the histone H1 molecule. *Eur. J. Biochem.* **77,** 45–51.

Hewish, D. R., and Burgoyne, L. A. (1973). Chromatin sub-structure. The digestion of chromatin DNA at regularly spaced sites by a nuclear deoxyribonuclease. *Biochem. Biophys. Res. Commun.* **52,** 504–510.

Hjelm, R. P., Kneale, G. G., Suau, P., Baldwin, J. P., Bradbury, E. M., and Ibel, K. (1977). Small angle neutron scattering studies of chromatin subunits in solution. *Cell (Cambridge, Mass.)* **10,** 139–151.

Hnilica, L. S. (1964). The specificity of histones in chicken erythrocytes. *Experientia* **20,** 13–14.

Ide, T., Nakane, M., Arzai, K., and Andoh, T. (1975). Supercoiled DNA folded by nonhistone proteins in cultured mammalian cells. *Nature (London)* **258,** 445–447.

Igo-Kemenes, T., and Zachau, H. G. (1977). Domains in chromatin structure. *Cold Spring Harbor Symp. Quant. Biol.* **42,** 109–118.

Igo-Kemenes, T., Greil, W., and Zachau, H. G. (1977). Preparation of soluble chromatin and specific chromatin fractions with restriction nucleases. *Nucleic Acids Res.* **4,** 3387–3400.

Imai, B. S., Yau, P., Baldwin, J. P., Ibel, K., May, R. P., and Bradbury, E. M. (1986). Hyperacetylation of core histones does not cause unfolding of nucleosomes: Data accords with disc structure of nucleosome. *J. Biol. Chem.* **261,** 8784–8792.

Inglis, R. J., Langan, T. A., Matthews, H. R., Hardie, D. G., and Bradbury, E. M. (1976). Advance of mitosis by histone phosphokinase. *Exp. Cell Res.* **97,** 418–425.

Jerzmanowski, A., and Maleszewski, M. (1985). Phosphorylation and methylation of *Physarum* Histone H1 during mitotic cycle. *Biochemistry* **24,** 2360–2367.

Judd, B. H., and Young, M. W. (1974). An examination of the one cistron: One chromomere concept. *Cold Spring Harbor Symp. Quant. Biol.* **38,** 573–579.

Keppel, F., Allet, B., and Eisen, H. (1977). Appearance of a chromatin protein during the erythroid differentiation of Friend virus-transformed cells. *Proc. Natl. Acad. Sci. U.S.A.* **74,** 653–656.

Kleinschmidt, A. M., and Martinson, H. G. (1981). Structure of nucleosome core particles containing uH2A (A24). *Nucleic Acids Res.* **9,** 2423–2431.

Kornberg, R. (1974). Chromatin structure: A repeating unit of histones and DNA. *Science* **184,** 868–871.

Kuo, M. T., and Plunkett, W. (1985). Nick-translation of metaphase chromosomes: *In vivo* labeling of nuclease-hypersensitive regions in chromosomes. *Proc. Natl. Acad. Sci. U.S.A.* **82,** 854–858.

Laemmli, U. K. (1985). "Higher-order Chromatin Loops; Evidence for Cell-cycle and Differentiation-dependent Changes." 10th Edward de Rothschild Sch. Mol. Biophys. Weizmann Inst. Sci., Jerusalem.

Laemmli, U. K., Cheng, S. M., Adolph, K. W., Paulson, J. R., Brown, J. A., and Baumbach, W. R. (1978). Metaphase chromosome structure: The role of nonhistone proteins. *Cold Spring Harbor Symp. Quant. Biol.* **42,** 351–360.

Lake, R. S. (1973). Further characterization of the F1-histone phosphokinase of metaphase-arrested animal cells. *J. Cell Biol.* **58,** 317–331.

Lake, R. S., and Salzman, N. P. (1972). Occurrence and properties of a chromatin-associated F1-histone phosphokinase in mitotic Chinese hamster cells. *Biochemistry* **11,** 4817–4826.

Langan, T. A. (1978). Methods for the assessment of site-specific histone phosphorylation. *Methods Cell Biol.* **19,** 127–142.

Langan, T. A. (1982). Characterization of highly phosphorylated subcomponents of rat thymus H1 histone. *J. Biol. Chem.* **257,** 14835–14846.

Langan, T. A., Zeilig, C. E., and Leichtling, B. (1980). Analysis of multiple site phosphorylation of H1 histone. *In* "Protein Phosphorylation and Bio-regulation" (G. Thomas, E. G. Podesta, and J. Gordon, eds.), pp. 70–82. Karger, Basel.

Langan, T. A., Zeilig, C., and Leightling, B. (1981). Characterization of multiple-site phosphorylation of H1 histone in proliferating cells. *Cold Spring Harbor Conf. Cell Proliferation* **8,** 1039–1052.

Lennox, R. W., Oshima, R. G., and Cohen, L. H. (1982). The H1 histones and their interphase phosphorylated states in differentiated and undifferentiated cell lines derived from murine teratocarcinomas. *J. Biol. Chem.* **257,** 5183–5189.

Levin, J. M., Jost, E., and Cook, P. R. (1978). The dissociation of nuclear proteins from superhelical DNA. *J. Cell Sci.* **29,** 103–116.

Levinger, L., and Varshavsky, A. (1982). Selective arrangement of ubiquitinated and D1 protein-containing nucleosomes within the *Drosophila* genome. *Cell (Cambridge, Mass.)* **28,** 375–385.

Lewis, C. D., Lebkowski, J. S., Daly, A., and Laemmli, U. K. (1984). Interphase nuclear matrix and metaphase scaffolding structure: A comparison of protein components. *J. Cell. Sci. Suppl.* **1,** 103–122.

McGhee, J. D., Rau, D. C., Charney, E., and Felsenfeld, G. (1980). Orientation of the nucleosome within the higher order structure of chromatin. *Cell (Cambridge, Mass.)* **22,** 87–96.

McGhee, J. D., Nickol, J. M., Felsenfeld, G., and Rau, D. C. (1983). Higher order structure of chromatin: Orientation of nucleosomes within the 30 nm chromatin solenoid is independent of species and spacer length. *Cell (Cambridge, Mass.)* **33,** 831–841.

McKnight, S. K., and Miller, O. L., Jr. (1976). Ultrastructural patterns of RNA synthesis during early embryogenesis of *Drosophila melanogaster*. *Cell (Cambridge, Mass.)* **8**, 305–319.

Marsden, M., and Laemmli, U. K. (1979). Metaphase chromosome structure: evidence for a radial loop model. *Cell (Cambridge, Mass.)* **17**, 849–853.

Martinson, H. G., True, R., Burch, J. B. E., and Kunkel, G. (1979). Semihistone protein A24 replaces H2A as an integral component of the nucleosome histone core. *Proc. Natl. Acad. Sci. U.S.A.* **76**, 1030–1034.

Matsui, S. I., Seon, B. K., and Sandberg, A. A. (1979). Disappearance of a structural chromatin protein A24 in mitosis: Implications for molecular basis of chromatin condensation. *Proc. Natl. Acad. Sci. U.S.A.* **76**, 6386–6390.

Matsui, S. I., Sandberg, A., A., Negoro, S., Seon, B. K., and Goldstein, G. (1982). Isopeptidase: A novel eukaryotic enzyme that cleaves isopeptide bonds. *Proc. Natl. Acad. Sci. U.S.A.* **77**, 1535–1539.

Matsumoto, Y., Yasuda, H., Mita, S., Marunouchi, T., and Yamada, M. (1980). Evidence for the involvement of H1 histone phosphorylation in chromosome condensation. *Nature (London)* **284**, 181–183.

Matsumoto, Y., Yasuda, H., Marunouchi, T., and Yamada, M. (1983). Decrease in uH2A (protein A24) of mouse temperature sensitive mutant. *FEBS Lett.* **151**, 139–142.

Matthews, H. R., and Bradbury, E. M. (1982). Chromosome organization and chromosomal proteins in *Physarum polycephalum*. *In* "Cell Biology of *Physarum* and *Didymium*" (H. C. Aldrich and J. W. Daniel, eds.), Vol. 1, pp. 317–369. Academic Press, New York.

Matthews, H. R., and Waterborg, J. H. (1985). Reversible modifications of nuclear proteins and their significance. *In* "The Enzymology of Post-translational Modifications of Proteins" (R. B. Freedman and H. C. Hawkins, eds.), Vol. 2, pp. 125–185. Academic Press, London.

Mende, L. M., Waterborg, J. H., Mueller, R. D., and Matthews, H. R. (1983). Isolation, identification and characterization of histones from plasmodia of the true slime mold *Physarum polycephalum* using extraction with guanidine hydrochloride. *Biochemistry* **22**, 38–51.

Mirzabekov, A. D., Chick, V. V., Belyavsky, A. V., and Bavykin, S. G. (1978). Primary organization of nucleosome core particle of chromatin: Sequence of histone arrangement along DNA. *Proc. Natl. Acad. Sci. U.S.A.* **75**, 4184–4188.

Mita, S., Yasuda, H., Marunouchi, T., Ishiko, S., and Yamada, M. (1980). A temperature sensitive mutant of cultured mouse cells defective in chromosome condensation. *Exp. Cell Res.* **126**, 407–416.

Mitchelson, K., Chambers, T., Bradbury, E. M., and Matthews, H. R. (1978). Activation of histone kinase in G2-phase of the cell cycle in *Physarum polycephalum*. *FEBS Lett.* **92**, 339–342.

Mitra, S., Sen, D., and Crothers, D. M. (1984). Orientation of nucleosomes and linker DNA in calf thymus chromatin determined by photochemical dichroism. *Nature (London)* **308**, 247–250.

Moss, T., Cary, P. D., Abercrombie, B. D., Crane-Robinson, C., and Bradbury, E. M. (1976a). A pH-dependent interaction between histones H2A and H2B involving secondary and tertiary folding. *Eur. J. Biochem.* **71**, 337–350.

Moss, T., Cary, P. D., Crane-Robinson, C., and Bradbury, E. M. (1976b). Physical studies on the H3/H4 histone tetramer. *Biochemistry* **15**, 2261–2267.

Mueller, R. D., Yasuda, H., and Bradbury, E. M. (1985a). Phosphorylation of histone H1 through the cell cycle of *Physarum polycephalum:* 24 sites of phosphorylation of metaphase. *J. Biol. Chem.* **260,** 5081–5086.

Mueller, R. D., Yasuda, H., Hatch, C. L., Bonner, W. M., and Bradbury, E. M. (1985b). Identification of ubiquitinated histones H2A and H2B in *Physarum polycephalum.* Disappearance of this protein at metaphase and reappearance at anaphase. *J. Biol. Chem.* **260,** 5147–5153.

Neelin, J. M., Callaghan, P. X., Lamb, D. C., and Murray, K. (1964). The histones of chicken erythrocyte nuclei. *Can. J. Biochem.* **42,** 1743–1752.

Nishimoto, T., Ajiro, K., Hirata, M., Yamashita, K., and Sekiguchi, M. (1985). The induction of chromosome condensation in tsBN2 a temperature-sensitive mutant of BHK21, inhibited by the calmodulin antagonist, W7. *Exp. Cell Res.* **156,** 351–358.

Olins, A. L., and Olins, D. E. (1974). Spheroid chromatin units (v bodies). *Science* **183,** 330–332.

Panyim, S., and Chalkley, R. (1969a). A new histone found only in mammalian tissues with little cell division. *Biochem. Biophys. Res. Commun.* **37,** 1042–1049.

Panyim, S., and Chalkley, R. (1969b). The heterogeneity of histones. I. A quantitative analysis of calf histones in very long polyacrylamide gels. *Biochemistry* **8,** 3972–3979.

Paulson, J. R., and Laemmli, U. K. (1977). The structure of histone-depleted metaphase chromosomes. *Cell (Cambridge, Mass.)* **12,** 817–828.

Pehrson, J., and Cole, R. D. (1980). Histone H1° accumulates in growth-inhibited cultured cells. *Nature (London)* **285,** 43–44.

Pienta, K. J., and Coffey, D. S. (1984). A structural analysis of the role of the nuclear matrix and DNA loops in the organization of the nucleus and chromosome. *J. Cell Sci., Suppl.* **1,** 123–135.

Pinon, R., and Salts, Y. (1977). Isolation of folded chromosomes from the yeast *Saccharomyces cerevisiae. Proc. Natl. Acad. Sci. U.S.A.* **74,** 2850–2854.

Rall, S. C., and Cole, R. D. (1971). Amino acid sequence and sequence variability of the amino-terminal regions of lysine rich histones. *J. Biol. Chem.* **246,** 7175–7179.

Rattle, H. W. E., Langan, T. A., Danby, S. E., and Bradbury, E. M. (1977). Studies on the role and mode of operation of the very lysine rich histones in eukaryotic chromatin. Effect of A and B site phosphorylation on the conformation and interaction of histone H1. *Eur. J. Biochem.* **81,** 499–505.

Richmond, T. J., Finch, J. T., Rushton, B., Rhodes, D., and Klug, A. (1984). Structure of the nucleosome core particle at 7 Å resolution. *Nature (London)* **311,** 532–537.

Rusch, H. P. (1980). Introduction. *In* "Growth and Differentiation in *Physarum polycephalum*" (W. F. Dove and H. P. Rusch, eds.), pp. 1–8. Princeton Univ. Press, Princeton, New Jersey.

Rusch, H. P., Sachsenmaier, W., Behrens, K., and Gruiter, V. (1966). Syhchronization of mitosis by the fusion of the plasmodia of *Physarum polycephalum. J. Cell Biol.* **31,** 204–209.

Schlepper, J., and Knippers, R. (1975). Nuclear protein kinases from murine cells. *Eur. J. Biochem.* **60,** 209–220.

Seale, R. L. (1981). Rapid turnover of the histone ubiquitin conjugate protein A24. *Nucleic Acids Res.* **9,** 3151–3158.

Shick, V. V., Belyavsky, A. V., Bavykin, S. G., and Mirzabekov, A. D. (1980). Primary organization of the nucleosome core particles. Sequential arrangement of histones along DNA. *J. Mol. Biol.* **139,** 491–517.

Simpson, R. T. (1978a). Structure of the chromatosome, a chromatin particle containing 160 base pairs of DNA and all the histones. *Biochemistry* **17,** 5524–5531.

Simpson, R. T. (1978b). Structure of chromatin containing extensively acetylated H3 and H4. *Cell (Cambridge, Mass.)* **13**, 691–699.

Smith, B. J., Walker, J. M., and Johns, E. W. (1980). Structural homology between a mammalian H1° subfraction and avian erythrocyte-specific histone H5. *FEBS Lett.* **112**, 42–44.

Stadler, J., Seebeck, T., and Braun, R. (1978). Degradation of the ribosomal genes by DNAse I in *Physarum polycephalum. Eur. J. Biochem.* **90**, 391–395.

Stein, A., and Bina, M. (1984). A model chromatin assembly system. Factors affecting nucleosome spacing. *J. Mol. Biol.* **178**, 341–363.

Suau, P., Kneale, G. G., Braddock, G. W., Baldwin, J. P., and Bradbury, E. M. (1977). A low resolution model for the chromatin core particle by neutron scattering. *Nucleic Acids Res.* **4**, 3769–3786.

Suau, P., Bradbury, E. M., and Baldwin, J. P. (1979). Higher-order structures of chromatin in solution. *Eur. J. Biochem.* **97**, 593–602.

Thoma, F., Koller, T., and Klug, A. (1979). Involvement of histone H1 in the organization of the nucleosome and of the salt-dependent superstructures of chromatin. *J. Cell Biol.* **83**, 403–427.

Uemura, T., and Yanagida, M. (1984). Isolation of type I and II DNA topoisomerase mutants from fission yeast: Single and double mutants show different phenotypes in cell growth and chromatin organization. *EMBO J.* **3**, 1737–1744.

Vidali, G., Boffa, L. C., Bradbury, E. M., and Allfrey, V. G. (1978). Butyrate suppression of histone deacetylation leads to an accumulation of multiacetylated forms of histone H3 and H4 and increased DNase I sensitivity of associated DNA sequences. *Proc. Natl. Acad. Sci. U.S.A.* **75**, 2239–2243.

von Holt, C., Strickland, W. N., Brandt, W. F., and Strickland, M. (1979). More histone structures. *FEBS Lett.* **100**, 201–218.

Waterborg, J. H., and Matthews, H. R. (1983). Patterns of histone acetylation in the cell cycle of *Physarum polycephalum. Biochemistry* **22**, 1489–1496.

Waterborg, J. H., and Matthews, H. R. (1984). Patterns of histone acetylation in *Physarum polycephalum;* H2A and H2B acetylation is functionally distinct from H3 and H4 acetylation. *Eur. J. Biochem.* **142**, 329–335.

West, M. H. P., and Bonner, W. M. (1980). Histone H2B can be modified by the attachment of ubiquitin. *Nucleic Acids Res.* **8**, 4671–4680.

Whitlock, J. P., Jr., Augustine, R., and Schulman, H. (1980). Calcium-dependent phosphorylation of histone H3 in butyrate-treated HeLa cells. *Nature (London)* **287**, 74–76.

Whitlock, J. P., Jr., Galeazzi, D., and Schulman, H. (1983). Acetylation and calcium-dependent phosphorylation of histone H3 in nuclei from butyrate-treated HeLa cells. *J. Biol. Chem.* **258**, 1299–1304.

Wilhelm, M. L., Wilhelm, X. F., Toublan, B., and Jalouzot, R. (1982). Accessibility of histone H4 gene of *Physarum polycephalum* to DNase-I during the cell cycle. *FEBS Lett.* **150**, 438–444.

Worcel, A., and Benyajati, C. (1977). Higher order coiling of DNA in chromatin. *Cell (Cambridge, Mass.)* **12**, 83–100.

Wu, R. S., Kohn, K. W., Bonner, W. M. (1981). Metabolism of uniquitinated histones. *J. Biol. Chem.* **256**, 5916–5920.

Yasuda, H., Matsumoto, Y., Mita, S., Marunouchi, T., and Yamada, M. (1981). A mouse temperature-sensitive mutant defective in H1 phosphorylation is defective in DNA synthesis and chromatin condensation. *Biochemistry* **20**, 4414–4419.

Yasuda, H., Mueller, R. D., Logan, K. A., and Bradbury, E. M. (1986). Identification of

Physarum polycephalum histone H1°; high levels of H1° in plasmodia and of phosphorylated H1° in sclerotia. *J. Biol. Chem.* **261,** 2349–2354.

Yasuda, H., Mueller, R. D., and Bradbury, E. M. (1987). Cell cycle phosphorylations of histone H1° and comparisons to histone H1 and H3. *J. Biol. Chem.* (submitted for publication).

Zeilig, C. E., and Langan, T. A. (1980). Studies on the mechanism of activation of mitotic histone H1 kinase. *Biochem. Biophys. Res. Commun.* **95,** 1372–1379.

Index